Physical Chemistry from Ostwald to Pauling

Physical Chemistry from Ostwald to Pauling

THE MAKING OF A SCIENCE IN AMERICA

John W. Servos

PRINCETON UNIVERSITY PRESS

PRINCETON, NEW JERSEY

Library of Congress Cataloging-in-Publication Data

Servos, John W. (John William), 1951–
Physical chemistry from Ostwald to Pauling : the making of a
science in America / John W. Servos.
p.   cm.
Includes bibliographical references.
1. Chemistry, Physical and theoretical—United States—History.
I. Title.
QD452.5.U6S47   1990   541.3′0973—dc20   89-70256

ISBN 0-691-08566-8 (alk. paper)

This book has been composed in Linotron Times Roman

Princeton University Press books are printed on acid-free paper,
and meet the guidelines for permanence and durability of the
Committee on Production Guidelines for Book Longevity of the
Council on Library Resources

Printed in the United States of America by Princeton University Press,
Princeton, New Jersey
10 9 8 7 6 5 4 3 2 1

For Catherine and William Servos

Contents

List of Figures

List of Tables

Preface

EVERY AUTHOR will be familiar with that most common and difficult question, "What is your book about?" During the several years I have worked on this project, I have experimented with several answers. Perhaps the simplest and most satisfactory is that it is an inquiry into how and why a new scientific discipline took root, grew, and flourished in a particular social setting. The discipline is physical chemistry; the setting is America in the decades around the turn of the century.

But such a bald statement demands some explication. Why study the history of entities as disorderly, ill-defined, and common as scientific disciplines? Why select a science as technically demanding and politically inert as physical chemistry? Why focus on the United States when it is well known that this science had its origins in Europe? And why end the story in the early 1930s, a time when some would say that things were just getting interesting?

Ruthlessly honest answers would demand the acknowledgment of chance. Like many if not most books, this one went through several lives and was shaped by all manner of accidental factors that now can hardly concern a reader. But as any student of science knows, just because the context of discovery differs from the context of justification, it does not follow that justifications are unimportant. With this in mind, let me explain, as best I can, the scope and design of this book.

It is about a discipline, and by that I mean a family-like grouping of individuals sharing intellectual ancestry and united at any given time by an interest in common or overlapping problems, techniques, and institutions. There is much ambiguity in this definition. It says nothing about size, for example, or about the degree to which individuals must be "related" in order to qualify as members of the same discipline. Yet to be more precise would be to exaggerate the exactness of the phenomenon. Disciplines may enroll a few dozen scientists or thousands. They may be tightly focused on a few questions and techniques, or they may be so diffuse as to challenge the skill of the best textbook writer. Some are happy families, with little controversy over methods and goals. Others are fractured into many research schools, each with a different agenda, each evolving its own traditions of thought and work, and each competing for resources and recognition. A single discipline may be all of these things at various points in its history.

Disciplines are difficult to define, but they are important in the intellectual and social life of scientists and scholars, especially those in academe. We publish in disciplinary journals; we work in departments that reflect present or past disciplinary contours; we take and teach courses bearing titles like "history of science" and "organic chemistry"; we identify ourselves at cocktail parties and in biographical directories by discipline; our applications for grants and fellowships are read and evaluated by peers, meaning other members of our discipline. With few exceptions, the more actively engaged we are in the production and transmission of knowledge, the more powerfully disciplines influence our behavior and self-images. We are, of course, more than members of a discipline, but our professional lives revolve around these entities; they help us define our ambitions, successes, and failures.

Despite or perhaps because of this ubiquity, disciplines have not enjoyed a favorable press in recent years. They have been described as purely political entities dedicated to serving the interests of small elites and as repressive institutions that stifle creative impulses and impose artificial limits on the growth of knowledge. For some proponents of "interdisciplinary" education, the discipline is something to be overcome.

Yet the discipline not only confines, it also liberates. The successful discipline affords its practitioners social and intellectual security, institutional support, and a sense of direction and opportunity. Although a small elite may exert great influence within a discipline, it would be a grave mistake to assume that its motives must be nefarious, its goals self-serving, and its influence retrograde. The discipline is not dysfunctional; it is functional. There is a nearly universal tendency for scholars, even those who set out to break disciplinary molds, to organize themselves into such units.

The reason is not far to seek. Disciplines not only lend structure and meaning to lives, they also bring order and significance to knowledge. To appreciate this, it is necessary only to try to imagine a world that ignores them. Any glimpse of unity that a schooling without specialties might afford could hardly compensate for its barrenness and sterility. Few if any of us could flourish on the boundless sea that is knowledge without categories. With no reason to drop anchor here rather than there, with no coordinates or landmarks to mark and communicate positions, a word like exploration would lose all meaning. Specialization, which is as important to science and scholarship as to pin production, would be impossible. Disciplines may shrink our horizons, but they compensate by giving us means by which we can make our knowledge productive.

Disciplines play important roles in all kinds of intellectual activity, but nowhere are they as important as in the sciences. Essential to the sustained accumulation of facts, the elaboration of ideas, and the transmission of technique, disciplines are at least partly responsible for giving modern science its cumulative and progressive character. As Thomas S. Kuhn has suggested, it would be surprising if some form of progress did not result from the concentration of effort on selected problems. Disciplines are lenses that focus individual effort. Scientific disciplines, with their textbooks, journals, abstracting services, review articles, societies, and powerful sanctions against amateurs, are especially powerful lenses. They are strikingly efficient in identifying soluble problems and in bringing resources to bear on such problems.

Of all the scientific disciplines one could write about, physical chemistry is perhaps not the first to come to mind. It is not as flamboyant as molecular biology or particle physics. Its practice does not command headlines or prompt Congressional inquiries; its concepts seldom attract popular attention; its story offers comparatively few opportunities to explore issues central to modern political history. Chemists find this discipline difficult to define; physicists sometimes look upon it as a trivial application of their subject; undergraduate chemistry majors tend to see it as the bane of their existence—a forbidding hurdle standing between them and a degree. All will grant the usefulness of physical chemistry and the virtues of knowing it, but few develop much affection for it.

Yet physical chemistry was not always dowdy. In the early twentieth century, it was nearly as chic and exciting as molecular biology is today. The names of its progenitors, Ostwald, van't Hoff, Arrhenius, and Nernst, were familiar to the scientifically literate. Their studies of solutions and chemical thermodynamics transformed scientists' understanding of chemical affinity. A generation later, their successors would effect another revolution by using quantum theory to generate new pictures of the molecule and chemical bond. Physical chemists' striking success in exploring the terrain between chemistry and physics inspired other scientists who were dissatisfied with traditional disciplinary boundaries and helped stimulate the growth of such other borderland specialties as biochemistry and geochemistry. Their science helped launch the high-technology industries of the day—petroleum cracking and nitrogen fixation. Its name could be invoked by a novelist like Sinclair Lewis as a symbol of the progressiveness, power, and difficulty of modern science.

One reason for writing about the history of physical chemistry, then, is simply because its story, while much less known than that of molecular biology or particle physics, has been no less important to the history of twentieth-century science.

But there is another reason as well. The fate of most successful disciplines is fragmentation into smaller and more cohesive specialties. Coalescing around a few tightly focused research schools, they expand and diversify until dismembered by the forces generated by their own growth. The name of the parent field may endure, but more for the convenience of educators and bibliographers than as a cohesive and vital category of scientific research.

Physical chemistry presents us with a poignant illustration of this process. Born out of a revolt against the disciplinary structure of the physical sciences in the late nineteenth century, it soon acquired all the trappings of a discipline itself. Taking form in the 1880s, it grew explosively until, by 1930, it had given rise to a half-dozen or more specialties that, more and more, were coming to serve as the principal reference frames for their members. Older physical chemists lamented the fragmentation of their science; younger ones, who now considered themselves primarily colloid chemists, kineticists, or crystallographers, celebrated the progress that accompanied specialization. Ironically, one of these descendants, variously called chemical physics, structural chemistry, or quantum chemistry, would perform the broad integrative functions that the founders of physical chemistry had aspired to fulfill. These developments—the coalescence of physical chemistry in the 1880s and the emergence of a new chemical physics in the 1930s—frame the book that follows. The first was associated most directly with the career of Wilhelm Ostwald, the second with that of Linus Pauling.

A few words must be said about what some readers may find to be a disturbing emphasis on the history of this discipline in America. Physical chemistry as a network of ideas is not American any more than it is German or French. And if this were a work of straightforward intellectual history, my concentration on American institutions and scientists would be unforgivable.

But disciplines are more than simply aggregates of disembodied ideas. They find leaders who are imbued not only with the norms of science but also with the values of national cultures; they draw on traditions of thought and activity that may vary from country to country, and within countries from locale to locale; they are propagated by journals and textbooks that are written in particular languages; in each nation and region they meet peculiar economic and social conditions, which may favor or hinder their development, or which may channel it along particular lines; and for reasons of convenience, and sometimes necessity, practitioners usually respect national boundaries when they organize themselves, either formally and informally. Even today, in an age of air travel and global telephone connections, distance produces subtle but significant variations in the practice of science. In earlier times, when long-distance communication was more laborious and western culture less homogeneous, opportunities for variation and divergence were greater and national and local styles more prominent.

For the historian, this dual nature of disciplines constitutes a dilemma. To concentrate on the universal by relating the history of a discipline without attention to national boundaries all but necessitates the sacrifice of the local details that may prove essential to understanding how and why the discipline flourished in particular places. The result may be a beautiful account of intellectual development, but one that provides little insight into the economic and cultural conditions that made such development possible. Alternatively, to focus exclusively on the history of a disciplinary community in one nation precludes any meaningful discussion of the content of the science. At best, we get sketchy summaries of conceptual developments; at worst, we lose all touch with ideas—the very things that give disciplines their *raison d'être*. In either case, the integrity of the historical subject—the discipline in its intellectual and conceptual totality—is destroyed.

I do not pretend to have solved this dilemma. It is impossible in the confines of a single volume to do justice to the history of a modern discipline in all its complexity. Yet, by what I hope are judicious compromises I have sought to retain the strengths and avoid the weaknesses of the two approaches outlined above. The origins of physical chemistry are in Europe, and in the first two chapters I describe those origins at some length. Subsequently, I allude to developments in Europe as they impinged on American scientists, elaborating only where it seems necessary to follow the work of selected research schools on this side of the Atlantic. The schools on which I focus are those of Arthur A. Noyes, G. N. Lewis, and Wilder D. Bancroft. Not only were these scientists among the most prominent American physical chemists of their generation, they were also critically important teachers and institution-builders. The study of their lives and labor, supplemented by a broader but shallower survey of the community in which they worked, is the best route I have discovered to explore both the expansion and diversification of physical chemistry in the early twentieth century and the path that led from the physical chemistry of Ostwald to the chemical physics of Pauling.

Acknowledgments

THIS PROJECT began when, more than a decade ago, I began to look into the history of Caltech's chemistry department. It has subsequently assumed the form of a seminar paper, a dissertation, several articles, and now a book. At every stage I have benefited from the sound advice, warm encouragement, and unfailing tact of my former teachers, especially Owen Hannaway and Robert Kargon.

Others have also made this a better book and me a better historian. Robert E. Kohler helped sustain my interest in this project both through his writings and his timely comments on drafts of several chapters. I do not share all of his views on the history of disciplines, but I have found no one whose work is as consistently stimulating. Gerald L. Geison, Michael S. Mahoney, Thomas S. Kuhn, and Charles C. Gillispie gave me friendship and generous support during the years I was privileged to work in Princeton's Program in the History of Science. They taught me more about writing, scholarship, and the history of science than they can possibly know. Jeffrey Sturchio and Walter Kauzmann, readers for Princeton University Press, gave generously of their time and thought and saved me from several errors and omissions.

Readers of the first chapter will recognize my debts to Robert Scott Root-Bernstein. While I ostensibly directed his doctoral dissertation on the Ionists, in truth he was my teacher. Larry Owens, through his unique gift for cultural history, has led me to see new dimensions in my subjects and their institutions. He proved equally resourceful in tracking down obscure books during the year he served as my research assistant. I also wish to thank JoAnn Morse, Geoff Sutton, Ted Porter, Peter and Pauline Dear, and John Carson, former students who did not share my interest in physical chemistry but whose lively intelligence, good conversation, and exciting work have consistently sparked my imagination.

The late R. E. Gibson generously shared his knowledge of the Geophysical Laboratory of the Carnegie Institution of Washington and vetted sections of this book that deal with the history of petrology. I am also indebted to David Cahan and Wolfgang Girnus for reading and commenting on an earlier draft of the first chapter. The late Robert S. Mulliken shared with me drafts of his autobiography, *The Life of a Scientist*, and gave me the pleasure of several hours of telephone conversation about his rich work and experience. Paul H. Emmet, Don M. Yost, Duncan MacRae, Ernest H. Swift, and Oliver R. Wulf graciously responded to queries about the early years of their careers as physical chemists. Farooq Hussein, Leon Gortler, Jeffrey Sturchio, P. Thomas

Carroll, Ron Doel, and John Heilbron have shared with me the results of their archival research and transcripts of interviews. Judy Goodstein and Deborah Cozort helped me navigate through the Caltech and MIT archives. Faye Angelozzi and Rhea Cabin have been superb secretaries and good friends. Laura Kang Ward, the manuscript's copyeditor, watched not only my commas and semicolons but also my German syntax and algebra. I was fortunate to have her help.

In addition to these personal debts, I owe thanks to several institutions. Sigma Xi generously afforded me travel funds that made my first trip to the Caltech archives possible. The Smithsonian Institution and my hosts at the National Museum of American History, especially Jon Eklund, gave me ideal conditions in which to write the dissertation upon which this book is based. The American Council of Learned Societies, the National Science Foundation, and the Princeton History Department provided grants that helped underwrite a year of uninterrupted research and writing. More recently, the Trustees of Amherst College awarded me a fellowship that made it possible to complete this book.

Portions of this book are taken, with adaptations, from articles that appeared in *Isis*, *Historical Studies in the Physical Sciences*, and the *Journal of Chemical Education*. They appear here by permission. Chapter 5 contains material that appeared, in different form, in *Chemistry and Modern Society: Essays in Honor of Aaron J. Ihde*, ed. John Parascandola and James C. Whorton (Washington, D.C.: American Chemical Society, 1983), and is reprinted with permission from the American Chemical Society, copyright 1983 American Chemical Society.

I am grateful, above all, to the members of my family, especially my wife, Virginia, who has listened patiently to several versions of this book.

List of Note Abbreviations

PPC	President's Papers, University of Chicago
PPMIT	MIT, Office of the President, 1897–1930 (AC 13), Massachusetts Institute of Technology
PPUC	President's Papers, University of California, Berkeley
RAM	Robert A. Millikan Papers, California Institute of Technology
SF	Simon Flexner Papers, American Philosophical Society
TWR	Theodore William Richards Papers, Harvard University
USGSLS	U.S. Geological Survey, Lake Superior Division Correspondence, State Historical Society of Wisconsin, Madison
USGSVH	U.S. Geological Survey, Lake Superior Division Correspondence, Charles R. Van Hise Letterbooks, State Historical Society of Wisconsin, Madison
VB	Vannevar Bush Papers, Massachusetts Institute of Technology
WAN	William Albert Noyes Papers
WDB	Wilder Dwight Bancroft Papers, Cornell University
WDB, CIW	Wilder D. Bancroft Files, Carnegie Institution of Washington
WJO	Winthrop John Van Orman Osterhout Papers, American Philosophical Society
WRH	William Rainey Harper Papers, University of Chicago
WRW	Willis R. Whitney Papers, Union College

JOURNALS AND OTHER FREQUENTLY CITED SOURCES

AC	*Annalen der Chemie*
ACJ	*American Chemical Journal*
AHES	*Archive for History of Exact Science*
AJS	*American Journal of Science*
ARPC	*Annual Review of Physical Chemistry*
BDCG	*Berichte der deutschen chemischen Gesellschaft*
BGSA	*Bulletin of the Geological Society of America*
BJHS	*British Journal for the History of Science*
BMNAS	*Biographical Memoirs of the National Academy of Sciences*
BPSW	*Bulletin of the Philosophical Society of Washington*
CR	*Chemical Reviews*
DSB	*Dictionary of Scientific Biography*
HS	*History of Science*
HSPS	*Historical Studies in the Physical Sciences*
IECNE	*Industrial and Engineering Chemistry, News Edition*
JACS	*Journal of the American Chemical Society*
JAMA	*Journal of the American Medical Association*
JCE	*Journal of Chemical Education*
JCP	*Journal of Chemical Physics*
JCS	*Journal of the Chemical Society*

JG	*Journal of Geology*
JPC	*Journal of Physical Chemistry*
KZ	*Kolloid-Zeitschrift*
OKEW	*Ostwalds Klassiker der exakten Wissenschaften*
PAAAS	*Proceedings of the American Academy of Arts and Sciences*
PM	*Philosophical Magazine*
PNAS	*Proceedings of the National Academy of Sciences*
PR	*Physical Review*
PSM	*Popular Science Monthly*
PZ	*Physikalische Zeitschrift*
RACR	*Review of American Chemical Research*
SM	*Scientific Monthly*
TAICE	*Transactions of the American Institute of Chemical Engineers*
TC	*Technology and Culture*
TES	*Transactions of the Electrochemical Society*
TFS	*Transactions of the Faraday Society*
TQ	*Technology Quarterly*
TR	*Technology Review*
ZPC	*Zeitschrift für physikalische Chemie*

Physical Chemistry from Ostwald to Pauling

CHAPTER 1

Modern Chemistry Is in Need of Reform

"MODERN CHEMISTRY is in need of reform." Or so claimed Wilhelm Ostwald at his examination for a master's degree in 1877.[1] Ostwald was then a twenty-three-year-old student at the University of Dorpat, a remote outpost of German scholarship in Russia's Baltic provinces. His claim was not the carefully considered manifesto of a revolutionary; rather it was a joint product of the intellectual exuberance of this son of a German cooper and of the requirements of an examination system that encouraged students to select and defend broad theses in their dissertations. Ostwald's blunt assertion may have raised few eyebrows among his teachers; in retrospect, however, it appears as an early sign of the urgent and driving desire to reshape his environment, intellectual and institutional, that ran as an extended motif through his career. During the late 1870s and early 1880s, while a student at Dorpat and a teacher at the Riga Polytechnic Institute, Ostwald confined himself to the rehabilitation of the theory of chemical affinity. In the mid-1880s, and especially after he was appointed to a professorship at the University of Leipzig in 1887, Ostwald widened his horizons and sought to set the whole science of chemistry on new foundations. In the 1890s his ambition to reform again expanded in scope, this time to encompass all the physical sciences through his energetics program. Finally, following his retirement from Leipzig in 1906, Ostwald devoted himself to philosophy and a variety of social causes, believing that the entire theory of knowledge required remaking and that social institutions and conventions as diverse as language, coinage, the printing industry, and methods of measuring time were in need of rationalization.[2]

The scope of Ostwald's ambitions was ever widening, but it was as a chemist that he made an enduring mark. His program for reforming chemistry, as it evolved in the 1880s and 1890s, may be simply stated: he sought to redirect chemists' attention from the substances participating in chemical reactions to the reactions themselves. Ostwald thought that chemists had long overemphasized the taxonomic aspects of their science by focusing too narrowly upon the composition, structure, and properties of the species involved in chemical processes. He recognized that this approach had considerable power, as amply demonstrated by the rapid growth and achievements of organic chemistry. Yet for all of its success, the taxonomic approach to chemistry left questions regarding the rate, direction, and yield of chemical reactions unanswered. To resolve these questions and to promote chemistry from the ranks of the descriptive to the company of the analytical sciences, Ostwald believed chemists would have to study the conditions under which compounds formed and decomposed and pay attention to the problems of chemical affinity and equilibrium, mass action and reaction velocity. The arrow or equal sign in chemical equations must, he thought, become chemists' principal object of investigation.

To shift the point of attack from the description and ordering of chemical species to the development of general laws of chemical change, Ostwald advocated the adoption of physical techniques—physics and chemistry would have to be joined together as they had not been since the first years of the nineteenth century. Like Comte, whose works he admired, Ostwald perceived a hierarchy among the sciences and a general pattern to their evolution. Every science, he thought, progressed through three overlapping but distinguishable stages. The first consisted of discovering and describing a set of phenomena; the second, of arranging those phenomena into orderly categories; and the third, of determining the general laws to which the phenomena were subject. Physics had entered the third stage; chemistry had not. For chemistry to do so, it would be essential that chemists borrow some of the methods that had made their colleagues in physics so successful. This meant not only adopting physical tools and canons of precision, but also a physical habit of thought. In particular, chemists had to focus on the quantifiable aspects of chemical phenomena and learn to fit mathematical expressions to their results. They had to relate their findings to the existing body of physical concepts, because chemistry and physics, as Ostwald saw them, were both parts of a more comprehensive physical science which ought not to encompass contradictions. And chemists had to learn, as had physicists, that their own subject was one rather than many. Divisions within chemistry, such as between organic and inorganic, might be transcended if the object of study became the reaction itself rather than the species of matter that participated in it. The study of the particular could not be abandoned, but it had to be subordinated to the search for the general.[3]

Ostwald aimed at bringing about changes in chemistry of a magnitude comparable to those effected by Lavoisier a century earlier, and, like Lavoisier's chemistry, the new science that Ostwald sought to create sailed for some time under several colors. Initially, Ostwald called it *allgemeine Chemie*, for in his view it would not be a new part of chemistry so much as a new basis for all existing parts of the science—not a branch of the tree, but the life sap of the entire organism. Although this term expressed the ambition of unifying chemistry under a set of general laws and principles, others who shared many of Ostwald's goals, like Walther Nernst, preferred the term *theoretische Chemie*, because it called to mind the aim of building a deductive science akin to theoretical physics. Both of these names enjoyed popularity, but by the late 1890s a third term had come into general use: *physikalische Chemie*. This label had long been applied to work on the borderland between physics and chemistry, a tradition with which Ostwald felt strong attachments, and Ostwald himself often used it interchangeably with the name *allgemeine Chemie*. It had the virtue of stressing the methods that the new science would employ. Chemistry might provide the material of study, but physics would afford the model and many of the means of investigation.[4]

Ostwald, and those who shared his goal of reforming chemistry, were largely successful in achieving their aims. They did not eliminate internal divisions within chemistry but did fashion tools and concepts that found application in every branch of the science. They did not close the divide separating chemists and physicists but did do much to narrow it. And if they fell short of making chemistry a fully deductive science, they nevertheless brought large and important elements of it within the domain of analytical treatment.

By the time Ostwald died in 1932, physical chemistry had attained both intellectual respectability and institutional expression. Three of the first ten Nobel laureates in chemistry had been physical chemists and a fourth had been closely identified with the specialty.[5] Physico-chemical institutes and chairs, journals and societies flourished in a half-dozen countries. Courses in physical chemistry were mandatory for chemistry majors at most major universities, and introductory textbooks of chemistry, which had been based on the properties and reactions of the elements, were coming more and more to be organized around physico-chemical principles. The subject had become a common feature in the curriculum of chemical engineers, biochemists, and geologists; and even organic chemists, who often had been skeptical of claims made on behalf of the new specialty, were beginning to adopt physico-chemical techniques. By 1930, many if not most scientists viewed physical chemistry as comprising the core of the science of chemistry. Ostwald, his confederates, and his students were largely responsible for shaping this perception and for effecting the achievements upon which the perception was based.

My dual aim in this chapter is to discuss the origins and nature of Ostwald's program for reforming chemistry and to consider some of the conceptual developments that made so many of his ambitions realizable. In subsequent chapters we shall examine how the new chemistry that Ostwald helped to create, and for which he was the chief spokesman, was brought to America, where it was interpreted and reinterpreted, and grew and prospered. To grasp the content and context of Ostwald's program, however, it will first be necessary to understand both the sources of his discontent with the chemistry he encountered as a student and the traditions that he and his co-workers drew upon in their early efforts to reform their science. To do this, we must look back to the period when modern chemistry first took form, paying particular attention to the associations chemistry had with other branches of knowledge.

CHEMISTRY AND THE CARTOGRAPHY OF SCIENCE

One of the most difficult and most important lessons of the history of science is that it is a grievous error to impose contemporary definitions of scientific subjects and disciplines upon the past. The categories into which scientific activity is divided are mutable. Every scientist in every field at every time must define for himself, or borrow from his teachers, certain assumptions regarding what is and what is not part of his domain of scholarship. He must situate his subject in relation to other fields and develop a mental picture of the terrain in which he works, identifying the landmarks and boundaries according to which he will develop and order his priorities as an investigator and teacher. Insofar as groups of scientists hold or inherit a set of common perceptions of some more-or-less narrow domain of natural knowledge, we may speak of them as members of a single discipline. Their perspectives may rarely if ever be identical, but when their mental maps are overlaid, one atop another, the points of identity or similarity are such as to allow them to speak with one another without need of interpreters. They form a community with a common vocabulary and a set of shared assumptions about priorities and methods, problems and acceptable forms of solution. Some lines may be blurred—the specific coordinates of landmarks and measures of elevation and distance may differ somewhat—but they will agree on general contours; when asked how to go from point A to point B, they will all understand the question and suggest similar forms of answers. Likewise, they will generally agree if asked to identify the subjects or disciplines contiguous to their own, although they may have but limited knowledge of the terrain of those neighboring fields.

These mental maps define in significant—although not absolute—ways the possibilities open to scientists as creative scholars. They also determine in large part the extent to which workers in one discipline may be influenced by or exert influence upon those in others. But the boundaries defining categories of scientific thought and research are constantly changing, and features that loom large to one generation of scientists may seem remote or trivial to their intellectual descendants. These changes may occur in many ways. Scientists are continually adjusting their mental maps in small ways to conform with the results of their own research and that of their colleagues. Individuals, through accidents of training or experience, may come to see their disciplines in new relations to their neighbors, and they may then perpetuate those new perceptions through the students they train or through the institutions they develop. Major groupings of disciplines may dissolve and others may take form as new ideas or techniques alter scientists' perspectives on nature. And from novel vantage points, areas of ignorance may be discovered that lead individuals to explore new territories and to survey anew their own domain and its borders with neighbors. Subjects and disciplines in science are as time-bound as species in biology, and chemistry represents no exception to this generalization.[6]

Chemistry was not among the classical sciences of antiquity; it was a label attached to an area of ignorance discovered in early modern times. Its earliest practitioners lacked the cohesiveness and shared perspective of astronomers or anatomists. Chemists were concerned with the names and properties of substances, their genesis and corruption, and the instruments whereby transformations could be effected, but they differed over the contours, external relations, and goals of this science—or, better, this group of sciences that we today give a single name.

As late as the eighteenth century, relations between chemistry and other branches of knowledge were ambiguous in the extreme. During the century separating the productive years of Newton and Lavoisier, chemistry occupied an uncertain borderland between the two great constellations of sciences, natural philosophy and natural history.[7] For some chemists, especially those working within a Newtonian tradition, chemistry was firmly allied with other segments of natural philosophy. Together with Newtonian natural philosophers they assumed that matter was constructed from homogeneous corpuscles and that the diversity apparent in matter was the product of the multitude of ways in which corpuscles might be grouped together to form units of higher organization. They also shared the natural philosopher's goal of explaining the behavior of corpuscular matter in terms of attractive forces akin to those that Newton had used to account for the motions of celestial bodies. Their aim was a chemistry of forces and mechanisms. Other chemists, who did not share such speculative tendencies, or who hid them better, emphasized the links chemistry had with natural history. Like botanists, they saw their science as being fundamentally about classification. External properties and, later, internal composition were means for systematizing nature's abundant minerals and compounds. Between these two extremes, represented respectively by a John Friend and an A. S. Marggraf, there were many chemists, like Torbern Bergman, who were open to influences from both traditions.

A cursory inspection of eighteenth-century textbooks illustrates these ambiguities.[8] Some authors classified chemistry as part of natural philosophy, others did not. Some saw it as a science in the dynamical Newtonian sense, others as a subject closely allied with the systematic sciences of botany and mineralogy and the practical arts that had long been associated with natural history: medicine, pharmacy, and metallurgy. At the end of the century, the British chemist William Nicholson grappled with the problem of giving his subject a coherent definition in his *Dictionary of Chemistry*. The result was an awkward compromise:

We might define it negatively by affirming that every effect which is not purely mechanical is chemical; . . . chemistry, as a science, teaches the methods of estimating and accounting for the changes produced in bodies, by motion of their parts amongst each other, which are too minute to affect the senses individually; and as an art, we should affirm that it consists in the application of bodies to each other, in such situations as are best calculated to produce those changes.[9]

Just as there was an absence of consensus over goals and assumptions among chemists, so too there was an absence of uniformity in training. The apothecary shop, the mining academy, the medical school, and the study of natural philosophy within or without university walls might all serve as background to a chemist's avocation, but each in its own fashion.[10] The awkwardness of Nicholson's definition and the ambiguities involved in assigning chemistry a position vis-à-vis other branches of knowledge are understandable in view of the disparate paths leading to the abstruse art.

The boundaries of chemistry became clearer in the nineteenth century both to its practitioners and to a wider public. Conceptual changes were essential to this clarification. During the first decades of the nineteenth century, chemists went a long way toward developing a shared set of allegiances to certain techniques, theories, and goals. At the same time, social changes that were affecting all of the sciences were having an especially profound effect upon chemistry, which, because of its numerous contacts with the practical arts, had both enormous potential for growth and great need of new institutions for the creation and diffusion of knowledge. By mid-century, if not earlier, chemistry had attained the stature of an independent science, and chemists in many parts of Europe could aspire to professional status.

The rapid development of chemical methods and theory was, of course, intimately associated with the work of Lavoisier and Dalton. Lavoisier's nomenclature gave chemists a common language, just as his writings gave them a common set of assumptions: the sum total of matter in a chemical reaction is constant—a point Lavoisier's work underscored rather than first established; changes involving combustion, calcination, and respiration are instances of a general class of oxidation reactions; an element is a substance that cannot be decomposed into simpler constituents. Dalton in turn provided chemists with an ontological foundation absent in the work of Lavoisier by identifying each of the simple substances with a different species of indivisible atom. Tentative as much of it was, the work of Lavoisier and Dalton formed landmarks that chemists of later generations could not ignore.

Nor could chemists fail to be influenced by one unintended but nonetheless significant consequence of their work: a shift in the arena of debate from chemical forces and mechanisms to chemical units. Lavoisier had stressed weight as the quality of primary concern to chemists: unlike the forces of chemical affinity, the weights of substances participating in chemical reactions could be quantified readily. Dalton took this approach one step further by suggesting that the relative weights of atoms themselves could be measured. The determination of atomic or combining weights and, later, the application of atomic theory to organic compounds dominated the research of several subsequent generations of chemists. This program represented a sharp departure from the Newtonian tradition in chemistry. Not only did Dalton's atomic theory differ from the earlier corpuscular tradition by positing as many kinds of atoms as there were simple substances, it also suggested that the quantification of atomic weights was a more fruitful line of investigation than the quantification of chemical affinities. Throughout the nineteenth century, chemists could and did object to the atomic doctrine, but most conducted their research within the context established by it. This outcome was inimical to the Newtonian vision of a chemistry of forces.[11]

While Lavoisier and Dalton provided chemists with methods and ontological assumptions that gradually became common property, the development of universities and scientific journals afforded chemists increasingly uniform training and effective methods of communication. Although institutional changes affected chemists in many parts of Europe, nowhere were they as dramatic as in the German states. Already by the beginning of the century, chemical journals and university chairs had begun to proliferate. The first chemical journal appeared in 1778; by 1800 a half-dozen of these periodicals existed.[12] Some were little more than trade journals catering to the interests of apothecaries or mineral analysts, but others sought to build broader readerships by including contributions on many aspects of chemistry and by reporting on research done abroad. Salaried positions for chemists also expanded rapidly in the last decades of the eighteenth century. German schools and academies employed 18 chemists in 1750, but 48 in 1800.[13] Especially noteworthy was a trend toward the creation of chairs in the philosophical faculties of the universities. Unlike their counterparts in medical schools or mining academies, the occupants of these positions served a clientele with diverse needs and expectations. Chair-holders were not necessarily men of *Bildung*, but they had the opportunity to be more than teachers of practical skills.[14]

Slowed somewhat by the political and social upheavals of the early nineteenth century, the expansion of the institutional framework for chemistry proceeded apace after 1820. Of special importance was the rise of the teaching laboratory and research school. Laboratories had long been attached to academic chairs, where they served the personal needs of the professor, but during the first half of the nineteenth century they gradually evolved into instruments for grooming research scientists. Decisive was the success of Justus von Liebig, who, in the 1820s at the University of Giessen, molded together the elements of the first modern research school: a teaching laboratory, a journal, a set of techniques competent to generate significant new knowledge, and a body of students who, for reasons that are not entirely clear, were ready to participate in the process of creation.[15] With Giessen as a model, the laboratory gradually became an essential part of chemical training at other German universities, a counterpart to the philologist's seminar.[16]

Developments in Germany had effects elsewhere. After mid-century, Germany set the pace in chemical research and education. Its universities became international centers for chemical training; its journals and chemical institutes were admired, envied, and emulated abroad. Although some national and regional peculiarities endured, the trend in chemical education during the nineteenth century was toward greater homogeneity.[17]

Contemporary definitions of chemistry reflected this convergence of chemists' assumptions regarding goals, methods, and training. By the 1830s, clarity had replaced awkwardness in its description. Chemistry was "the science which treats the composition of substances and their relations to one another."[18] It was distinguished from physics, itself just emerging from natural philosophy, insofar as physics dealt with general laws applying to all matter and the forces operating upon mattter, whereas chemistry dealt with the behavior and constitution of specific substances.[19]

The common emphasis on the problems of chemical composition proved exceedingly fruitful during this period, as is best demonstrated by the spectacular advances made in the chemists' understanding of organic compounds. But something was lost as well as gained by this clear demarcation of chemistry's domain. As definitions of chemistry and physics grew more precise, connections between chemistry and fields such as heat and electricity became more tenuous.[20] The physicalist goal of understanding the forces and mechanisms involved in chemical change receded to the background as problems of analysis and—with the growth of organic chemistry—synthesis came to the forefront. By mid-century, the chemist and the physicist in leading European universities worked in different institutes, used different instruments, and measured different properties. Indeed, they even spoke different languages, for whereas the chemist needed only arithmetic to express weight relations, the student of physics was becoming ever more dependent upon the higher mathematics. Lothar Meyer, an acute mid-century observer, lamented these developments in his textbook of 1864, *Modern Theories of Chemistry*:

It cannot be denied that, by the acceptance and development of the atomic theory, chemistry became more and more estranged from the nearly related science of physics. The provinces of the two sciences were more sharply defined, each discipline pursued its own path and the portions common to both remained untouched, unless, as was frequently the case, they were appropriated by chemistry alone. Almost daily new relations between chemical and physical phenomena were discovered; but even the greatest discoveries made by the application of physical methods to chemical research did not, since the aim of each had become different, serve to reunite the now severed sciences.

It was now most important for chemists to prepare, study, and classify as large a proportion of those compounds the existence of which was predicted by the atomic theory. Thus chemistry assumed more and more the form of a descriptive natural science, in which theoretical speculations, . . . , became now of secondary import only.[21]

CHEMISTRY IN THE MINOR KEY: PHYSICALIST TRADITIONS IN NINETEENTH-
CENTURY CHEMISTRY

Chemistry developed largely in isolation from physics during the middle decades of the nineteenth century, but the physical chemistry of the 1880s was
not entirely lacking a heritage. As Meyer suggested, some chemists did pursue
research that may be called physico-chemical. They did not belong to a single
school of thought, nor did they develop a coherent body of doctrine; rather,
they studied a variety of topics, some of which would later be incorporated
into the physical chemistry of Ostwald and his contemporaries. Although the
scattered and diffuse nature of this material makes generalization hazardous,
much of the physico-chemical work of the mid–nineteenth century may be
classified into three broad categories: the development of physical instruments
and their application to the study of chemical composition, research on the
relation between physical properties and chemical composition, and the study
of the physical principles that governed the processes of chemical change.

Scientists whose work fell into the first two categories held aims that often
were relevant to those interested in conventional topics in analytical and organic chemistry, and several occupied positions of influence in the universities
of Germany. Isolated neither intellectually nor geographically, they were often well-known and respected by their contemporaries. Robert Bunsen, Hermann Kopp, and Hans Heinrich Landolt were three prominent chemists whose
work may be placed in these categories: Bunsen's in the first, Kopp's and
Landolt's in the second.

Of these figures, Bunsen had the most dramatic success. After building an
impressive record of research in organic chemistry at the University of Mar-
burg, Bunsen moved to Heidelberg in 1852, where he presided over the chemical institute for nearly forty years. He owed his reputation to experimental
skill and ingenuity in the creation of new instruments and techniques for chemical analysis. With a carbon-zinc battery of his own devising, Bunsen was able
to isolate relatively large samples of magnesium and the metals of the rare
earths, substances hitherto little studied because they could not be obtained in
workable quantities or stable form. His collaboration in 1859 with a colleague
in physics, Gustav Kirchhoff, resulted in the development of a second analytical tool: the spectroscope. The product of cooperation between a chemist and
a physicist when this was uncommon, spectrum analysis soon proved itself the
most sensitive of techniques for the detection of trace quantities of elements.[22]

Bunsen's labors on instruments such as these gave him a broad knowledge
of contemporary physics, and he is said to have stressed the value of physical
study in his lectures. But the bridge Bunsen threw up between chemistry and
physics was a narrow one. It was designed for the exchange of fact and technique, not for commerce in theory. Eschewing theory in his lectures and
publications, Bunsen discouraged students from straying far from the realm of

demonstrable fact.[23] His work led to the development of many valuable physico-chemical instruments, but it was directed toward the advancement of the purely chemical art of analysis.

Like Bunsen, Kopp and Landolt were accomplished experimental scientists rather than profound or original theoreticians. Their research focused on relations between the composition and physical properties of compounds, especially the organic compounds, whose makeup fascinated many of their contemporaries. Kopp, a student of Liebig and later a professor at Giessen and Heidelberg, hoped to demonstrate that all physical properties were simple functions of chemical composition. There was a key, he suspected, with which one could predict the physical properties of a substance from its composition and, conversely, identify a substance of unknown composition by its physical characteristics. Between 1839 and the late 1860s, Kopp studied the molecular volumes, boiling points, and heat capacities of a variety of compounds and defended the idea that these properties are essentially additive. The physical properties of a molecule, he suggested, are basically the sum of the physical properties of its atomic components. Thus, for example, Kopp maintained that lengthening an organic compound by adding methylene groups increases its boiling point by regular increments. The principles he formulated to describe these relations were completely empirical and, as it turned out, never entirely exact.[24]

Landolt, like Kopp, was the student of a grand master (Bunsen), held a series of distinguished positions (Bonn, Aachen, and Berlin), and sought to establish connections between physical properties and chemical composition. The property Landolt chose to investigate most fully was the refractive power of dissolved substances. In the early 1860s, just as Kopp was nearing the end of his productive years, Landolt began to study the influence of dissolved hydrocarbons on the transmission of light. He soon concluded that molecular refractivity is an additive property, that the refractive power of a molecule is the sum of the refractivities of its atoms. Although many exceptions were soon discovered, the generalization proved sufficiently accurate to give the measurement of molecular refractivities a place in the analysis and standardization of sugars, alkaloids, terpenes, camphors, and perfumes. But Landolt, like Bunsen and Kopp, was no theoretician. He never carried his studies further than was necessary to establish molecular refractivity as a useful analytical tool.[25]

Some, including several of the leaders of physical chemistry at the end of the century, have emphasized the continuities between, on the one hand, the work of Bunsen, Kopp, Landolt, and others interested in relating chemical and physical properties and, on the other, the dynamic physical chemistry of the 1880s and 1890s. Ostwald, for instance, repeatedly sought to garb himself in Bunsen's mantle. He used a portrait of Bunsen as the frontispiece of the first issue of the *Zeitschrift für physikalische Chemie*, edited his collected works, and wrote a popular biography of the grand master.[26] To be sure, Ostwald and his associates could and did draw upon the techniques and data of these chemists, but they also stood to gain much by identifying themselves with great names from the past. Like the proponents of any young enterprise, the physical chemists of the 1880s discovered that a distinguished ancestry might soften skepticism from without and assuage doubts from within.[27] Nevertheless the discontinuities between these older traditions and the new physical chemistry are more significant than the continuities. The invention of new instruments and analytical tools was not the principal goal of Ostwald and his colleagues, nor was the creation of a physical and quantitative basis for classifying chemical species. Rather, their chief aim was to resolve the problem of chemical affinity, and here the preceding generation offered fewer candidates suitable for canonization.

In this aim of understanding the chemical reaction itself, Ostwald had to look back to a third group of physico-chemical investigators, those who had been interested in studying the "hows" and "whys" of chemical change rather than identifying and fixing the relations of the "whats." With a few exceptions, these scientists were outsiders whose work was alien to the concerns of their contemporaries. They often came from countries or regions distant from established centers of chemical research. Few were influential teachers; most failed to establish enduring research schools. Ironically, most of the work of these outsiders had its antecedents in the contributions of a man who was, in his own time, most decidedly an insider, Claude Louis Berthollet— pupil of Macquer, colleague of Lavoisier, minister to Napoleon, and doyen of the French scientific community.[28]

Berthollet, at about the time Dalton was formulating his atomic theory, was following a very different line of reasoning. Born in 1748, Berthollet grew up during an era when Newtonian concepts of short-range attractive and repulsive forces dominated chemical discourse in France. His teacher, P. J. Macquer, had sought to specify rules governing the workings of elective affinities; his colleague, Guyton de Morveau, had gone so far as to quantify affinities by measuring the cohesive forces by which metals adhered to mercury.[29] Both believed chemical affinities to be absolute and both flirted with the idea put forward by Buffon that chemical attraction could be explained by an inverse-square law. Berthollet brought this Newtonian tradition to a halt, not by rejecting it, but by extending it in ways that made further development impracticable.

The immediate occasion for Berthollet's work on affinity was his visit to the Natron Lakes of Egypt in 1799, where he saw clear evidence that certain reactions did not proceed to completion in the presence of large quantities of their products. He recognized the implications quickly: chemical affinities were not absolute; mass as well as attractive force played a role in chemical combination and decomposition. In memoirs published during the next two years, and in his *Essai de statique chimique* of 1803, Berthollet developed this idea.[30] Adopting the Newtonian goal of explaining chemical phenomena through reference to forces, Berthollet postulated that chemical affinity was a force analogous to if not identical with gravity. Just as Newton had found the force of gravity to be proportional to the mass of an object, so too Berthollet suggested that the force of affinity was dependent on the mass of the reacting substance. In the competition between two substances to form a combination with a third for which they have unequal affinities, a large quantity of the substance with weaker affinity may possess an attractive force equal to or greater than the attractive force of a small quantity of the other substance. Nor were affinity and mass the only factors affecting chemical reactions. The physical state of the participants in a reaction also might affect the degree to which they exercise their affinities. Most reactions take place in solution, and if one of the participating species is extracted from solution, its effective concentration or active mass decreases. Insoluble and volatile substances remove themselves from solution and, to the degree to which they do so, decrease their active masses. This, Berthollet argued, helped explain why most reactions proceed to an end point at which all the reactants are converted into products. Changes of state—vaporizations and condensations—could remove products from the arena of reaction as quickly as they were formed, thus allowing the reaction to run until all reactants were exhausted. If it were not for such perturbing influences, Berthollet surmised, the outcome of chemical reactions would not appear to be so final and absolute.[31]

Berthollet's conception of affinity had an important corollary. Since, according to his view, each particle of matter attracts every other (although differences in this attractive force were acknowledged to be possible at close ranges because of differences in the sizes and shapes of corpuscles), combinations between particles in variable proportions are likely. Berthollet asserted that solutions were no less the product of chemical combination than compounds of definite proportions. Homogeneity was the distinguishing characteristic of chemical species. Indeed, he went further by suggesting that solutions should be considered the archetypical form of chemical combination; compounds of definite proportions, although common among solids and gases, were exceptions—the products of special circumstances or the chemist's intervention.

Berthollet's complex and sophisticated chemistry of forces enjoyed only a brief season of prosperity. His attack on the older conception of chemical affinity won support, but by 1820 his own theory had been relegated to the realm of the intriguing but speculative. The reasons are not far to seek. For one thing, he had made the tactical error of linking concepts that might best have been kept separate. Berthollet's supposition that variable rather than definite proportions were the prevailing mode of chemical combination became a lightning rod for opposition. His adversaries drew reinforcement from Dalton's new atomic theory, which lent the principle of definite proportions a suitable theoretical basis. A long and heated exchange of experimental results between Berthollet and the advocates of fixed proportions led many observers to conclude that Berthollet had seriously overstepped the limits of what could be claimed for his theory of mass action.[32] Equally significant was a flaw in Berthollet's program for quantifying chemical forces. He could offer no general method for measuring the absolute or relative affinities of substances. And because his system made the outcome of a chemical reaction dependent on both mass and affinity, it was impossible to predict the end points of chemical reactions. His elegant reasoning seemed barren of practical consequences.[33]

In the broadest sense, Berthollet's ideas fell victim to the inadequacy of techniques for measuring affinities and the poor fit between his conception of variable proportions and the emerging atomic theory. By deprecating the importance of definite proportions, Berthollet was de-emphasizing the significance of chemical composition at the very time its study was coming to dominate the interests of chemists. By making solutions paradigms of chemical combination, he was running counter to a rising tide of interest in the composition of gases and solids. Those committed to the study of composition could point to a growing battery of experimental methods for advancing their program; Berthollet had no corresponding set of methods for quantifying forces. Although his ideas received attention in textbooks, they led to little in the way of experimental work during the first half of the nineteenth century.

In the second half of the century, however, a number of scientists were led to reconsider Berthollet's ideas. Some were motivated by purely chemical concerns. Heinrich Rose of Berlin found Berthollet's discussion of mass action valuable in understanding solubility effects so important in analytical chemistry. The English chemist Alexander Williamson, while studying the mechanism whereby ether was formed from alcohol, called for a re-examination of Berthollet's program in papers published in 1850 and 1851. Not long afterwards, in France, Marcellin Berthelot and Pean de St. Gilles developed a method for studying the equilibrium involved in the esterification of acetic acid by ethanol.[34] Others were drawn to Berthollet's writings through the study of physical problems. In the late 1850s and 1860s, work on the behavior of gases at high temperatures suggested that mass played a role in the dissociation of diatomic and polyatomic molecules into their atomic constituents.[35]

Two Norwegians, Cato Guldberg and Peter Waage, consolidated and extended this work on mass action in a series of papers that began to appear in 1864.[36] Friends since youth and linked through marriage to sisters, Guldberg and Waage existed on the periphery of the European scientific community. Guldberg taught mathematics at the Royal Norwegian Military Academy; his brother-in-law was a chemist at the University of Christiana. They were steeped in the writings of Berthollet and were familiar with the research done on rates of reaction by Berthelot and St. Gilles, but they themselves were practically unknown to scientists outside Scandinavia and remained so until the 1880s.

It is not clear how Guldberg and Waage became interested in the problem of chemical affinity, but it is certain that by 1867 they were convinced that chemistry should become, like mechanics, a science of forces and their effects. The first step toward making it such would be the development of a mathematical theory of chemical affinity.[37] Their starting point was not far removed from where Berthollet had left off sixty years earlier. They postulated that the affinity between substances might best be described as a chemical force that was proportional to the product of the active masses of the substances entering into a chemical reaction. And, like Berthollet, they adopted the view that reactions do not always run to completion in the sense that all reactants are transformed into products. Rather, some chemical processes result in a condition of equilibrium, a state in which the force of the forward reaction is precisely balanced by that of the reverse reaction. Unlike Berthollet, Guldberg and Waage developed a mathematical expression to describe the state of equilibrium and tested it using the techniques and data of Berthelot and St. Gilles.

Their first formula, presented in their paper of 1864, was littered with arbitrary constants; it could not have made much impression even on those chemists able to read the Norwegian in which it was written. Selecting a double decomposition reaction of the form

$$A + B = A' + B',$$

they described equilibrium in the following terms:

$$k(p - x)^a(q - x)^b = k'(p' + x)^{a'}(q' + x)^{b'},$$

where the constants k, k', a, a', b, and b' all depended on the nature of the reaction, p and q were the initial masses of the reactants (equivalent to the initial concentrations of A and B), p' and q' were the initial masses of the products, and x stood for the amount of each product formed in the reaction.[38] Unwieldy as this formula was, it constituted the first general expression of equilibrium to prove effective in accounting for experimental data. In later papers they refined the equation to a form more familiar today:

$$kpq = k'p'q'.$$

In a reaction at equilibrium, the product of a constant, k, and the concentrations of the reactants is equal to a second constant, k', times the concentration of the products.[39]

Guldberg and Waage were well aware of the potential significance of their work, but they also recognized that few chemists were prepared to share their outlook. "Investigations in this field," they wrote in 1867, are doubtless more difficult, more tedious, and less rewarding than those which now engage the attention of most chemists, i.e., the discovery of new compounds. It is our opinion, however, that nothing can so soon bring chemistry into the class of the truly exact sciences as just the researches with which this investigation deals . . . a branch of chemistry which, since the beginning of the century, has unquestionably been far more neglected than it deserves.[40]

Their papers, however, led to no immediate upsurge of interest in mass action. In 1869 the Danish chemist Julius Thomsen devoted a section of a memoir to their theory; and in 1873 the German physicist August Horstmann found that their law could be applied to the dissociation of gases.[41] But a dozen years passed before Ostwald systematically tested Guldberg and Waage's conclusions.

The linguistic and geographical isolation of these scientists and their free use of mathematics militated against the rapid incorporation of their work into chemists' discourse. Equally important, the principle of mass action at first appeared to have but few uses. Guldberg and Waage had been able to apply their equation to only a handful of reactions, like esterification, in which both the reactants and products were present in measurable quantities at equilibrium.[42] Most chemical processes do not fit this description. Nor did Guldberg and Waage's results throw much light on the nature of chemical affinity. They accepted it as a given, an attractive force inherent in matter and specific in its deployment.

By focusing on mass, Guldberg and Waage were taking a detour around a larger question—the origins of that power which drew certain substances into combination. Some of their predecessors and contemporaries, however, had and were confronting the issue directly. It had long been known, for example, that the passage of an electric current through a liquid conductor is accompanied by chemical changes; a long line of chemists and physicists had studied this phenomenon during the years following the invention of the Voltaic pile in 1800. No less a chemist than Berzelius, impressed with the power of the electric current to decompose the most durable of compounds, was led to conclude that chemical affinity was a manifestation of electrical forces between the particles of matter. Likewise, it had been recognized at least since the time of Lavoisier and Laplace that most chemical reactions are accompanied by the evolution of heat. But efforts to equate chemical affinity with the force of electrical attraction or to gauge it by measuring the thermal effects of chemical change met with failure in the nineteenth century. Berzelius's electrochemical theory could not explain how such electronegative elements as chlorine could substitute freely for electropositive hydrogen in organic compounds; nor could it explain Faraday's finding that the same quantity of electricity is always required to decompose an equivalent of any compound regardless of the force of the affinities to be overcome. Attempts by Julius Thomsen and Marcellin

Berthelot during the 1850s and 1860s to use thermal effects as a measure of affinity likewise were embarrassed by recalcitrant facts. Both believed that every chemical change completed without the interposition of external energy will always tend to the production of that body liberating the maximum quantity of heat—an idea that Berthelot called the principle of maximum work. Most spontaneous chemical reactions are exothermic and hence fit his generalization; but some absorb heat, and efforts to find external sources of energy, while often ingenious, were not fully persuasive.[43]

By 1870, sufficient evidence was in print to link mass, electricity, and heat to chemical affinity. But few, if any, chemists had full command of the literature on all of these topics, and no one had developed a conceptual framework into which the evidence could be incorporated. Guldberg and Waage had proved that mass could alter the position of chemical equilibria and had found ways to quantify its effects, but they did not define affinity itself. It was known by all chemists that electric currents could counteract the power of affinity under certain circumstances, but it was impossible to equate electrical attraction and affinity in any simple way. And Thomsen and Berthelot, while showing that the heat evolved in chemical reactions might serve as a rough index to the affinity between reacting substances, had not been able to prove that this was a precise and universal relation. Affinity studies did not represent a coherent set of problems, nor did they command the attention of a network of scientists in close communication with one another.

The geographical remoteness and personal styles of these chemists did little to foster productive contacts. Guldberg and Waage left no trace of a research school; working in Oslo, they could hardly have been expected to attract many advanced students. Julius Thomsen's position in Copenhagen was hardly much better, and he was, in any case, reluctant to supervise students.[44] Berthelot had disciples in Paris and enjoyed the attention of his contemporaries, but was renowned as much for his authoritarian ways as for his wide-ranging knowledge; those who disagreed with him did so at hazard to their careers.[45] The prosperous research schools of the mid–nineteenth century were to be found in Germany, and none of those took the study of chemical affinity as central to its research.

Poor communication among chemists and the weakness of the institutions in which affinity studies were prosecuted together comprise only part of the reason why such research lacked coherence. Equally important was the isolation of chemists from physicists. Tools with which chemists might have clarified and unified their findings already existed in the concepts of work and energy and the principles of the conservation and dissipation of energy. These were sufficient to provide a theoretical basis to the study of chemical affinity and its relations with heat and electricity. But they were applied to chemistry only slowly. Thermodynamics was largely in the hands of physicists. The few chemists who sought to wield this tool, like Thomsen and Berthelot, did so in a clumsy manner. And when physicists themselves began to apply thermodynamics to chemistry in the 1870s, their results were, for the most part, ignored by chemists.

The work of the American J. Willard Gibbs is the outstanding example of how the thermodynamics of the 1870s could be tapped to clarify the relations between chemical, electrical, and thermal phenomena. Working in splendid isolation at Yale University, Gibbs used the two laws of thermodynamics and his own concept of chemical potential (obtained by differentiating the energy of a system with respect to the mass of any one of its component substances) to develop a system of chemical thermodynamics that was very nearly complete. His equations defined the condition of equilibrium in homogeneous and heterogeneous systems and expressed both the driving force of chemical reactions and the electromotive force of galvanic cells in terms of changes in chemical potential. Using his system, it was possible to derive expressions describing the effects of gravitational and electrical forces, mass, temperature, and pressure on systems at equilibrium. His work, in short, contained precise answers to questions that chemists were only beginning to formulate.[46]

But if Gibbs's memoirs show how fruitful thermodynamics could be when applied to chemistry, their reception bespeaks the distance that separated chemists and physicists in the 1870s. Publishing in the little-known *Transactions of the Connecticut Academy of Arts and Sciences*, Gibbs barraged the leading scientists of Europe with reprints.[47] The papers were read and appreciated, at least in part, by some of the physicists who received copies—in particular, by James Clerk Maxwell and his students. But they were utterly neglected by chemists on his mailing list, a group that included Thomsen, Berthelot, Kopp, and Bunsen. Gibbs's style was that of a theoretical physicist: concise, mathematical, and abstract. Both the structure of his thought and his system of notation were unfamiliar. Moreover, he had made little effort to describe the experimental implications of his work; chemists, in turn, were unprepared to appreciate those implications for themselves. As a result, Gibbs's ideas had negligible influence on chemistry until after 1890, and even then his work was cited more often than it was studied.

During the mid–nineteenth century, then, chemists and physicists were following several lines of experimental and theoretical investigation related to the study of chemical processes. It should be emphasized, however, that much of this work assumes significance only with the benefit of hindsight; its consequences became manifest only toward the end of the century. The memoirs of scientists like Guldberg, Waage, Thomsen, and Gibbs were as foreign to the central concerns of contemporary chemists as those figures were distant geographically from the citadels of chemical research. In the 1860s and 1870s, the surest avenue for chemical advance seemed to be the one followed by Kekulé, Kolbe, Baeyer, Wislicenus, and Hofmann, that is to say, work on composition, structure, and synthesis.

In the 1880s this view was challenged by a group of chemists who asserted that their science, through its emphasis on composition and structure, threatened to become narrow and sterile. The indiscriminate multiplication of examples of the synthetic art was an unending task, they argued, since an indefinitely large number of compounds would always remain to be discovered. And such generalizations that resulted from this work formed a poor substitute for a set of principles yielding predictions regarding all chemical change. These chemists integrated earlier work on chemical affinity and on the relation between physical and chemical properties into a new framework in the hope that the product might form the basis of a science of general chemistry.

AMBITIONS AND IDEAS CONVERGENT: OSTWALD, VAN'T HOFF, AND ARRHENIUS

Jacobus Henricus van't Hoff, Svante Arrhenius, and Wilhelm Ostwald founded the discipline of physical chemistry. They did not invent the name, nor did their ideas lack precursors; their theories were not without flaws, and their conception of physical chemistry was not without ambiguities and internal tensions. Nevertheless, these three chemists created the discipline. Etymologically, discipline is derived from the Latin *disciplina*, a word meaning instruction or training—not so much in a body of doctrine as in a mode of conduct or in a set of methods that could be applied to one or several lines of inquiry. Hence, ancient schools of philosophy and rhetoric were known as *disciplinae*. When applied to a science in modern times, the word implies a certain coherence of aims and techniques maintained through transmission from teacher to student. The content of the science may change over time, but there is a genetic link among practitioners, the methods they use, and the questions they ask. To maintain this continuity, institutions are necessary: schools in which techniques are practiced and honed; societies in which workers meet and renew their ties; journals in which results are exposed to public scrutiny. Van't Hoff, Arrhenius, and Ostwald created physical chemistry in this sense. They systematized their predecessors' work and developed new concepts and techniques that served as growing points for subsequent research. They brought physical chemistry to the attention of their contemporaries, both through unremitting propaganda and through impressive demonstrations of its utility. And, together with their students, they founded the laboratories and journals through which the new discipline grew and prospered. They gave coherence and continuity to what had been diffuse and sputtering traditions of study.

The members of this triumvirate emerged from obscure beginnings: each grew up in a community on the periphery of German scholarship and science. Van't Hoff, born in 1852, was a native of Rotterdam and an organic chemist by training. A dreamy young man with a taste for Comte's philosophy and

Byron's poetry, he had exhausted the educational resources of his native country by the time he was twenty and then had gone to Germany and France for schooling in the intricacies of organic chemistry. At the age of twenty-two he wrote a startlingly original essay on the three-dimensional structure of the carbon atom. It was a fundamental contribution to stereochemistry, but it earned him much controversy and few immediate rewards. Jobless for nearly two years following its publication, van't Hoff resorted to teaching chemistry and physics at a veterinary school before finally landing a position in 1878 as professor of chemistry, mineralogy, and geology at the newly constituted University of Amsterdam.[48] Arrhenius, seven years van't Hoff's junior, received all of his training in Sweden, working under both chemists and physicists at the University of Uppsala and the Hogskola in Stockholm, an appendage of the Swedish Academy of Sciences. His doctoral dissertation had been received with indifference by the examining board at Uppsala, and his future as a scientist appeared bleak until the third member of this group, Ostwald, exercised his influence to gain him a position.[49] Ostwald himself was, as we have seen, the product of the German community in Russia's Baltic provinces. After study at the University of Dorpat, Ostwald accepted a position at the Riga Polytechnicum, and in 1887 was called to a chair at Leipzig. He later claimed that his early years on the periphery of German scientific life were a decisive factor in his career, saying that if he had studied at a larger and better-known university, "I probably would have fallen wholly under the influence of the chemical greats of the time, of a Baeyer for example. These men devoted themselves completely, without exception, to organic chemistry. . . . I, too, undoubtedly would have become an organic chemist."[50]

Neither Ostwald nor Arrhenius became organic chemists, and van't Hoff, although beginning his career in organic research, did not long remain satisfied with it. Instead, these men took up a constellation of problems relating to chemical affinity: the effects of mass and temperature on chemical equilibrium, the phenomenon of osmotic pressure, and electrolytic conduction. At the outset, their studies and perspectives were as little connected as those of earlier workers on these topics. Gradually, however, their paths converged until their lines of thought were at times duplicating and at times complementing one another. In the mid-1880s they discovered, much to their surprise, that the work they had been doing fit together to form a whole. Each had been led to adopt a similar outlook on his science; indeed, each had independently come to the conclusion that an understanding of the behavior of matter in solutions was the key to a science of general chemistry.

Ostwald, as we have noted, was already keen to effect reforms in chemistry when a student at Dorpat in the 1870s. In his earliest papers he adopted the position that the study of affinity should constitute the primary task of chemists. The proximate cause of his early and lively interest in this subject appears to have been two of his teachers at Dorpat, Karl Schmidt and Johann Lemberg.

Schmidt, a former pupil of Liebig, Wöhler, and Heinrich Rose, was the ordinarius in chemistry; Lemberg his principal assistant. Both were interested in geological chemistry. Schmidt confined his research largely to the analysis of mineral waters, but his younger colleague was far more interested in the natural processes whereby minerals took form and decomposed.[51] Seventy-five years earlier, geological phenomena had spurred Berthollet to study the influence of mass on chemical reactions, and Lemberg, in his teaching, found ample opportunity to discuss Berthollet's concepts of chemical equilibrium and mass action. "He emphasized," Ostwald later recalled, "that there was no absolutely insoluble material, no absolutely complete reaction, indeed, nothing absolute in nature."[52]

Encouraged by Lemberg, Ostwald made mass action the subject of his dissertation research. During his studies, Ostwald read a paper written in 1869 by Julius Thomsen, "On Berthollet's Theory of Affinity." It was a formative encounter. Not only did Thomsen discuss the work of Guldberg and Waage, thus alerting Ostwald to the neglected papers of these Norwegians, but he also showed how thermochemical data could be used to measure the relative affinities of certain pairs of acids and bases. Different acids, on neutralization with the same alkali, evolve different amounts of heat. When one acid displaces another from its compound with a base, the decomposition of the salt is accompanied by a thermal change. If the heat of neutralization of the free acid is greater than that of the acid bound by the salt, then heat is evolved; if less, then heat is absorbed. Knowing the value of the thermal effect, it is then possible to calculate the degree of decomposition of the salt and the ratio of the affinity constants of the two acids for the base. Thomsen's method was of considerable importance because Guldberg and Waage, while developing an equation describing the relationship between mass and affinity, had failed to provide guidance on how to apply their expression to a wide variety of reactions. The problem their work presented was that of measuring the concentrations of participants in a reaction at equilibrium when traditional methods of chemical analysis displace the equilibrium by removing a species from solution. Thomsen believed that he had overcome this difficulty by measuring concentrations indirectly. His technique permitted the equilibrium condition to be studied without altering the composition of the solution. More important, it suggested to Ostwald that other physical methods might be used for the same purpose.[53]

The chemical laboratory at Dorpat lacked equipment for precise calorimetric measurements, but Ostwald turned this deficiency into an advantage. He conceived the idea that changes in volume accompanying chemical reactions might also be used to evaluate affinity constants. A series of experiments on the systems Thomsen had studied showed this to be the case. Moreover, Ostwald found that his technique was far simpler and more general than the delicate and time-consuming method of Thomsen. Later, in his doctoral dissertation, Ostwald used changes in the refractive index of solutions for the same purpose, and still later took up the measurement of reaction velocities. His aims were to correlate the values of affinity constants derived by a variety of methods and to settle on the most accurate, general, and simplest techniques for obtaining such data. Although he was measuring some of the same physical properties that researchers like Kopp had studied earlier, he was making these measurements with the problem of affinity clearly in view.[54]

Ostwald published the results of this work in a series of papers during the late 1870s and early 1880s, but even so he had to struggle after earning his degree. His energy and originality could not entirely compensate for a provincial background; his research was not on a topic that was likely to catch the attention of the directors of chemical institutes of major universities. For three years he taught in a Realschule while working as an assistant to the professor of physics at Dorpat, Arthur von Oettingen. In 1881, he marginally improved his situation by accepting a professorship at the Riga Polytechnic Institute, a young institution designed to meet the needs of an increasingly industrial city.

Although not much of a prize, Ostwald made the best of this opportunity. He commenced work on a textbook that would synthesize existing work on topics situated between chemistry and physics, the *Lehrbuch der allgemeinen Chemie*. He also continued his research on chemical affinity and began to attract students to his laboratory. It is testimony to his effectiveness as a teacher and his growing reputation as a scholar that the rector of the Riga Polytechnicum obtained funds to build a new laboratory for his use.[55] Recognition also began to come from abroad. The British chemist M. M. Pattison-Muir singled out Ostwald's work for special praise in a review of research on chemical affinity in 1879, and during the 1880s references to Ostwald began to appear in German textbooks. The 1883 edition of Lothar Meyer's *Die Modernen Theorien der Chemie*, for instance, gave generous attention to Ostwald's physical methods of determining affinity constants.[56] Ostwald's growing reputation, his productivity in the laboratory, and the appearance of the first volume of his textbook together prompted the offer of a chair at Leipzig University in 1887. To be sure, it was the second chair of chemistry (the first belonged to the organic chemist Wislicenus), and it brought with it the task of supervising the laboratory studies of pharmaceutical students. Others, including van't Hoff and Landolt, had already declined to accept the call. But for Ostwald the offer was a godsend. The Russian government was beginning to obstruct the careers of German-speaking academics in the Baltic region; this program of Russification would eventuate in a purge of professors of German origin.[57] More important, Leipzig offered Ostwald a stage on which he could make the most of his ample talents as a research director and propagandist. Leipzig was a major university with a distinguished faculty and a commitment to research—everything Riga was not. He accepted the invitation with alacrity.

By 1887 Ostwald's career was well under way. But despite the professional success, his research was not living up to its original promise. He exhibited energy at the laboratory bench and the writing desk but had not advanced much beyond his original insight, that a variety of physical methods could furnish affinity constants. Ostwald had not discovered a single technique for measuring affinities that was at once reliable, efficient, and general. He had come to doubt the accuracy of his optical method, and measurements of specific volume, while an improvement on Thomsen's calorimetry, were still cumbersome. His contributions, although valuable, hardly scratched the surface of the problem of chemical affinity. Nor had he been able to give much thematic unity to his mammoth textbook of general chemistry. But answers to Ostwald's needs were just becoming available when he moved to Leipzig. They came from unexpected quarters: Holland and Sweden.

Ostwald had confronted the problem of chemical affinity in a direct and bull-like fashion, but his two future associates, van't Hoff and Arrhenius, approached it obliquely. Of these three scientists, van't Hoff was the deepest thinker and had the most complex personality and style. He was of a retiring disposition and could be taciturn with colleagues and students. He had none of Ostwald's facility of expression. Indeed, his writings resemble lecture notes: a profusion of headings and subheadings, short paragraphs, terse sentences, and abrupt transitions from subject to subject. He did not collaborate in the laboratory as did Ostwald and Arrhenius; indeed, he seems to have been jealous of his solitude. Nor was he one to form conclusions during the free give-and-take of conversation—his style was to brood over questions, for hours, days, and sometimes years.[58] He did a great deal of experimental work, but published only that portion of it that bore directly on his theoretical concerns.[59] He was, above all, a theoretician who took pride in his imagination and his ability to treat broad classes of phenomena by means of a few economical symbols and words. He died of tuberculosis just short of his sixtieth birthday—as if in confirmation of Thomas Mann's correlation of creative instinct and consumption.

Van't Hoff's work does not lend itself to easy summary. Both his career and his research appear to fall into three discrete stages. The first extended from his work on the asymmetric carbon atom in 1874 to the publication of the second volume of his textbook on organic chemistry, *Ansichten über die organische Chemie*, in 1881. During these years he worked as a private tutor, taught at the veterinary school in Utrecht, and was appointed to a professorship at the University of Amsterdam. The second phase of his career was spent entirely at Amsterdam, where he studied the rates of chemical reactions and applied the laws of thermodynamics to the equilibrium condition, to the problem of affinity, and to the behavior of solutions. In the final stage of his career, lasting from 1896 until his death in 1911, van't Hoff held a research appointment at the Prussian Academy of Sciences in Berlin and worked toward unraveling the complex solubility relations of the compounds found in the salt beds of Stassfurt, the richest source of potash in western Europe.

Convenient as this periodization is, it obscures continuities in van't Hoff's development as a scientist. Running as a thread through all of his work was the theme that much—though not all—of chemistry could be reduced to physics. He appears to have absorbed this attitude as a young man, when, prior to his graduate work in chemistry, he had read the writings of Auguste Comte. Comte had stressed the primacy of mathematical physics as a model for the development of other sciences and had pointedly criticized the qualitative or, at least, nonmathematical character of contemporary chemistry. Van't Hoff, according to his student and biographer, Ernst Cohen, kept these lessons before him throughout his career.[60] He acquired training in mathematics and physics before leaving Holland for Germany; he consistently sought to frame questions that were amenable to mathematical treatment; and he repeatedly looked to the work of physicists for guidance in the treatment of chemical problems. Although it would be an exaggeration to ascribe van't Hoff's choice of research topics to Comte's influence, Comte nevertheless appears to have inspired the young Dutchman with a sense of direction and purpose.

The first problem to which van't Hoff applied himself was that of clarifying the relationship between the constitutional formulae and properties of organic compounds. The structural theory he studied as a student was a flat, two-dimensional affair. The existence of two optically active forms of compounds with identical constitutional formulae had suggested to several chemists that the spatial arrangement of atoms in molecules might affect their physical and chemical properties, but none had visualized how and why these differences existed; the shapes of molecules were generally taken to be inaccessible to study. In 1874 van't Hoff, twenty-two and without a publication to his credit, wrote a concise and remarkably imaginative paper in which he used the evidence of optical activity to develop a three-dimensional model of the carbon atom. The affinities of carbon, he argued, are directed toward the corners of a tetrahedron, with the carbon atom itself at the center. If this model were correct, then carbon atoms linked to four different univalent groups might exist in two mirror-image forms, and all carbon compounds that rotated the plane of polarized light when in solution should possess an asymmetric atom. Van't Hoff showed that existing evidence was entirely consistent with this theory and further demonstrated that when compounds lost their asymmetry they also lost their optical activity. In brief, he demonstrated that a simple hypothesis regarding the spatial arrangement of atoms could explain the properties of a host of particular substances.[61]

In his second major work, the *Ansichten über die organische Chemie* (1878–1881), van't Hoff sought to extend his treatment of the relationship between constitutional formulae and physical and chemical properties by quantifying the forces between atoms.[62] The effort was a failure in almost every sense. Like Berthollet, he began by assuming that chemical affinity was a force akin to and perhaps identical with universal gravitation. Unlike Berthollet, however, van't Hoff tried to develop equations to describe the magnitude of this force as it was exerted between atoms and as it was affected by atomic size and shape. It was a hopeless task, not least of all because van't Hoff had no knowledge of the absolute size or mass of atoms and hence had to couch his analysis entirely in terms of hypothetical quantities. The book appears to have had few readers, and van't Hoff himself later described it as ''hardly worth knowing.''[63] But it was of value to its author, for, in writing it, van't Hoff came to recognize an important deficiency in chemistry—the absence of a theory explaining why the rate of chemical change varied markedly depending upon the substance involved. It was not at all clear, for instance, why it was easier to oxidize certain substitution products of methane, such as methanol, than methane itself, or why there was a direct correlation between the ease of oxidation and the number of oxygen atoms already present in a molecule. This question was the germ of his next book, the *Études de dynamique chimique*.

As the title implies, the *Études* was more than a study of the theory of reaction rates, although this did comprise the first half of the book. Rather, it was a broadly conceived examination of many of the same problems that Berthollet had treated in his *Essai de statique chimique*, approached from a new perspective—that of the kinetic theory of matter and thermodynamics rather than that of gravitational forces and static equilibria.

The *Études*, whether viewed from the perspective of the principles it enunciated or from that of the methods used in developing those principles, is one of the richest books in the history of chemistry. Van't Hoff's analysis of rates of reaction, for example, was simplicity itself. He began by classifying reactions into categories based on the number of molecules participating. Then, using data that was for the most part already available in the literature, he showed how, for each category, a simple equation could express the relation between the rate with which the concentration of reactants changes and the elapse of time. His general expression for a reaction involving n molecules, present in the proportions in which they take part in the reaction, was

$$- dC/dT = kC^n,$$

where C is the concentration, T is the temperature, and k represents what van't Hoff called the velocity constant. Using this relation, together with empirical data on the rate of a particular reaction, it became possible in principle to draw conclusions regarding the mechanism of the change.[64]

Important as this was to the study of chemical kinetics, van't Hoff hardly paused to consider the consequences, for in writing the *Études* his original purpose was overshadowed by a far grander scheme. In the last half of the book, van't Hoff, in rapid-fire style, attacked one major problem after another:

What is the nature of chemical equilibrium? How do changes in temperature, pressure, and mass affect the position of equilibrium? How may affinity be defined and quantified? And how does affinity relate to the thermal and electrical effects that often accompany chemical change? His results constituted a unification of data, techniques, and ideas that had been evolving separately for many years; they were cemented together by van't Hoff's ample intelligence and imagination and liberal use of thermodynamics.

Van't Hoff was not the first scientist to apply the principles of thermodynamics to chemistry. With the exception of the much-neglected Gibbs, however, van't Hoff's predecessors had been tentative in their use of the tool or had applied it to rather narrowly defined problems. Among the papers to which van't Hoff had access, a memoir written in 1873 by August Horstmann of Heidelberg was the most significant.[65] Horstmann, a student of Bunsen, Clausius, and Landolt, had used the Clausius-Clapeyron equation for the vaporization of a liquid to develop an expression describing the equilibrium condition in the case of gaseous dissociation:

$$Q/T + R \ln(p_1/p_2) + C = 0.$$

The pressures of dissociated molecules, p_1 and p_2, are dependent on the absolute temperature T and the heat of dissociation Q (R and C are constants). His result was understood by his colleagues to be a significant contribution to the debate over Avogadro's hypothesis in that it explained why the density of a gas is not constant at low and moderate temperatures.[66] Important as this finding was, however, it came as the answer to a particular question; Horstmann did not press much further in his application of thermodynamics and neither did the readers of his paper, with the exception of van't Hoff. The reasons are not hard to fathom. It took considerable knowledge of physics and not a little creativity to fashion links between the general and abstract laws of thermodynamics and the experimental data of chemistry. Few of Horstmann's fellow chemists had the advantage of having studied under both a Clausius and a Bunsen.

Van't Hoff did take Horstmann's analysis further. He was able to do so in part because of his training in physics and mathematics, and in part because he defined chemical phenomena in such a way as to make them accessible to thermodynamic treatment. The law of mass action, for example, had not been linked with thermodynamics by Guldberg and Waage. They had understood affinity to be a force and the equilibrium condition to be a static balance of forces. In the *Études*, van't Hoff gave a different interpretation to both concepts. He defined equilibrium in dynamic terms. In any reversible change, equilibrium results when the rates of the forward and reverse reactions are equal. The same expression that Guldberg and Waage had used could still be applied:

$$kpq = k'p'q',$$

where p and q are the masses of the reactants, p' and q' are the masses of the products, and k and k' are constants. But these constants were now understood by van't Hoff to be velocity constants rather than affinity coefficients. To state the principle in the terms van't Hoff used in the *Études*, at equilibrium,

$$C_2{}^{n_2}/C_1{}^{n_1} = k_1/k_2 = K,$$

where C_1 and C_2 are the concentrations of the two systems, n_1 and n_2 are the numbers of molecules in each of the systems taking part in the reaction, k_1 and k_2 are the velocity constants of the forward and reverse processes, and K is what van't Hoff called the equilibrium constant.[67] One could determine the equilibrium constant of a reaction either by measuring the concentrations of the reactants and products at equilibrium or by gauging the rates of the forward and reverse processes.

Van't Hoff was not the first to state the law of mass action; Guldberg and Waage had done this. He was not the first to give the equilibrium condition a dynamic interpretation; Williamson had done so in 1850, and others, including Horstmann, had thought of it in these terms. Nor was he the first to suggest ways to measure equilibrium constants; Ostwald and others had directed their work toward this goal. Nevertheless, van't Hoff was the first to draw these results together and to show how they reinforced one another.

To this point in the *Études*, van't Hoff's treatment of chemical change rested upon kinetic reasoning and experimental findings. He found these resources inadequate, however, to his next goal—that of showing how the position of equilibrium in a chemical reaction depends upon temperature.[68] Here he had recourse to thermodynamics, and in particular to Horstmann's paper of 1873. Without bothering to provide a derivation, van't Hoff simply adapted Horstmann's equation,

$$Q/T + R \ln(p_1/p_2) + C = 0,$$

for use as a general expression of the equilibrium condition. He did so by taking Q to represent the heat of the reaction rather than the heat of dissociation, and by substituting the natural logarithm of the equilibrium constant K for the term $R(\ln p_1/p_2)$, so that

$$Q/T + \ln K + C = 0.$$

The introduction of the equilibrium constant into Horstmann's equation involved an unstated assumption—that an analogy could be drawn between the vapor pressures of gases and the concentrations of solutes in solution, or

$$R(p_1/p_2) = C_2{}^{n_2}/C_1{}^{n_1}.$$

Relying upon this analogy, it was then a simple matter to introduce the equilibrium constant, since, according to his statement of the law of mass action,

$$C_2{}^{n_2}/C_1{}^{n_1} = k_1/k_2 = K.$$

With a general thermodynamic expression in hand by which he could describe the equilibrium condition of any chemical system, it was then a straightforward matter to develop an equation describing how the equilibrium constant should change with temperature. Van't Hoff took the derivative of his general equation with respect to the absolute temperature and then simplified to reach the following result, later called the equation of the reaction isochore, since it assumed that the reaction took place at constant volume:

$$d(\ln K)/dT = Q/2T^2.$$

This was a powerful expression indeed. It made it possible, for instance, to calculate the value of Q, the heat of a reaction, from observations of the equilibrium such as those Ostwald was making. Moreover, by its use van't Hoff was also able to show that the principle of Berthelot and Thomsen—all reactions proceed with the evolution of heat—is not valid except at absolute zero. Instead, reactions obeyed what van't Hoff called the principle of mobile equilibrium: "Every equilibrium between two different conditions of matter [systems] is displaced by lowering the temperature, at constant volume, toward that system the formation of which evolves heat." Hence, the heat of reaction was not a direct measure of affinity.[69]

If heat was not a direct measure of affinity, what was? As in the case of his treatment of chemical equilibrium, van't Hoff approached an answer by defining his terms in such a way as to make the problem accessible to thermodynamic treatment. Thus he proceeded by defining affinity in terms of its effects—in particular, the work that it performed. And the work of affinity, he argued, was equivalent to the quantity of work done by a chemical process when it took place in a reversible way. By following a series of reversible changes by which one system was transformed into another and returned to its original state, van't Hoff was then able to derive an expression for the work of affinity, A, that, although not general, was valid for systems in which there is a temperature P at which $A = 0$:

$$A = Q(P-T)/P,$$

where Q is the heat produced by the transformation, P is the absolute temperature at which $A = 0$ (the transition point), and T is the absolute temperature at which the reaction occurs.[70]

In the final pages of the *Études*, van't Hoff went on to suggest ways in which the value of A might be calculated from vapor pressures, as in the case of the affinity of a salt for its water of crystallization, or from osmotic pressures, as in the case of the affinity of a solute for a solution. Most important, however, he noted that his expression for the work of affinity could also be understood to represent the electrical work that a chemical change is capable of producing.[71] Helmholtz, in 1882, had shown that a chemical change that occurs in such a way as to do electrical work is reversible: the electromotive force produced by the reaction is the same as that required to reverse it. By taking his expression for the work of affinity and differentiating with respect to temperature, van't Hoff was able to obtain an equation for electrical work that Helmholtz had developed by other means in his paper of 1882:

$$A = Q + T(dA/dT).$$

The electrical work of a reversible galvanic cell, A, is equal to the heat of the reaction Q plus the product of the absolute temperature T and a temperature coefficient, dA/dT. And since the electrical work is proportional to the electromotive force in such cells, the determination of an electromotive force could be considered a measure of the work of affinity. Van't Hoff had, in short, presented a set of expressions that related heat, electromotive force, and chemical affinity.

In a paper published shortly after the *Études*, van't Hoff took one additional step.[72] By combining the equation for the electrical work of a reversible cell, $A = Q + T(dA/dT)$, and his equation of the reaction isochore, $d(\ln K)/dT = Q/RT^2$, and integrating, van't Hoff obtained

$$RT \ln K = -A.$$

The circle was closed. When measuring the electromotive force produced by a reversible reaction, one was at the same time measuring the equilibrium constant, and vice versa. Both kinds of measurement in turn might be used to determine the work of affinity.

By any standard, the *Études* was an important book. Van't Hoff had broken new ground in the study of reaction kinetics and had given the law of mass action a new interpretation. Using the work of Horstmann as a starting point, he had greatly enlarged the scope of chemical thermodynamics and had developed expressions for both the effect of temperature on equilibrium and the work of affinity. He had disposed of the principle of Berthelot and Thomsen by showing that the heat evolved in a chemical reaction is a direct measure of the work of affinity only at absolute zero, and he had stated his own principle of mobile equilibrium. Finally, he had demonstrated that the electromotive force of a reversible chemical process was a measure of the work of affinity. That most of this had already been done by J. Willard Gibbs does not detract from van't Hoff's achievement, for, like all of his contemporaries in chemistry, van't Hoff worked in ignorance of Gibbs's papers.

Important as the *Études* appears in retrospect, it was little appreciated during the years immediately following its publication.[73] In part this neglect may be ascribed to the size and naiveté of its audience: the circulation of the *Études* was not very large, and few of the chemists who perused it could have been prepared for its contents. But in part it must have been due to the style in which van't Hoff presented his results. As in the case of the *Ansichten*, he had failed to highlight his theoretical findings effectively. He had moved quickly from one problem to another, making little effort to ease transitions for his readers or to summarize his conclusions, and several of his most important equations were presented with little or no attempt at derivation. He postponed proof of his equation of the reaction isochore, for instance, to subsequent papers published in Dutch and Swedish journals.[74] These tactics detracted seriously from the force of his arguments.

To understand the reception accorded the *Études*, however, it is equally important to note that van't Hoff's generalizations could be applied to particular problems only with the expenditure of considerable creative effort. A great gulf separates the statement of a thermodynamic expression from its application to specific cases. Van't Hoff had presented several examples of how that gulf might be bridged, but much work remained to be done. It was, for example, notoriously difficult to specify exactly what processes occurred in a Voltaic cell and to eliminate impurities and secondary reactions so as to obtain constant and reliable values for the electromotive force. Likewise, as Ostwald had already discovered, determining the equilibrium constant of a reaction by measuring the concentrations of reactants and products usually entailed difficult and time-consuming procedures subject to many sources of error. Moreover, not all reactions lent themselves to such measurements.

In addition to these many experimental difficulties, there was also a major conceptual obstacle confronting those who would follow van't Hoff's path, that is, ignorance of the state in which most chemical reactions occur—solution. Many of van't Hoff's results were premised on an analogy between the behavior of gases and solutions. He had, for example, used this analogy in moving from Horstmann's equation for the heat of dissociation of a gas to his own general equation of chemical equilibrium. But to reason by analogy was quite different from providing direct evidence. The laws of thermodynamics, from which Horstmann's equation was derived, had been developed by considering gases that underwent reversible cycles of heat transfer involving isothermal and adiabatic expansions and compressions. Van't Hoff's problem was finding a means of making solutions accessible to the same sort of analysis. This required that he think of solutions in a new way, not as purely chemical systems, but as physical systems with properties suited to thermodynamic manipulation. Once again, as in the case of the *Ansichten*, van't Hoff's attempt to treat a subject systematically had led him to a fresh set of questions.

Van't Hoff resolved many of these new problems by focusing on the phenomenon of osmotic pressure. He first mentioned osmosis in the final chapter of the *Études*, where he suggested that the measurement of osmotic pressure across a semipermeable membrane might constitute a means of determining the affinity of a dissolved substance for water. After completing this chapter, van't Hoff discussed the subject with Hugo de Vries, professor of plant physiology at Amsterdam. Botanists and physiologists had been studying osmosis for some time, and de Vries, who himself had been conducting experiments on osmotic phenomena, was well acquainted with their results; he referred van't Hoff to the work of a German botanist, W.F.P. Pfeffer, who had already published fairly extensive data on the osmotic pressures of dilute solutions of cane sugar. Pfeffer's data showed quite clearly that osmotic pressure was proportional to concentration or, in other words, inversely proportional to the volume of the solution in which a given amount of solute is contained. In addition, de Vries informed van't Hoff of his own discovery of isotonicity: there is no net transfer of solvent across a semipermeable membrane that separates two different solutions having the same osmotic pressure.[75]

The results of Pfeffer and de Vries were important to van't Hoff because they indicated that osmotic pressure was not the result of specific chemical attractions between the semipermeable membrane and the components of a solution, as some chemists maintained. Rather, they suggested that both the membrane and the solvent were passive during osmosis. Van't Hoff concluded that osmosis was a physical rather than a chemical process; it could be explained by an hypothesis analogous to the kinetic theory of gas pressure. The osmotic pressure of dilute solutions is due to the bombardment of the semipermeable barrier by dissolved molecules of solute. It was inversely proportional to the volume of the solute, just as gas pressure was inversely proportional to volume, and for the same reasons. Just as important, he soon realized that by treating the osmotic pressure of a solution in the same manner as the elastic pressure of a gas, it became possible to fully explore the similarities between gases and liquids through the application of thermodynamics. Thus, according to his treatment, the dilution of a solution or the movement of solvent across a semipermeable membrane became analogous to the reversible expansion of a gas. Consequently, these processes were brought under the scope of the second law of thermodynamics. Van't Hoff pursued the implications of this idea in a series of papers that appeared in 1886 and 1887.[76]

In these memoirs, which were fully as rich and original as the *Études*, van't Hoff accomplished three major results. Each was derived from the laws of thermodynamics by applying the cyclic-process methods of Carnot and Clausius to reversible transformations involving solutions. First, van't Hoff showed that it was possible to develop analogues of the laws of Boyle, Gay-Lussac, and Avogadro for dilute solutions. The equation summarizing these gas laws, $PV = RT$, was applicable to matter in the dissolved state provided that the solution was dilute and P was taken to represent osmotic pressure. Even the constant R had the same value for gases and dilute solutions. In other words, the properties of a given volume of solute were identical to those of a gas occupying the same volume. The analogy between ideal gases and dilute solutions, as van't Hoff remarked, was so close as to be almost an identity.[77] Second, he presented thermodynamic derivations of both the law of mass action and the equation of the reaction isochore, thus shoring up results he had already presented in the *Études*. Third, he demonstrated that several of the effects that solutes had upon solutions were proportional to the osmotic pressure and hence to the concentration. These effects included the lowering of freezing point, the elevation of boiling point, and the decrease of vapor pressure. In doing this, van't Hoff provided theoretical justification for a number of principles that chemists had already developed on an empirical basis, such as Raoult's law, which related the molecular weight of a solute to its effect on the freezing point of a solution. One consequence of this was that when measuring any one of these effects, one was at the same time making a determination of all of the others. In addition to unifying hitherto disparate phenomena under a single principle, this result promised experimental utility, since measuring osmotic pressure was far more difficult than measuring freezing point or vapor pressure.

Van't Hoff's analogy between gases and solutions had given rise to a powerful theory of solution, but as has usually been the case in chemistry, the fit between theory and experimental data was far from perfect. Van't Hoff himself recognized two major shortcomings in his treatment. The first was a limitation similar to one that physicists encountered when applying the gas laws, for just as those laws described only the behavior of ideal gases, so the laws of solution were strictly true only for solutions approaching an ideal, infinitely dilute state.[78] This limitation was inherent in the methods van't Hoff used in developing his theory. In order to carry through his thermodynamic derivations, he had found it necessary to make several simplifying assumptions, positing for instance that the volume of the solute is negligible by comparison with the volume of the solution. Van't Hoff appreciated that predictions based upon such assumptions would be more or less accurate depending on the dilution of the system under study.

This difficulty was foreseeable, but the second was entirely unanticipated. When he checked predictions regarding osmotic pressure, freezing point, and other properties of dilute solutions against data available in the literature, he found that attenuated solutions of electrolytes showed considerable deviations from what his theory required. Their osmotic pressures were consistently higher and their freezing points consistently lower than they ought to have been. In order to resolve these discrepancies, van't Hoff introduced an empirically determined factor, i, into his equations.[79] Thus for solutions of electrolytes the equation $PV = RT$ became $PV = iRT$, the value of i depending upon the compound in solution. He had no explanation of why this correction factor was necessary in the case of solutions that conducted an electric current and not in the case of nonconducting solutions, such as those of sugar in water. Nor could he explain why the value of i should differ depending on the electrolyte in question. Answers to these questions came to him on 30 March 1887 in a letter from his Swedish colleague, the twenty-eight-year-old Svante Arrhenius.[80]

Arrhenius had much in common with van't Hoff and Ostwald. Like them, he came from a family of modest but by no means humble circumstances.

Whereas van't Hoff was the son of a physician and Ostwald the son of a cooper, Arrhenius's father was an administrator at the University of Uppsala, responsible for the management of income from university-owned lands.[81] Like his older colleagues, Arrhenius had begun his university studies at rather backward and provincial institutions: Uppsala and the Stockholm Hogskola. Years later, Arrhenius would refer to his fellow Scandinavians Guldberg and Waage as "natural philosophers."[82] The same term might well be applied to some of Arrhenius's teachers and to the young Arrhenius himself, for the boundary between physics and chemistry was not drawn as sharply in the schools of Sweden in the 1870s and 1880s as in the universities of Germany. As a result, Arrhenius, like van't Hoff and Ostwald, acquired a better command of physics during his student years than was typical of a chemist trained in Germany. The physics was not always up-to-date, but it would prove sufficient to his needs.

Arrhenius took an indirect path to his work on solutions. While looking for a dissertation topic as a student in Stockholm, he recalled that his former chemistry professor at Uppsala, P. T. Cleve, had emphasized the impossibility of ascertaining the molecular weight of substances like cane sugar, which could not be volatized without decomposition. Arrhenius, who was then studying electricity under the tutelage of the physics professor at Stockholm, Eric Edlund, thought he might be able to solve the problem. He had noticed that the molecular conductivity of an electrolyte was lowered when some of the conducting solution was replaced by a poor conductor like alcohol. Arrhenius thought it might be possible to determine the molecular weight of cane sugar by comparing its effect on conductivity with the effects of other nonconductors of known molecular weight. His experiments did not meet this purpose, but they did reveal that the conductivity of solutions varied with concentration in interesting ways. By the time he finished his dissertation in 1883, his original aim had been overshadowed by his realization that conductivity measurements offered a powerful technique for examining the nature of solutions themselves.[83]

Arrhenius's dissertation, published in 1884, was divided into two parts.[84] In the first, he described a method for determining the conductivities of extremely dilute solutions and presented the results of measurements performed on a large number of electrolytic solutions at various concentrations. From these data Arrhenius extracted one very important principle: whereas the conductivity of a solution decreased with attenuation, the molecular conductivity increased, tending toward a maximum value as the solution approached infinite dilution. Diluting an electrolytic solution seemed to enhance the conducting capacity or activity of each molecule of the electrolyte.

In the second part of his dissertation, Arrhenius sought to explain this puzzling phenomenon. He did so by developing a theory of electrolytic solutions that, in many respects, was a synthesis and extension of ideas first proposed by Alexander Williamson and Rudolph Clausius. Williamson, in developing an account of the process of etherification in 1850, had developed a model in which molecules in solution were constantly exchanging atoms and radicals with one another. At any given moment, according to Williamson, a certain undetermined fraction of the molecules of a solute like hydrochloric acid would therefore be decomposed into their constituents. Williamson, interested in the mechanisms of ordinary chemical reactions, believed these constituents to be electrically neutral atoms, but in 1857 Clausius modified Williamson's idea to explain an electrical phenomenon. Noting that the application of the smallest possible current to an electrolytic solution resulted in the migration of ions, Clausius argued that a certain small portion of the molecules of an electrolyte are decomposed into positive and negative ions even in the absence of an electric current. Like Williamson, he envisioned a dynamic equilibrium between the decomposed and undecomposed species.[85]

Arrhenius accepted Clausius's conclusion and extended it by relating molecular conductivity to the equilibrium between electrolytic molecules and their decomposition products. He did this by assuming that an electrolyte in solution is divided into active and inactive portions, the active portion being that which conducts electricity and the inactive portion being that which does not. In order to explain his results, Arrhenius further proposed that, on dilution, an increasing proportion of inactive molecules are transformed into the active state. The ratio between the number of ions present in an electrolyte and the number of ions it would contain if it were entirely in the active state he called the activity coefficient, α.[86] Arrhenius suggested that this coefficient was proportional to the molecular conductivity: the more dilute the solution, the greater the molecular conductivity and hence the larger the activity coefficient. Most important, Arrhenius went on to argue that the electrical and chemical activity of solutions were related. Using data from the first part of his dissertation, Arrhenius showed what he thought to be a direct proportionality between the strength of an acid and its activity coefficient.[87] He concluded that if the electrically active portion of a solution was identical with its chemically active mass, then it would be possible to predict reaction rates, affinity constants, and heats of reaction through the use of conductivity data. This he sought to demonstrate in the final sections of his thesis.

Arrhenius's dissertation was strikingly ambitious, but it was of the sort that teachers often call undisciplined. Many of his conclusions rested upon flimsy experimental supports. He was measuring conductivities of solutions far more dilute than any hitherto studied, and some of his generalizations depended upon only a handful of examples—the link he postulated between the strengths of acids and their activity coefficients was based, for instance, on the study of only five acids. Nor was Arrhenius entirely clear in his definition of essential concepts. The active portion of an electrolyte, he told his readers, "is probably a compound of the inactive part and the solvent. Or possibly inactivity may be caused by the formation of molecular complexes. Or again the difference between the active and inactive parts may be purely physical."[88] Such ambiguities could not have inspired much confidence among the members of his examination committee; they judged the dissertation to be undeserving of honors, in effect telling Arrhenius that he was qualified to teach in secondary schools but not in the university.

Arrhenius's career might well have come to a premature end had he not sent a copy of his thesis to Wilhelm Ostwald, whose work Arrhenius had come to admire during the course of his doctoral research. Ostwald reacted enthusiastically to the dissertation, apparently because he saw promise in the use of conductivity measurements to determine affinity constants.[89] He visited Sweden during the summer of 1884, met Arrhenius, and later helped arrange a traveling fellowship for the young Swede. Arrhenius was supported by that fellowship, working in Germany, when in February 1887 he received reprints of van't Hoff's papers of 1886. The effect was immediate. Arrhenius's dissertation had put him in command of a large fraction of the literature of physical chemistry and had made him an authority on electrolytic solutions. He was in an ideal position to resolve the difficulties that van't Hoff had encountered in applying the laws of solutions to electrolytes. Within a few weeks of obtaining van't Hoff's memoirs, Arrhenius had devised a theory explaining the need for the correction factor, i, in van't Hoff's equations. He described this theory in the celebrated letter to van't Hoff of 30 March 1887 and developed it at greater length in articles published later that year.

The most important of these articles appeared in the first volume of the *Zeitschrift für physikalische Chemie*. It at once put a capstone on the theory of solutions developed by van't Hoff and suggested new opportunities for research that would keep scores of scientists occupied for years to come. Like van't Hoff, Arrhenius assumed that osmotic pressure was the result of impacts made by the particles of substances in solution on the walls of the containing vessel.[90] If the osmotic pressure of electrolytic solutions was consistently higher than that of nonconductors, then it must be because they produce a greater number of particles in solution. The dissociation of molecules of electrolyte into smaller constituents would produce this effect. His own earlier work and that of Clausius pointed to the existence of ions in solution, but neither had envisioned the degree of dissociation to be so large as the deviations in osmotic pressure would seem to require. Nevertheless, Arrhenius proceeded under the assumption that ionic dissociation was the cause of these deviations. If the fraction of molecules that dissociate could be measured, he argued, then it would be possible to determine the total number of particles in such solutions and hence the osmotic pressure.[91]

Arrhenius accomplished this by relating van't Hoff's coefficient, i, to his own activity coefficient, α, by means of the equation

$$i = 1 + (k - 1)\alpha,$$

where k represents the number of ions into which an electrolyte dissociates.[92] The activity coefficient, according to this new treatment, came to be synonymous with the degree of dissociation, the ratio of dissociated molecules of solute to the sum of dissociated and undissociated molecules. In extremely dilute solutions, nearly all of the electrolyte dissociates, α approaches unity, and the value of i approaches that of k. So, as a solution of a binary electrolyte such as KCl ($k = 2$) approaches infinite dilution, α should approach a value of 1, and van't Hoff's coefficient should approximate a value of 2, leading one to expect an osmotic pressure twice as high as that produced by an equal concentration of a nonelectrolyte. In somewhat more concentrated solutions, in which the degree of dissociation would be considerably less than total, the exact degree of dissociation could also be evaluated, since Arrhenius had shown in his dissertation that α was proportional to the molecular conductivity, a measurable quantity.

Arrhenius was able to test his hypothesis by comparing the values obtained for i by means of conductivity data with values computed by means of measurements of freezing-point depression, a method van't Hoff had already shown to provide such information. Using solutions of ninety substances, Arrhenius demonstrated that the values arrived at by these two methods were nearly identical in the vast majority of cases. The evidence for his dissociation theory was impressive.

Arrhenius marshalled additional supporting evidence by noting the many properties of electrolytic solutions that were additive or nearly so: the heat of neutralization of strong acids and bases, specific volume and specific weight, specific refractivity, together with all those properties that van't Hoff had shown were proportional to osmotic pressure. In each instance, the property of the solution as a whole was the sum of the properties of its parts. In dilute solutions of electrolytes, in which dissociation was nearly complete, the parts consisted of the molecules of the solvent and the independent positive and negative ions of the electrolyte; undissociated molecules of the electrolyte could be ignored. Thus, for example, equivalent quantities of strong acids and bases always generate the same heat during neutralization regardless of the nature of the acids and bases. Arrhenius explained this by arguing that these electrolytes are highly dissociated in solution, giving rise to large concentrations of hydrogen and hydroxide ions. Neutralization is in fact a process in which water is formed from these ions. The heat of neutralization is therefore always the heat generated by the single reaction $H^+ + OH^- = H_2O$, a reaction accompanied by the evolution of 13,600 calories of heat irrespective of the source of the ions.[93]

Strong acids and bases also afforded Arrhenius examples of the linkage of electrical and chemical activity. In his dissertation he had pointed out that substances that are electrically active in solution, such as acids and bases, are also chemically active. Now, since he had established the connection between electrical activity and dissociation, it followed that there also was a connection between dissociation and chemical activity. The process of reaction in solution seemed to be related in many cases to the dissociation of molecules into their constituent ions. This relationship would later prove to have many significant ramifications.

The work on solutions that van't Hoff and Arrhenius completed during the mid-1880s had at least one avid reader—Ostwald. His own studies of mass action and his labors on the *Lehrbuch der allgemeinen Chemie* had given him an intimate knowledge of the phenomena his colleagues were investigating. By 1887, he had befriended both van't Hoff and Arrhenius and had begun to incorporate their results into his textbook and his research program. In particular, Ostwald was intrigued by Arrhenius's suggestion that electrical and chemical activity were related. After reading Arrhenius's dissertation in 1884, Ostwald had begun a series of investigations into the conductivity of weak organic acids, substances for which he had already determined affinity constants on the basis of other physical measurements. Finding that molecular conductivities were very nearly proportional to affinity constants, as Arrhenius had predicted, Ostwald then set out to find a general expression connecting the concentration of an electrolyte and its molecular conductivity. Initially, he approached the problem by collecting massive amounts of data, which he then sought to fit to an empirical equation.[94] Shortly after Arrhenius announced his theory of electrolytic dissociation, however, Ostwald appears to have realized that it ought to be possible to apply the law of mass action to the equilibrium between the molecules of a dissolved electrolyte and their ions. Such an application might yield a rational expression for the variation of molecular conductivity with dilution. The result, published in January 1888, came to be known as Ostwald's dilution law.[95] It linked the law of mass action and the theory of electrolytic dissociation in the form of the equation

$$\alpha^2/(1 - \alpha)v = k,$$

where v represents the volume containing one gram–molecular weight of solute and k is a constant. The relation to the law of mass action is immediately clear if one considers α as the portion of a binary electrolyte that exists as ions and $1 - \alpha$ as the portion that is undissociated. The active mass of each species will be proportional to its concentration, or inversely proportional to its volume. Letting v represent the volume, then the active mass of each ion is α/v, and the active mass of the undissociated electrolyte is $(1 - \alpha)/v$. According to the law of mass action, the active mass of the products stands in a constant relation to the active mass of the reactants, or

$$\frac{(\alpha/v) \, (\alpha/v)}{(1 - \alpha)/v} = k.$$

By simplification this yields the dilution law:

$$\alpha^2/(1 - \alpha)v = k.$$

Knowing van't Hoff's work, it was a simple matter for Ostwald to derive this expression directly from the laws of solution. And since Arrhenius had shown that the ratio of the molecular conductivity of a solution at a finite concentration Λ to its molecular conductivity at infinite dilution Λ_0 yielded values for the degree of dissociation, it was possible to substitute the term Λ/Λ_0 for α:

$$(\Lambda/\Lambda_0)^2/[1 - (\Lambda/\Lambda_0)]v = k.$$

If valid, the dilution law would make it feasible to calculate the degree of dissociation of a dissolved electrolyte over a broad range of concentrations using only two experimental values: the substance's molecular conductivity in an infinitely dilute solution and in a solution of one finite concentration. Both values were relatively easy to determine.

Several months after announcing the dilution law, Ostwald published the results of his experimental tests of the expression.[96] In experiments with over two hundred weak organic acids, he found values of k to remain constant over a wide range of dilute to moderate concentrations. It was an impressive confirmation of his principle and of the theory of electrolytic dissociation that stood behind it.

In the span of a few years, van't Hoff, Arrhenius, and Ostwald had developed a powerful and highly suggestive theory of matter in the state of solution, effectively simplifying and relating a large body of previously isolated facts. According to van't Hoff, dilute solutions were fundamentally similar to gases. They obeyed the laws of Boyle, Gay-Lussac, and Avogadro, with osmotic pressure substituted for gaseous pressure. According to Arrhenius, the anomalous behavior of electrolytic solutions resulted from a dissociation of molecules of solute, the degree of dissociation varying with the concentration of the solution and the nature of the electrolyte. The ions that resulted from dissociation carried unit charges of electricity and were primarily responsible for both electrical and chemical activity. The degree of dissociation of a specific electrolyte was open to calculation through measurement of electrical conductivity or any of the properties that van't Hoff had shown to be proportional to osmotic pressure. With the addition of Ostwald's dilution law, it became possible to forecast with some degree of confidence the degree of dissociation of many electrolytes over a range of concentrations and hence to make quantitative predictions about their chemical and physical properties.

THE NEW CHEMISTRY OF THE IONISTS

Eight years after Ostwald published his dilution law, he looked back on the achievements of the 1880s and asserted that "in the history of science there are few examples of such concentrated progress."[97] With characteristic exuberance he went on to suggest that what he, van't Hoff, and Arrhenius had accomplished was commensurable with the attainments of Lavoisier and his co-workers. A skeptic might retort that the history of science knows few examples of such concentrated arrogance, but to dismiss Ostwald's claim would be a mistake. Ostwald was not as penetrating a thinker as Lavoisier, but together the complementary contributions of Ostwald, Arrhenius, and van't Hoff did not suffer by comparison with those of their predecessor of the late eighteenth century. Their memoirs on the theory of solution and van't Hoff's *Études* had made a broad range of phenomena accessible for the first time to quantitative treatment and, as a result, had shown how seemingly disparate properties of matter were in fact closely related. They had introduced greater exactitude into the definitions of many of the working concepts of chemistry and had made important changes in the instruments of their science. And, like Lavoisier, they had opened up many new opportunities for research.

Important as these particular contributions were, they nevertheless constituted but a part of a more thoroughgoing reform of chemistry begun by these three scientists. Their work of the 1880s embodied a new mode of doing chemistry and suggested a new conception of the internal arrangement and external relations of the science. The subject that Ostwald and his colleagues had studied as young men was an experimental science that looked within for concepts and methods. Its practice did not require the use of mathematics beyond arithmetic, since there was little need of the calculus in the study of composition and structure. It had a strong theoretical component, but chemical concepts were largely qualitative and often articulated poorly with prevailing ideas in physics. That chemical and physical theories did not always fit occasioned little surprise, for they were not always expected to; after all, physicists were concerned with matter in the abstract, whereas chemists had to deal with the perverse peculiarities of particular substances. Phenomena on the borderland between the two sciences, such as those of electrochemistry, were treated on an *ad hoc*, haphazard basis when they were treated at all. And in chemistry, unlike physics, the possibility of a specialty devoted to theory hardly existed. In the first place, divisions internal to chemistry, such as those between the study of organic and inorganic substances, seemed unbridgeable. Structural theory, for instance, was of little value to inorganic chemists, and the periodic table told organic chemists little about carbon and its compounds that they did not already know. Second, chemists were committed to the notion that their science would grow and prosper through experiment and inductive reasoning. Nature was met in the laboratory and not at the writing desk.

The Ionists, as Ostwald, van't Hoff, and Arrhenius soon became known, did not reject this heritage whole. But they did touch and alter every aspect of it: the balance between experiment and theory, the role of mathematics in chemistry, the relations between chemistry and other sciences, and the structure or topography of chemistry itself.

Permeating their work, for example, was a new attitude toward the laboratory, according to which it was the place where ideas were tested rather than where they were to take shape. This outlook is exemplified by van't Hoff, who, among those who did not know him, had a reputation for being a lazy or clumsy man at the bench. This reputation was undeserved. If he published few experimental papers during the 1880s and seemed to rely on the data of others to support his arguments, it was not because he could not or did not do experimental work himself; rather, it reflected his conviction that experimental studies above and beyond those needed to evaluate an idea were extravagant and wasteful. As his friend and co-worker James Walker noted, van't Hoff was preeminently a theorist who took pride in his ability to recognize interesting patterns in scanty data.[98] To a lesser extent this was true also of Arrhenius, whose dissertation rested on such sparse experimental evidence. Even Ostwald, who had enormous enthusiasm and capacity for research in the laboratory, was immune to the traditional goals of the laboratory chemist—the discovery of new compounds and isolation of new elements. He and his associates preferred the hunt for new ideas and new means of winning them. Some called their science theoretical chemistry, and not without justification, for although the Ionists did not forsake experimental work, they did subordinate it to their theoretical pursuits.[99]

Just as the laboratory held a different position in the science of the Ionists than it had hitherto occupied in chemistry, so too did mathematics. It would be an exaggeration to say that they made chemistry a mathematical science. None of them were sophisticated mathematicians, and their papers did not make extraordinary mathematical demands of readers. Never did they use more than the basic techniques of the differential and integral calculus. Nevertheless, the Ionists perceived far greater opportunity for the use of mathematics in chemistry than did the vast majority of their fellow chemists, and they took some of the first steps toward mathematizing parts of their science. Each of them appreciated the economy and precision of expressing complex relations in abstract terms; they all knew how to fit data to equations; and on occasion they deduced new and interesting relationships by manipulating the terms of their equations. Van't Hoff, for instance, could not have arrived at his equation of the reaction isochore entirely on the basis of experimental data and chemical intuition. Mathematics did not dominate their science, but it played an essential role in it.

The Ionists' emphasis on theory and their mathematical literacy were associated with another characteristic that set them apart from contemporary chemists—their common conviction that chemistry and physics could and should be unified. The behavior of particular substances or multicomponent systems might be more complicated than that of matter in the simple or abstract state, but there should be no contradiction between the principles used to describe each case. Where there were discrepancies, the structure of one or both sciences was faulty. In the *Études*, van't Hoff showed utter confidence in his opinion that Berthelot's principle of maximum work was incorrect except at a temperature of absolute zero; his certainty derived from a firm belief that reversible chemical processes fell under the jurisdiction of the laws of thermodynamics. Likewise, when Arrhenius posited the existence of active and inactive species in electrolytic solutions, he drew great support from Clausius's demonstration that since Ohm's law applied to such solutions, ions must exist in them even in the absence of an electric current. In these instances, van't Hoff and Arrhenius both rejected ideas that had won widespread acceptance among chemists: the notion that the heat evolved in a chemical change was a direct measure of affinity, and the belief that oppositely charged particles could not be formed in a solution without the application of an external force. They did so not solely because experimental evidence conflicted with these ideas—there were ways of interpreting the evidence to save the theories—but rather because these beliefs stood in opposition to fundamental physical principles.

The Ionists believed in the unity of science and they were willing to act accordingly. Oftentimes this led them to alter chemical doctrines in order to bring them into conformity with physical theory. Nevertheless, they were not out-and-out reductionists. Physical theory had much to offer, but it was not sufficiently developed to allow for an easy and painless explanation of all chemical problems. The behavior of Voltaic cells, for instance, was not well understood, and the measurement of the electromotive force of reactions in such cells was a challenging problem, fraught with interpretive and practical difficulties. Nor did physical theory seem to offer much guidance on how to move from a theory of dilute aqueous solutions to a more comprehensive understanding of the properties of all solutions. The Ionists often looked to physics for models and tools, but they did not imagine that chemistry would become merely an annex to physics. Indeed, on occasion they were prepared to assert the priority of chemical explanations over physical. In 1896, for example, Ostwald published a long and critical history of electrochemistry in which he showed how, in his view, a chemical theory of the origin of galvanic electricity came to triumph over the physicists' contact theory.[100] The point here is that although the Ionists thought the sciences to be one, they by no means believed that the physicists' understanding of nature was in every respect more general and valid than that of the chemists. A reconciliation of chemical and physical theories would require compromises of both parties; the Ionists cast themselves in the role of mediators.

The work of the Ionists revealed unanticipated opportunities in the borderlands between chemistry and physics, but it was rich in implications for chemists and other scientists as well. The great majority of reactions that chemists studied in academic laboratories took place in solution, as did most industrial processes; hence the theory of solution held potential for application in every branch of chemistry, pure and applied. Likewise, their work on reaction kinetics, equilibrium, and affinity promised to yield information about the outcomes and mechanisms of all reactions, those involving organic as well as inorganic compounds. Chemical changes themselves and the physical factors affecting them were the Ionists' focus of study rather than the properties, composition, and structure of particular classes of compounds. Consequently the Ionists could imagine transcending many of the divisions between organic and inorganic chemists—scientists who, because of their preoccupation with composition, were bound by the possibilities inherent in the study of one or another kind of substance. The Ionists' work contained the germ of a science of general chemistry that would be defined not by the presence or absence of carbon or any other element, but by a set of principles and techniques applicable wherever chemical change occurred. And it followed that in those other sciences in which chemical processes played a role, the work of the Ionists should find application. Solutions, for example, were the media of most natural processes, among them respiration and other metabolic changes, the action of drugs, anesthetics, and toxins, and the formation of minerals, ores, and salt deposits. Already prior to 1887 the consideration of these phenomena had led a number of botanists, physiologists, and geologists to conclude that a better understanding of solutions was a desideratum for advance in their own sciences.[101] The new theory of solution therefore promised to become a common reference point for scientists in all of these fields, a possibility that Ostwald, van't Hoff, and Arrhenius were quick to recognize.

Implicit then in the work of the Ionists was nothing less than a new mental map of chemistry and its borders with other sciences. Surveying from the vantage point of their theories of affinity, equilibrium, and solution, the Ionists discerned broad avenues linking chemistry and physics and new routes for commerce between chemistry and the biological and geological sciences. And from their perspective the domain of chemistry itself expanded and acquired a new integrity. Territories lying on the margins of chemistry, like electrochemistry and chemical thermodynamics, were revealed to be of unexpected scope and resources. Older and better-explored fields, like analytical and inorganic chemistry, assumed surprising new contours; the class reactions of quantitative analysis, for example, turned out to be tests of the presence of ions rather than integral molecules. Even the topography of the large and rich province of organic chemistry was revised on their atlas, both because they perceived a possibility of opening new paths to connect it with other regions of chemistry and because, in the greatly enlarged realm of their science, the study of organic composition and structure inevitably shrank by comparison with the whole.

By 1887 Ostwald, van't Hoff, and Arrhenius had each developed pictures of chemistry that, while not identical in every respect, were sufficiently similar so as to permit easy communication and mutual understanding. Although they had come to this shared outlook by independent paths, there were common factors in their backgrounds that had made this convergence possible. For one thing, each came from a country on the periphery of German science and scholarship—places where they could stay in touch with what was being done in Germany in fields like thermodynamics and electrical theory while maintaining independence from many of the assumptions that guided German research in chemistry. Second, as young men, each had given physics serious study: Ostwald as an assistant to von Oettingen at Dorpat, van't Hoff as a student in the Netherlands, and Arrhenius as a worker in the physical laboratories at Uppsala and Stockholm. The breadth of their educations was not unrelated to the comparative backward state of the scientific institutions and communities in which they worked; specialization had not proceeded as far in Latvia, Holland, and Sweden as in Germany or France. In addition, at early stages in their careers each of the Ionists had come to identify the relations between the physical and chemical properties of matter as a critical area of research. They initially had different reasons for doing so: Ostwald saw such correlations as a means of determining affinity constants, van't Hoff wished to better understand the rates of chemical reactions, Arrhenius sought links between electrical and chemical activity. Nevertheless, their individual research programs soon brought them into contact with both a common literature and each other.

If Ostwald, van't Hoff, and Arrhenius had come to a common picture of science in the 1880s, it was not a picture that was entirely congruent with the viewpoints of many of their contemporaries, either in physics or in chemistry. Each of the Ionists had studied physics, used physical ideas and methods in their work, and maintained personal contacts with members of the physics community, but they were not of physics. They worked in chemical institutes and published in chemical journals. Their results, for the most part, were not directed to questions that physicists were asking, and their means of attaining those results fell short of the rigor that physicists were trained to expect. Van't Hoff's reliance on artificial cyclic processes in proving his theorems, the way that both van't Hoff and Arrhenius jumped back and forth between kinetic and thermodynamic modes of argument, and the Ionists' common use of *ad hoc* assumptions in proofs and calculations—all of this seemed amateurish, especially after Gibbs's austere and penetrating papers became known, for there physicists found many of the results of the Ionists stated in a far more elegant, analytical form. Some physicists, like Max Planck, Gustav Wiedemann, and G. F. FitzGerald, suspected the Ionists of practicing physics without a license; their suspicions were intensified in the nineties when Ostwald embarked on a crusade to axiomatize the physical sciences by grounding them all in thermodynamics.[102]

Nor were the perspective and results of the Ionists immediately conformable with the views held by chemists. Organic chemists naturally resented the implication that their research on composition and structure was so much *Brot-*

studium.[103] Others felt that the ideas of the Ionists conflicted with chemical intuition.[104] That stable salts should spontaneously decompose in solutions at room temperature—as if of fright, as one critic put it—could seem outlandish, as could the idea that water played an essentially passive role in the behavior of dilute solutions. Many chemists were ill at ease with the style of the Ionists; it was simply too theoretical, too physical, too alien. It involved a leap, for example, to go from a chemistry that usually required no more than two decimal points to the properties of one-ten-thousandth-molal solutions, or to move from the familiar notions of mass and volume to the relatively foreign concepts of internal energy, maximum work, and entropy. Even those chemists who were favorably disposed toward the Ionists sometimes had difficulty comprehending their discourse. William Ramsay, for instance, defended their conclusions in England for more than a year before he grasped the distinction between an atom and an ion. And, lacking any knowledge of the calculus, he never became adept at handling chemical thermodynamics.[105]

If there were conceptual difficulties in integrating the new physical chemistry and existing sciences, so too were there institutional and practical obstacles. A training in physical chemistry would have to be very different from preparation in physics or in any of the established branches of chemistry. Students would have to be at home in the chemical laboratory and be skilled in the analytical techniques of that science. But in addition to having a command of the reagents and reactions that were every chemist's stock in trade, they would also have to acquire a working knowledge of the calculus, thermodynamics, and portions of electrical and kinetic theory. Likewise, a well-equipped laboratory for physical chemistry would look different from those of the chemist or physicist. In addition to the standard glassware and instruments found in the analyst's laboratory, the physical chemist would also want apparatus for measuring electrical conductivities and other physical properties, such as viscosities and refractivities; equipment for thermochemical and electrochemical investigations; and rooms that were free of electrical, thermal, and vibrational disturbances. Adapting existing laboratories to these needs could be an expensive affair.

The poor fit between the views of the Ionists and those of their contemporaries, together with their special educational and research requirements, created motive for the development of institutions that would have a measure of independence from institutions of either chemistry or physics. The outcome was paradoxical—scientists committed to the unity of knowledge found themselves building a new specialty.

CHAPTER 2

Physical Chemistry from Europe to America

THE HISTORY of physical chemistry in the 1880s is, in large part, a story of how the viewpoints and investigations of van't Hoff, Arrhenius, and Ostwald converged, and of how, during that process, there emerged both a powerful ensemble of new ideas and techniques and a new mental map of chemistry and its borderlands. But disciplines are more than mental constructs; they are also social institutions. They require ideas, but also money, jobs, recruits, and the various agencies (textbooks and journals, research schools and laboratories, associations and societies) necessary for the sustained growth and smooth transmission of knowledge and technique. Disciplines may arise from the imaginations of a few scholars, but they can flourish only by interdigitating themselves into dense networks of educational, industrial, and professional organizations.

This is the complex task that physical chemists began to undertake in the closing years of the 1880s. The translation of ideas into institutions required skills different from those at which van't Hoff, Arrhenius, and even Ostwald had previously excelled. Fields opened up by the introduction of thermodynamics into chemistry and by the theory of electrolytic dissociation had to be surveyed in detail. The Ionists had to make their ideas relevant to the interests and pursuits of others, both within and outside science, and this entailed a new emphasis on applications. And since growth was accompanied by conflicts, the Ionists were compelled to defend their ideas in public arenas. Whereas convergence and rapid conceptual advance were dominant motifs in the 1880s, growth, diversification, and persuasion became principal themes in the history of physical chemistry during the next two decades.

THE DIFFUSION OF PHYSICAL CHEMISTRY

In reaching out to the world outside their laboratories, the Ionists deployed talents that could not have been predicted. Van't Hoff, despite his retiring nature, gradually grew into a public role. By the mid-1890s, his research school at Amsterdam was receiving a stream of students from all parts of Europe and America. There followed an appointment to a distinguished professorship at the Prussian Academy of Sciences in Berlin, a Nobel Prize, speaking tours on both sides of the Atlantic, and a decade of intensive work on the physical chemistry of the rich Stassfurt salt beds. By the time he succumbed to tuberculosis in 1911, this one-time teacher at a veterinary school had become world-renowned as the preeminent chemical theorist of his era.[1]

Arrhenius, while less influential as a research director, displayed the broad interests of a natural philosopher and an unsuspected knack for popularization. Successively a lecturer, professor, and rector at the Hogskola of Stockholm, Arrhenius ended his career at the Nobel Institute, which had awarded him its chemistry prize in 1903. While continuing to make contributions to the study of salts in solution, Arrhenius made forays into geology, astronomy, and biology. In several papers he attributed ice ages and other great fluctuations in climate to changes in the carbon-dioxide content of the atmosphere; in a book on cosmic physics, he suggested that life on earth might have originated through colonization of the planet by spores from space; in another book, this one on immunochemistry, he argued that reactions between toxins and antitoxins might be treated by the law of mass action. Although winning more controversy than acceptance, Arrhenius's speculative ideas opened veins that have not been entirely exhausted.[2]

Yet none of the Ionists proved so able as propagandists and institution builders as Ostwald. To borrow the language of the economist, physico-chemical research up until 1887 was replete with inefficiencies in the communication and evaluation of data and ideas. Ostwald made a market in that information. The journal he organized, the *Zeitschrift für physikalische Chemie*, became the showcase for research in physical chemistry from its inception in 1887 until its international character was destroyed by world war. Coedited by van't Hoff, the journal brought together on its editorial board Horstmann, Landolt, and Lothar Meyer of Germany, Ramsay of England, Raoult and Berthelot of France, and Thomsen, Arrhenius, Guldberg, and Waage of Scandinavia—in short, most of those throughout Europe whose work, in whole or part, could be woven into the new physical chemistry.[3]

While his *Zeitschrift* served colleagues, Ostwald's books helped introduce this new discipline to a younger generation. Gifted with the ability to write lucid and graceful prose with little revision, Ostwald composed a shelf-full of textbooks: introductory works on general (1889), analytical (1894), and inorganic chemistry (1900)—all written from the standpoint of the new theory of solutions; a translation of the papers of J. Willard Gibbs (1892); a manual of laboratory techniques (1893); a massive history of electrochemistry (1896); and several installments of a new edition of his *Lehrbuch der allgemeine Chemie* (1891–1902). Widely translated, these texts gave hundreds of young students their first introduction to the new science and the opportunities it presented.

No less important than this profusion of textbooks, papers, reviews, and journals was Ostwald's laboratory at Leipzig. Housed in what was once an agricultural institute, the facilities were physically unimpressive. One visitor described the building as an old pile . . . in every way unfitted for the carrying on of those delicate experiments which brought Ostwald to the forefront of scientific workers. Research was carried on under countless difficulties; the light was bad, the rooms unventilated, the heating effected by means of stoves difficult to regulate and producing dust which caused much injury to the finer instruments; no precautions had been taken in laying the foundations to ensure the deadening of vibrations; thus many experiments were ruined; the lack of space precluded the use of telescopes for reading scales, and altogether it would have been difficult to construct a laboratory worse adapted for physico-chemical investigations.[4]

Not only were the physical conditions unsatisfactory, but Ostwald's appointment entailed the supervision of novices in analytical and pharmaceutical chemistry—a chore beneath the dignity of Wislicenus, the organic chemist who headed Leipzig's principal chemical institute. In addition, since Ostwald's predecessor, Gustav Wiedemann, had been much more a physicist than a physical chemist, there was no tradition of physico-chemical research at Leipzig upon which Ostwald could build; he had to equip a laboratory, win the respect of his colleagues, and recruit assistants and students while at the same time attending to his own investigations and to his literary and editorial activities.

Ostwald threw himself into these tasks with his customary energy. During his first months in the new position, Ostwald hired three assistants: to one, Julius Wagner, he gave responsibility for instruction in analytical chemistry; the second, a skillful organic chemist by the name of Ernest Beckmann, was given charge of the pharmaceutical students; and the third, a young physicist recommended by Arrhenius, stepped into a role as Ostwald's principal assistant in physical chemistry. His name was Walther Nernst. At first Nernst was a lieutenant without troops; two students took the laboratory course in physical chemistry during the fall of 1887, and only one enrolled the following spring. But Ostwald's ability as a lecturer and his growing reputation as an investigator brought thirteen workers into the laboratory by the end of the second year, and soon thereafter the total reached thirty, the capacity of the facility. The number of students taking laboratory instruction in physical chemistry stabilized at that level until 1897, when the completion of a fine new building allowed Ostwald to increase enrollments yet further.[5]

In the laboratory, Ostwald directed his research and that of his growing band of disciples toward testing, extending, and applying the laws of solution and the principle of electrolytic dissociation. The dilution law was the first fruit of this labor. Later, Ostwald developed a theory of acid-base indicators based upon the principles of mass action and electrolytic dissociation, made contributions to the study of crystallization, and developed a qualitative theory of catalysis for which he would win a Nobel Prize in 1909. Nernst took up the problem of the diffusion of ions in solution, stated principles describing the effects of salts on the solubility of other salts, and, building upon the work of van't Hoff and Arrhenius, contrived both a model and a quantitative theory of electrochemical action in galvanic cells. Others studied the processes occurring at electrodes during electrolysis, the electrochemical behavior of amalgams, and the colligative properties of solutions.

As Ostwald's research group grew, so too did a cosmopolitan spirit in his laboratory. A colloquium was begun at which students, instructors, and visitors discussed their current research. Twice a month, on Sunday evenings, Ostwald held an open house for his pupils and co-workers at which he displayed his own landscape paintings and impromptu string quartets and trios performed. His daughter later recalled meeting students from all parts of Europe and America at these soirees, and some from Japan and Egypt. In fact, Ostwald's laboratory held a special appeal for foreigners. The Scottish chemist James Walker noted that even when he was with Ostwald in the late 1880s, his fellow students "were of varied nationality with seldom more than one example of each nation. As far as my recollection goes, there was only one genuine German research student . . ."[6] Later, as class sizes grew, the proportion of foreigners remained high, English-speakers being the most common. Shortly after Ostwald's new institute opened in 1897, the story began to circulate that his assistants had forgotten their German, "although they had not yet learned English."[7]

At the center of the laboratory, at least until philosophical pursuits and other outside interests eclipsed his chemical concerns in the late 1890s, stood Ostwald. He set a frenetic pace. Paul Walden, his assistant and successor at Riga, was especially impressed by his labile energy. He was, Walden wrote,

> omnipresent in the laboratory. No sooner did you see him in lively conversation with his assistants than, stepping into the student workrooms, would you encounter him walking quickly from table to table, giving instructions on the construction of apparatus, explaining analyses and experiments currently in progress, suggesting new procedures and equipment, etc. Each student's work held interest for him; for each and every scientific ailment he had a cure. His personal laboratory was no sanctuary; students and assistants were permitted to enter it freely, and often even called on him at home.[8]

Another of his students, George Jaffé, who worked with Ostwald in 1899, best recalled his role in the weekly meetings of the research colloquium:

> Every piece of research was treated there more than once. The first time a topic was presented Ostwald himself, or one of his assistants, exposed the line that was to be followed. The second time, as a rule when the research was well under way, the researcher himself reported on his work. In this stage he mostly had to expose the difficulties which he had met. Finally, a report on the finished paper was given. . . . Ostwald himself could exert his influence even on investigations which he did not

conduct himself, . . . everybody knew what everybody else was doing. There was no affectation of mysteriousness in the place, rather a spirit of what Ostwald called "brotherly openness." . . .[9]

The most important product of this institute was physical chemists, scores of young scientists with personal knowledge of Ostwald's outlook and techniques or, as increasingly became the case as the years passed, of the methods of the brilliant young scientists who worked as his assistants: Nernst, Max Le Blanc, Robert Luther, Georg Bredig, Max Bodenstein, and Herbert Freundlich, among others. On the occasion of his fiftieth birthday, Ostwald counted 147 pupils who had achieved independent scientific success, 34 of whom were professors.[10] These, his former students, became the principal vectors of the dissemination of the views and techniques of the Ionists.

By the time Ostwald retired from his chair at Leipzig in 1905, his ambition of reconstituting chemistry had found partial fulfillment. In Germany, chairs in physical chemistry existed at Berlin, Göttingen and Leipzig; Ostwald's students held lower-ranking positions at many other universities and *technischen Hochschulen*; the *Zeitschrift für physikalische Chemie* was flourishing; and a national society for the application of the principles of physical chemistry to industry had been established. In Great Britain a number of Ostwald's friends and students, William Ramsay and James Walker most prominent among them, had succeeded in gaining positions of influence at English and Scottish universities. Arrhenius and van't Hoff had laid foundations for study and research in the field in Sweden and the Netherlands. Even in France, long hostile to influences from across the Rhine, the influence of the Ionists was slowly making itself felt.

Yet these achievements, impressive as they may be, must have seemed disappointingly modest to Ostwald, who, after all, had set out not merely to graft physical chemistry onto the existing traditions of chemistry, but to graft chemistry onto his new science. "Physical chemistry," he had told readers of his new journal in 1887, "is the chemistry of the future."[11] Yet throughout Europe there were those who wanted no part of such a future. At home, Ostwald confronted powerful and entrenched organic chemists who, not without reason, viewed Ostwald and his more extreme disciples as vandals in the cathedral. Dominant at most of Germany's twenty-one universities, these chemists proved reluctant to share their institutes with those who dismissed their intricate and beautiful science as so much compound-making.[12] They found natural allies in ministers of education, who in the decades around the turn of the century were coming to realize that the public purse could not afford to underwrite a chair and an institute in every specialty at every university. As a result, these ministries were more and more interested in controlling growth by checking the duplication of facilities and concentrating resources selectively.[13]

Such policies had little impact on organic chemists, who already controlled the chemical institutes at most universities and who could hardly be deposed. But it did restrict the multiplication of chairs and institutes in new fields, among them physical chemistry. In 1904, the twenty-one universities of Germany supported only four institutes of physical chemistry: Leipzig, Göttingen, Giessen, and Freiburg, and one of these (Freiburg) was directed by an untenured professor. Five other universities gave sections (Abteilungen) of their chemical institutes to physical chemists, but these were occupied by docents or untenured professors rather than by chair-holders. Not only were institutes of physical chemistry few in number, they were also poorer than other chemical institutes. At Leipzig in 1895/96, Ostwald worked with a budget less than half that available to his colleague in organic chemistry, Wislicenus; at Göttingen fifteen years later, appropriations for the institute of physical chemistry were one third those allotted the organic institute. According to Jeffrey Johnson's recent survey of the budgets of all German chemical laboratories during this period, the disparity between the income of institutes in physical chemistry and those in other fields actually increased between 1892/93 and 1910/11. Only in the *technischen Hochschulen*, institutions with faster growth rates than the universities and more open and egalitarian structures, did physical chemistry flourish. By 1910, physical chemists held chairs at all but four of Germany's eleven technical colleges, and together the technical colleges employed nearly as many physical chemists at all ranks as did Germany's twenty-one universities.[14]

In Britain, it was not organic chemists and state officials but an indigenous school of inorganic chemists that offered stiff resistance to the claims of the Ionists. Led by Henry Armstrong, professor of chemistry at the Central Technical College of London and long-time Secretary of the Chemical Society, opposition centered around the theory of electrolytic dissociation. The fierce debates, skillfully chronicled by the historian R.G.A. Dolby, may have attracted some adventuresome young chemists to Ostwald, but they wrapped the physical chemistry of the Ionists in an aura of controversy that many English dons found unsavory.[15] Not until after World War I would chairs in physical chemistry be created at Oxford or Cambridge. Had it not been for the generous support of William Ramsay, the distinguished professor of chemistry at University College London and devoted admirer of Ostwald, Ionists in Britain would have found themselves in desperate straits. Opening his laboratory to young chemists returning from the institute of Ostwald, Ramsay helped launch the careers of many of those who would later be numbered among Britain's most eminent physical chemists.[16]

In France, proponents of the Ionists' views found themselves enmeshed in bewilderingly complex political and intellectual struggles. The lingering bitterness of the war lost in 1871, the nonconvertibility of German degrees into French teaching positions, and the insularity of French intellectual life all militated against visits by French students to Ostwald's or other German institutes of physical chemistry. Consequently, only a handful of chemists active in France had personal knowledge of work underway in physical chemistry at foreign institutes. French textbooks ignored or slighted the works of the Ionists, and when a position for a physical chemist was created at the Sorbonne in 1898, it was taken as a recommendation that candidates have had little contact with Ostwald, Arrhenius, or van't Hoff.[17] It is indicative of the frosty reception given the physical chemistry of the Ionists that the first French-language journal of physical chemistry, begun in 1903, was organized not by a native of France, but by a professor at the University of Geneva.[18]

Isolated from foreign scientists, the French scientific community was also divided within.[19] French students of chemical thermodynamics, for example, had to contend with the still-powerful Marcellin Berthelot, professor at the Collège de France and permanent secretary of the Académie des Sciences. Although the Ionists sought to integrate Berthelot into the prehistory of their science, in France Berthelot was a part of the present rather than the past. Until entombed in the Panthéon in 1907, he steadfastly held to his principle of maximum work, according to which purely chemical reactions always tend to yield the product that liberates the greatest amount of heat. Younger colleagues who, following Gibbs and van't Hoff, criticized this doctrine, did so at peril to their careers.[20]

Yet French scientists could not entirely ignore the study of solutions or the implications of thermodynamics for chemistry. In Paris, a few young scientists, fortified by the intensive training in physics and mathematics that was integral to education at the *grandes écoles*, discovered and began to explore the writings of that native of a sister republic, J. Willard Gibbs.[21] Independently of contacts with the Ionists, several developed into something like physical chemists, although chemical physicists might be a better term, since their approach was generally far more abstract and formal than that of the Ionists.[22] At the provincial science faculties, fast becoming centers of industrial science, scientists more concerned with practical electrochemistry than theoretical thermodynamics took up the ideas and techniques of the Ionists with an enthusiasm altogether lacking in Paris. By 1900, laboratories in such cities as Nancy and Toulouse offered students a training in physical chemistry that was probably indistinguishable from that offered at the *technischen Hochschulen* of Germany.[23]

Throughout Europe it was, with a few notable exceptions, new institutions that proved most hospitable to the Ionists and their disciples, at least prior to World War I. In Germany, physical chemists enjoyed a far stronger position in the rapidly expanding *technischen Hochschulen* than in the universities. In Britain, it was such relatively new foundations as University College London and the University of Liverpool that offered students of Ostwald their initial opportunities, rather than Oxford or Cambridge. In France, the most vigorous development occurred at such provincial faculties as Nancy, Toulouse, and

Bordeaux rather than the Sorbonne or the *grandes écoles* of Paris. Elsewhere, Arrhenius, van't Hoff, and their students found homes in the recently created Nobel Institute of Stockholm and the young University of Amsterdam. The Ionists had begun their careers on the margins of European scientific culture; most of their students would do likewise.

America, where graduate education was barely older than the ideas of the Ionists and where universities were expanding at a rate unmatched by any European nation, would prove an especially fertile ground for this new discipline to take root and grow. Yet opportunities are no guarantee of achievement. Although unencumbered by the traditions and entrenched competitors facing the Ionists in Europe, American physical chemists confronted both the advantages and disadvantages of their nation's comparative backwardness. Instead of elbowing their way into existing laboratories and institutes, American physical chemists had to build them; instead of asserting themselves against powerful intellectual rivals, they had to create traditions of research and scholarship in a country that had long proved resistant to both.

OSTWALD AND THE AMERICANS

Nowhere were Ostwald's ambitions rewarded as richly as in the United States. Over forty American chemists worked with him in Leipzig prior to his retirement (see Table 2.1), and by 1906 members of this group had advanced to the rank of full professor at many of the premier institutions of higher education in America, among them Harvard, the Massachusetts Institute of Technology, Cornell, Wisconsin, Stanford, Columbia, and Johns Hopkins. A census conducted in 1901 by the American Chemical Society reported that there were over 500 students enrolled in physical chemistry courses in 39 institutions in the United States.[24] Over half were studying at colleges with Leipzig alumni on their faculties. The results of a 1946 questionnaire answered by over 130 leading American chemists provides one measure of their impact; three of Ostwald's American students (G. N. Lewis, A. A. Noyes, and T. W. Richards) were named among America's six most influential teachers of chemistry.[25]

Ostwald's American students distinguished themselves through research as well as teaching. The *Journal of Physical Chemistry*, the first periodical devoted to the subject outside Germany, began publication at Cornell in 1896; by 1906 it had printed over 300 research papers, almost all written by Americans and Canadians and over one quarter by former Leipzig students. Between 1896 and 1906 an additional 145 articles treating physico-chemical topics appeared in the *Journal of the American Chemical Society*, of which Leipzig alumni wrote approximately one third.[26] A substantial number of papers by Americans also appeared in other domestic publications, such as the *American Chemical Journal*, as well as in the major foreign periodical, the *Zeitschrift für physikalische Chemie*. In recognition of the presence of American contrib-

TABLE 2.1
Americans Who Worked in Ostwald's Laboratory

Names	Prior Education	Years at Leipzig	Later Positions
M. Loeb	Harvard/Berlin	1889	Clark/NYU
A. A. Noyes	MIT	1888–90	MIT/Caltech
E. Buckingham	Harvard/Strassburg	1890–91 and 1892–93	Bryn Mawr/USDA/NBS
W. D. Bancroft	Harvard/Strassburg	1890–92	Harvard/Cornell
J. E. Trevor	Cornell	1890–92	Cornell
C. S. Palmer	Amherst/Johns Hopkins	1892–93	U Col/industry
H. C. Jones	Johns Hopkins	1892–94	Johns Hopkins
H. M. Goodwin	MIT	1892–94	MIT
A. J. Wakeman	Yale/MIT	1892–94	Yale/Herter Lab/Ag. Res. Station
C. W. Coggeshall	Grinnell/Harvard	1892–95	Harvard/industry
J.L.R. Morgan	Rutgers	1892–95	Stevens/Brooklyn/Poly/Columbia
L. Kahlenberg	Wisconsin	1893–95	Wisconsin
O. F. Tower	Wesleyan	1894–95	Wesleyan/Western Reserve
T. W. Richards	Haverford/Harvard	1895	Harvard
E.H.S. Bailey	Yale/Strassburg/Illinois Wesleyan	1895	Kansas
W. R. Whitney	MIT	1894–96	MIT/GE
S. L. Bigelow	Harvard/MIT	1895–98	U Mich
G. A. Hulett	Princeton	1896–99	U Mich/Princeton
E. C. Sullivan	U Mich/Göttingen	1896–99	U Mich/USGS/Corning Glass
W. G. Smeaton	Toronto	1898–1902	U Mich
A.W.C. Menzies	Edinburgh	1898	Glasgow/Chicago/Oberlin/Princeton
J. White	Johns Hopkins	1898	Nebraska/Rose Poly
P. T. Walden	Yale	1899–1900	Yale
A. W. Young	Cornell/Stanford	1899–1900	Stanford

H. Schlundt	Wisconsin	1899–1900	Wisconsin/Missouri
H. W. Foote	Yale	1899–1900	Yale
C. W. Foulk	Ohio State	1899–1901	Ohio State
G. N. Lewis	Nebraska/Harvard	1900–1901	Philippines/MIT/U Cal
M. Sherrill	MIT	1900–1901	MIT
A. A. Blanchard	MIT	1900–1902	MIT
F. G. Cottrell	U Cal/Berlin	1901–1902	U Cal/US Bur. of Mines/industry
G. W. Heimrod	Harvard	1901–1905	Rockefeller Inst.
H. W. Morse	Stanford	1902	Harvard/U Cal/industry
G. V. Sammet	MIT	1902–1905	industry
W. C. Bray	Toronto	1902–1905	MIT/U Cal
W. V. Metcalf	Oberlin/Johns Hopkins/Würzburg	1903–1905	industry/Fisk
C. G. Fink	Columbia	1903–1907	GE/Columbia
J. P. Mitchell	Stanford	1904–1905	Stanford
W. H. Sloan	Stanford	1904–1905	San Francisco Board of Health/Stanford
S. C. Lind	Washington & Lee/MIT	1904–1905	MIT/U Mich/US Bur. of Mines/U Minn
H. F. Sill	Princeton/Göttingen	1904	Carnegie Tech/Westinghouse
E. B. Spear	Manitoba/Toronto	1904	MIT/industry
A. B. Lamb	Tufts/Harvard	1904–1905	NYU/Harvard
J. W. McBain	Toronto	1904	Bristol/Stanford

Sources: R.G.A. Dolby, "The Transmission of Two New Scientific Disciplines from Europe to North America in the Late Nineteenth Century," *Annals of Science* 34 (1977): 392; *American Men of Science: American Chemists and Chemical Engineers*, ed. Wyndham D. Miles (Washington, D.C., 1976); C.J.S. Warrington and R.V.V. Nicholls, *A History of Chemistry in Canada* (Toronto, 1949); and H. F. Sill to Andrew S. Draper, 23 April 1904, LAS, Box 1.

utors Ostwald had by 1906 appointed two of his former American pupils to the editorial board of the *Zeitschrift*, A. A. Noyes of MIT and T. W. Richards of Harvard.

Chemistry was growing rapidly in the United States at the turn of the century, but physical chemistry was expanding far more rapidly than the science as a whole. While only 5 percent of all articles published in the *Journal of the American Chemical Society* in 1896 dealt with physical chemistry, 15 percent did so ten years later, and by the mid-twenties over 25 percent of the papers published in that journal treated topics in physical chemistry (see Table 2.2). Although these figures may be taken only as rough indices of the distribution of American chemical research by field, they err—if at all—on the side of underestimating the growth of physical chemistry, since physical chemists developed their own journal in the 1890s, years before other specialists did likewise. Physical chemistry, one observer noted in 1926, "now seems about to swallow up chemistry proper."[27]

Although it is easier to discuss the quantity of research done by a group of scientists than the quality of that work, a group including such figures as T. W. Richards, A. A. Noyes, G. N. Lewis, and Willis R. Whitney was not lacking in talent. Three of Ostwald's American students (Lewis, Noyes, and Richards) received the Davy Medal, the highest honor of the Royal Society of London; four others (Cottrell, Fink, Sullivan, and Whitney) were awarded the Perkin Medal by the Affiliated Chemical and Electrochemical Societies of America for accomplishments in applied chemistry. T. W. Richards was the 1914 Nobel laureate in chemistry. Six members of the group were elected to the presidency of the American Chemical Society (Bancroft, Lamb, Lind, Noyes, Richards, and Whitney), and eleven were elected to membership in the National Academy of Sciences (Bancroft, Bray, Cottrell, Hulett, Lamb, Lewis, Lind, Noyes, Richards, Sullivan, and Whitney). Twenty-eight re-

TABLE 2.2
Articles on Physical Chemistry Published in the *Journal of the American Chemical Society* by Five-year Periods, 1894–1926

Years	Number of Articles on Physical Chemistry	Total Number of Articles	% of Total on Physical Chemistry
1894–1898	37	674	5
1899–1903	69	615	11
1904–1908	155	862	18
1909–1913	201	943	21
1914–1918	274	1,304	21
1919–1923	438	1,580	28
1924–1926[a]	339	1,248	27

[a] Three-year period.

ceived stars beside their names in the various editions of *American Men of Science*, signifying that they had been named by their peers to be among the leading scientists in America.

Statistics such as these show that physical chemistry expanded rapidly in America during the decades following 1890 and suggest that Ostwald's pupils played an important role in promulgating the specialty. But who were Ostwald's American students? Where did they come from? Why did they go to Leipzig and how were they received upon returning to the United States? Answers to these questions are important if we are to understand why this specialty prospered in the United States. No less important is the question of what Ostwald's students shared besides a common teacher. What did it mean to be a physical chemist in late-nineteenth-century America?

The family backgrounds of the Americans who worked in Ostwald's laboratory are most notable for their essential sameness. More than half were from New England and the Middle Atlantic states; none came from the states of the former confederacy and only a handful from west of the Mississippi. Three were born in Ontario but attended college in the United States and later settled there. The great majority were of British ancestry; in many cases the family's voyage to America occurred in the seventeenth century. Noyes's ancestors settled in Newbury, Massachusetts, in 1633; Bancroft's forebears landed in Massachusetts in 1632. Lamb, Lind, and Cottrell also were descended from colonists of the 1630s. It need hardly be said that most were raised as Protestants, although those whose personal papers survive betrayed no strong religious convictions. Their fathers were lawyers, bankers, merchants, and farmers. Most appear to have been accustomed to living comfortably; even those who traveled to Germany on scholarships were from fairly prosperous families. One or two might be described as men of independent means. In short, their family circumstances were quite typical of those entering professional careers in late-nineteenth-century America.[28]

The first of these Americans to enter Ostwald's institute made their way to his laboratory by circuitous paths. Americans in the 1880s and 1890s, of course, had ample access to information about German universities. Magazines, newspapers, and scientific periodicals occasionally published accounts of living conditions in German university towns; hundreds of professors, especially chemists, had recent personal experience of graduate study in German universities; and, as is always the case among students, there was an effective grapevine by which graduates of American colleges could learn of conditions in Germany from other graduates a year or two their senior. In addition, major German university towns often had American clubs, associations of students from the United States and Canada, which served as sources of up-to-date academic gossip for those recently arrived in Europe. Prior to the mid-1890s, however, none of these sources could provide much information about the new professor at the second chemical institute at Leipzig. Ostwald lacked a cadre of American disciples to furnish him with fresh recruits and his specialty and laboratory were too young to have attracted much attention on the other side of the Atlantic.

Chance brought Ostwald his first American students. Many found his institute a refuge from the organic chemistry they had gone to Germany to study. Morris Loeb obtained a doctorate at Berlin under the great dye chemist, A. W. von Hofmann, before visiting Leipzig and joining Ostwald's still-small band. Wilder D. Bancroft entered Ostwald's laboratory following work in organic chemistry under C. Loring Jackson at Harvard and Rudolph Fittig at Strassburg. Edgar Buckingham followed a similar path. The experience of Arthur Amos Noyes, whose student-day letters to his friend Harry Manley Goodwin still exist, may well have been typical of that of his compatriots. He embarked for Europe in 1888 in the company of several recent graduates of MIT. All had studied organic chemistry, and Noyes had already done a little research on the structure of organic compounds under the guidance of his teacher at MIT, L. M. Norton. Their goal now was to study under Adolph von Baeyer, the world-famous chemist who, only five years previously, had solved the problem of the structure of indigo. When their boat docked in Antwerp, however, they received a letter from von Baeyer telling them that there would be no room for them in his Munich laboratory. After sampling courses at Heidelberg and Bonn, they settled in Leipzig.[29] There, a year of work on an intractable problem in stereoisomerism dampened Noyes's enthusiasm for organic research. As he explained it:

> last May [1889] after the Spring vacation, I went to work at my old Arbeit in organic chemistry under Wislicenus, although it had already begun to look rather hopeless, a state of affairs by no means unusual in organic work. Still I plugged away at it until about the middle of the term, when it had become evident that it was only a waste of time to continue it. In the meantime I had been hearing a course of lectures on theoretical and physical chemistry by Prof. Ostwald, and had become much interested in the subject, especially as it is for the most part new to me; so it didn't take me very long to decide to give up organic work and begin on a physical chemical Arbeit which I did at the beginning of this term last October [1889]. The Arbeit which I have been and still am at work on is an investigation of the influence of one salt on the solubility of another in water; . . . I have already got a number of important results, and don't think I could possibly have found a subject that would be more interesting to me . . .[30]

Noyes's high hopes for his second *Arbeit* were not disappointed; his dissertation gave valuable experimental support to Nernst's solubility-product principle. By the summer of 1890, Noyes had successfully defended his thesis and had passed exams in chemistry, physics, and mathematics, although not without considerable cramming to compensate for his weak preparation in the calculus and chemical thermodynamics.[31]

Frustration and discouragement shook Noyes's commitment to organic chemistry: frustration over an unrewarding research project and discouragement over the prospect of attaining significant results in a field in which so much had been achieved and so many were working. By contrast, there was the novelty of Ostwald's lectures and the opportunity to do important research; together these made physical chemistry seem irresistible to the young and ambitious American.

The rapidity with which van't Hoff, Arrhenius, and Ostwald erected a theoretical structure covering much of the data of physical chemistry could not help but create a sense of excitement and momentum. At the same time, the structure in many places was still skeletal, and this preserved a sense that opportunities for future work were still available. There was an abundance of experimental and theoretical research problems and many of them could be resolved using simple equipment, by chemists with relatively little training. As an American physicist put it:

> During the last few years the scientific advances made in physical chemistry have been marvelously rapid, and especially has this been the case in those lines which relate to solution and electrolysis. Questions present themselves to the investigator from all sides of the subject, and their solutions are often within the reach of limited apparatus and limited time on the part of the investigator.[32]

Important too was Ostwald's personality. More than a few world-wise Europeans were taken by his enormous energy and broad intelligence. William Ramsay described him simply as "the most brilliant man I ever met."[33] Small wonder that, to inexperienced Yankees, he seemed invincible.

> My "old man," I mean, Prof. Ostwald, is also a very pleasant man; and he is a terror, I tell you, in knowledge. He has got not only chemistry, but physics down cold. He goes into our Physical Colloquium, and criticizes the new Arbeits with more authority than the physics professor himself, and in the chemical society he often gets into long discussions with Wislicenus . . .[34]

Nor was this impression peculiar to Noyes. Harry Clary Jones, writing ten years after he left Leipzig, recalled similar characteristics:

> Ostwald always impressed the writer of this review as having a control of physical and chemical literature that was simply wonderful—matters which other men of far maturer years knew where to find, Ostwald had always at his finger's end. This seemed to apply to almost every phase of Physics, Inorganic Chemistry, and Physical Chemistry. A personal characteristic which made a lasting impression on us all was the absolute lack of selfishness in our teacher. Investigation after investigation was suggested to us, and when complete we were told to publish it as if it were of our own thinking. . . .[35]

Omniscience and generosity in a teacher, a record of stunning achievement, and broad opportunity in a subject—a serious student could ask for little more.

Of course, not all Americans found Ostwald and Leipzig irresistible. By the turn of the century, Ostwald's institute had become crowded and Ostwald himself was spending less and less time with his students and leaving more and more of the instructional duties to his assistants. And students now could weigh the fading charms of Leipzig against the excitement available at new institutes of physical chemistry. All this was reflected in the experience of Irving Langmuir. Langmuir, a 22-year-old graduate of the Columbia School of Mines, arrived in Germany in the summer of 1903 undecided whether to work under Ostwald at Leipzig or under Nernst, Ostwald's former assistant who had won his own institute at Göttingen. He visited both and found the choice a difficult one. He much admired Ostwald; the summer after his freshman year in college he had made a bicycle tour of the Catskill Mountains and packed but two books: Darwin's *Origin of Species* and Ostwald's *Grundriss der allgemeinen Chemie*. Yet when he arrived at Leipzig, Langmuir discovered that Ostwald spent most of his time writing books. Langmuir thought the institute's laboratory and library were magnificent and enjoyed the city's music and theater, but Leipzig was also noisy and full of Americans, and its citizens spoke their own language sloppily. Most disturbing, Langmuir was given a cold reception by other students at Ostwald's institute; even his fellow Americans offered him little help in getting settled. Ostwald's gradual drift away from the laboratory and the growth of his research school had destroyed the family-like ambience of the "old pile of bricks" that once had housed his institute. At Göttingen, by contrast, Langmuir found a small but friendly American colony, a beautiful countryside, and a people who spoke pure Hanover German. The laboratory was poorly equipped, but he thought Nernst was "a splendid professor," and he was greatly impressed with the quality of lectures on physics and mathematics. At the end of the summer Langmuir decided to seek his degree at Göttingen.[36]

The first Americans to study physical chemistry in Germany—Loeb, Noyes, Bancroft, and Buckingham—did not leave America with the intention of entering a new discipline, but later American students, like Langmuir, by and large did. Most came from colleges at which they had been exposed to some coursework in the subject; many had contact with one or more of Ostwald's earlier students before leaving for Europe. They went to Germany with clearer views of their goals and often with letters of recommendation from Ostwald's former co-workers that smoothed entry into Ostwald's institute at a time when competition for positions was becoming intense.[37] Many sampled the fare at two or more institutes of physical chemistry. Moving freely from university to university was a hallowed tradition among German students; Americans, anxious to see as much as possible of the country and its famous professors, were even more peripatetic. A large number visited both Leipzig and Göttingen, as did Langmuir; a few spent a semester or two in van't Hoff's laboratory at Berlin. The great majority of those who remained to obtain degrees, however, continued to settle in Leipzig.

Graduates of Harvard and the Massachusetts Institute of Technology were especially prominent among the Americans who studied physical chemistry at Leipzig. Indeed, seventeen of Ostwald's American students went to Leipzig from these institutions. It is not surprising that Harvard was a source of substantial numbers of students. Harvard attracted many of America's ablest undergraduates and had recently reaffirmed its leadership as a center of learning and research in America after having been challenged in the 1870s by Johns Hopkins. Moreover, its chemistry department, even in the 1880s, possessed a chemist unusually sensitive to the relations between chemistry and physics— Josiah P. Cooke. In the nineties Cook's most illustrious pupil, T. W. Richards, went to Leipzig himself and thereafter influenced several others to do likewise.[38]

That so many graduates of MIT should have undertaken study at Leipzig is noteworthy. In the 1890s, MIT possessed neither the reputation for scholarship nor the climate for research that Harvard enjoyed. Its chemistry department was of only local repute; its students for the most part were from modest homes and usually were oriented toward subjects more practical than theoretical chemistry. Nevertheless, the Institute was growing rapidly in the late nineteenth century, and as it grew it experienced a great need for teachers to handle the chemistry courses and laboratory exercises that formed part of every student's program of study. Like many American colleges during this period, MIT looked to its own graduates for help; faculty inbreeding was not yet a recognized danger.[39] In 1890, MIT hired Noyes together with Henry P. Talbot and Augustus H. Gill as instructors. Talbot and Gill, like Noyes, had graduated from the Institute in the 1880s, served as laboratory assistants, and then gone to Germany. All received doctorates from Leipzig in 1890, although Gill and Talbot, despite having attended Ostwald's lectures, completed their theses under Wislicenus.[40] From their positions within the growing chemistry department, these young instructors worked to effect substantial changes in the undergraduate chemistry curriculum.

Noyes and Talbot were especially responsible for developing a strong program in physical chemistry at the Institute. They made their impact by introducing changes in an established course. MIT had for many years required that all students enrolled in the chemistry program take a course entitled ''Theoretical Chemistry,'' the syllabus of which included topics drawn from all parts of the subject: organic structural theory and stereochemistry, affinity theory, thermochemistry, and the periodic table.[41] The idea seems to have been to segregate speculative or recently developed concepts from the hardcore of facts and laboratory procedures studied in other courses. Noyes and Talbot, assigned responsibility for the course soon after their return from Germany, altered the syllabus so that 45 out of 75 hours of classwork were given to the phenomena of solution theory and chemical change. Fifteen hours were devoted to the study of chemical equilibria alone. In addition to classroom instruction, 30 hours of laboratory work were also required. Discussion of structural theory was transferred to separate courses in organic chemistry.[42] By the mid-1890s, MIT's course in theoretical chemistry had been transformed for all intents and purposes into an introductory course in physical chemistry.

The efforts of Noyes and Talbot to reshape this course and to win greater attention for the principles of mass action and electrolytic dissociation in other courses did not meet with unanimous praise. The extra work seemed an imposition to many students, and some faculty members thought such studies too abstract to meet the needs of engineers.[43] Nevertheless, the changes were popular among those students and teachers who were oriented more to the basic sciences than to engineering. Through their advocacy of curricular reforms and their emphasis on research, Noyes and, to a lesser degree, Talbot earned reputations as reformers and leadership in the party of those seeking to make MIT more than just a good engineering school. They also won talented recruits for physical chemistry. Noyes, in particular, was successful in spotting able students in his classes and grooming them for future careers in research. The best he would engage as assistants in his own work on solubility effects and reaction rates. Although some of these assistants later left chemistry, such as C. G. Abbot, who went on to a distinguished career as an astrophysicist and Secretary of the Smithsonian Institution, others followed Noyes's footsteps to Leipzig: first his friend H. M. Goodwin, later his students and co-workers Willis R. Whitney, Miles Sherrill, A. A. Blanchard, G. V. Sammet, and S. C. Lind.

The upper-level course on theoretical chemistry, also sometimes called "chemical philosophy," was common to a number of American colleges, and the technique Noyes employed of altering the design of this established course was also used by several other Leipzig alumni. Richards at Harvard and Morgan at Columbia both inherited courses on chemical philosophy and transformed them into courses on physical chemistry. Others, such as Trevor and Bancroft at Cornell, Tower at Wesleyan, and Buckingham at Bryn Mawr, established new courses under the name of physical chemistry. By 1900, the principles of solution theory, mass action, and chemical thermodynamics were called theoretical chemistry in some chemistry departments, chemical philosophy in others, and physical chemistry in still others.

The number of names under which these topics were taught gradually dwindled after the turn of the century, so that by the outbreak of World War I physical chemistry was by far the most common appellation. Nevertheless, the multiplicity of titles in use prior to the war raises the question of whether and to what extent Ostwald's American students were in agreement over the content and goals of their field. That each worked at Leipzig under Ostwald's direction for a year or two does not necessarily imply that their views of chemistry were thereby significantly shaped. In what measure did they accept and take away what was offered them by Ostwald? How did their common—though not simultaneous—sojourn in Leipzig shape their views of chemistry, and did it result in a consensus regarding beliefs and purposes after their return to America?

The record here is unambiguous. Most of Ostwald's American pupils were deeply influenced by their time at Leipzig. For most this influence was enduring. In their research, most passed their careers working within the conceptual limits defined by their teacher. As teachers themselves they emulated Ostwald, not so much in matters of style—for Ostwald's personality was unique—as in matters of content. Their courses and textbooks conformed to the mental map of science developed in Ostwald's own voluminous writings: a map in which physical chemistry occupied a central position, a natural crossroads of intellectual commerce. They commonly evinced loyalty to Ostwald's views regarding the goals of chemistry, differing with him only after Ostwald came to view himself more a philosopher than a chemist.[44] In these major ways, and in many lesser ways, Ostwald's American students made their debts obvious. They were a cadre of true believers, convinced, as was Ostwald, that physical chemistry was the chemistry of the future. They had enthusiasm, talent, and that personal knowledge of a master's technique which is so important in a laboratory science.

Like other scientists of their generation—the experimental biologists of the 1890s who criticized traditional morphology, or the astrophysicists who sought to transcend the limits of positional astronomy, or the geophysicists who complained of the guesswork called geology—physical chemists defined themselves in opposition to older traditions. And like these other young Turks, physical chemists wished to make their parent science analytic rather than descriptive, quantitative rather than qualitative, precise and definite rather than speculative. Fundamental among their shared beliefs was the conviction that the laws governing chemical change should constitute the focus of a chemist's research.

> The physical chemist studies the reaction and not the end products. . . . [He studies the reaction] qualitatively and quantitatively, with special reference to such factors as initial concentration, temperature, solvent, pressure, electrical stress, and time. He does more than this. He correlates his facts and draws conclusions from them, so that it becomes possible to generalize from one reaction to all reactions.[45]

This emphasis on general laws, Ostwald's students argued, differentiated physical chemists from other chemists. Many thought the gulf separating their field from organic chemistry was especially wide.

American organic chemists held little of the power in the chemical community that their German colleagues possessed. America had no dyestuffs industry worthy of the name and hence there was but limited demand for organic specialists. By contrast, American agriculture and industry had great need of technicians to perform quality-control tests on the ores, metals, drugs, bulk chemicals, fertilizers, and other products that were being produced and consumed in ever-growing quantities. Not surprisingly, analytical and agricultural chemists dominated most academic departments in the United States.[46] Nevertheless, Ostwald's American pupils were hardly less vitriolic than their teacher in their criticisms of organic chemistry. "Too many chemists," Morris Loeb wrote in 1889, "hurry in their studies toward the El Dorado of carbon synthesis, striving to obtain at the earliest moment tangible results in the shape of new substances, be they dyes, drugs, or merely triumphs of synthetical art. . . . Like the miners of '49 the specialist in organic chemistry has but one thought"—that thought being to discover "a rich nugget." Both Loeb's concept and the literary figure were borrowed directly from Ostwald.[47]

Some of Loeb's colleagues were more charitable (or diplomatic) and conceded that other branches of chemistry did contain, in addition to a vast accumulation of details, some generalizations of a higher order. S. L. Bigelow, among others, recognized this and claimed that it was the physical chemist's province to make a specialty of those generalizations. Physical chemistry, he wrote in his textbook of 1912, "stands in the same relation to the subdivisions of the science of chemistry in which philosophy stands toward all the sciences. Its main object is to unify thought within the science of chemistry; therefore, it might well be named, the 'philosophy of chemistry.' "[48] Bigelow was here indulging in rhetoric that Ostwald first used when he sought to establish his new field as an *allgemeine Chemie*, a science that would serve as a framework for all chemical knowledge, extending to cover the principles of analytical, inorganic, and even organic chemistry. Like his teacher, Bigelow did not clearly distinguish between the set of all the laws and principles of chemistry and that subset of generalizations dealing with phenomena on the borderland of physics and chemistry. The physical chemist's task, Bigelow claimed, was to make chemistry a true science by specifying and systematizing the laws governing all chemical phenomena.

Although Bigelow's rhetoric was reminiscent of his teacher's views on *allgemeiner Chemie*, the substance Bigelow provided in his text more closely resembled the narrower research program in physical chemistry that Ostwald had charted in his Leipzig laboratory. Its central chapters dealt with osmotic pressure and solution theory, electrolytic dissociation, and the applications of the law of mass action; no attempt was made to discuss principles truly drawn from other branches of chemistry, such as recent generalizations regarding coordination compounds or older principles of organic structural theory. The disparity between claims and substance found in Bigelow's textbook was characteristic of the work of many of Ostwald's American students.

Ironically, the science that was to systematize the great principles of chemistry was characterized during its early years in America by a concern with specifying the final decimal point which to outsiders seemed to border on the obsessional. The reasons are not hard to find. The remarkable series of theoretical advances that van't Hoff, Arrhenius, Ostwald, and Nernst made between 1884 and 1891 had opened up many lines of experimental work. Exploring the ramifications of Arrhenius's electrolytic dissociation theory was perhaps the most obvious and important of them.

The dissociation theory had been proven to apply to weak electrolytes in dilute aqueous solutions, but by the late nineties it was becoming increasingly clear that more concentrated solutions of such electrolytes (concentrations greater than, say, 0.5 molar) as well as solutions of certain strong electrolytes failed to behave according to predictions. Shortly before the turn of the century, chemists began to investigate nonaqueous solutions and here too found the dissociation theory wanting. These anomalous findings presented the Ionists and their followers with riddles that demanded the utmost in experimental precision. The problem of reconciling the behavior of strong electrolytes with the dissociation theory was especially vexatious. The ionization theory was demonstrably successful when applied to dilute solutions of weak electrolytes, so it seemed briefly that the key to understanding the behavior of strong electrolytes might be found if only solutions were attenuated sufficiently. Later, workers such as Noyes came to suspect that patterns might emerge if solutions of strong electrolytes were studied at extreme temperatures. Others sought to trace anomalous results to systematic errors in techniques for determining the dissociation of solutes. Such factors as the purity of materials and constancy of experimental conditions were of great importance to these studies and rendered results susceptible to dispute. Even the most common procedures were called into question as investigators sought to find possible sources of error. Research more notable for the virtuosity of experimental techniques than the significance of conclusions was often the outcome.[49]

Problems associated with the electrolytic dissociation theory attracted greater attention from American physical chemists than any other topic prior to World War I, but not all of Ostwald's former students saw the strenuous prosecution of research on solutions as an urgent desideratum. The goal of understanding chemical change was a very broad one; it could be pursued in many ways. Wilder D. Bancroft, who advanced rapidly to a professorship at Cornell after founding the *Journal of Physical Chemistry* in 1896, stands out for his vigorous promotion of alternative research strategies. He believed that those engaged in the intense scrutiny of dilute aqueous solutions were neglecting more promising research opportunities. The physical chemist, he thought, was in the dilemma of finding his field of vision daily growing smaller. "A tenth-normal solution," he wrote in 1905, "is now considered a concentrated one and some people are so extreme as to maintain that we can not expect agreement between theory and experiment for anything except infinitely dilute solutions.'' Instead of losing themselves in the study of ''slightly polluted water,'' Bancroft argued that physical chemists should look to develop a better knowledge of heterogeneous equilibria, catalysis, and colloidal behavior—topics that Bancroft thought were richer in practical applications and more accessible to study. Bancroft's *Journal* became the principal outlet of those, such as Louis Kahlenberg of Wisconsin, who were skeptical of the work of the ''dilute school.'' Nevertheless, although Bancroft differed with many of his colleagues over the most promising avenues for the growth of physical chemistry, the differences were over means and not ends. Bancroft was interested in finding more effective ways to approach the study of chemical change; his research fell within a framework shared with those studying dilute solutions.[50]

Another common ground for Ostwald's American students was their conviction that physical chemistry was what may be termed a donor specialty. That is, they believed that problems awaiting solution in other branches of pure and applied science might be solved through application of laws and techniques derived from physical chemistry. This view was entirely consonant with the position that it was the primary task of physical chemists to explore the general principles governing chemical change. Since they believed their field to have made substantial progress toward this goal, they quite naturally saw possibilities for widespread application of their knowledge. Of course, it was nothing new for scientists seeking support to invoke utility as a justification for their work. But Ostwald's disciples made the case for the usefulness of physical chemistry with uncommon vigor, and they demonstrated a willingness to pursue applications which belied the possibility that their commitment was primarily rhetorical. Ostwald and his American pupils saw their field as being relevant to other branches of chemistry, other sciences, and industry.

Within chemistry, physical chemists could point to the bearing the solution theory had upon the field of analytical chemistry. The qualitative analysis of inorganic substances provided the most dramatic and convincing example of the explanatory power of the electrolytic dissociation theory. Although analytical chemistry had attained a high degree of practical development during the first three quarters of the nineteenth century, the theory underlying many of its most common procedures and typical reactions was little understood. The electrolytic dissociation theory provided a framework in which these processes might be interpreted, showing that most analytical tests were actually ion-specific, rather than molecule-specific. Thus, all solutions containing copper salts gave positive results when tested for the presence of copper, no matter what the nature of the specific salt, because solutions of all copper salts contained dissociated copper ions. Much of the rest of analytical chemistry was also found to rest upon the properties of ions in solution.[51]

Approaching analytical chemistry from this viewpoint had obvious pedagogical advantages; ''if he [the student] knows 100 different reactions he also knows 10,000. He does not have to learn every possible case, therefore, but simply the standard reactions of the various kinds of ions."[52] Ostwald was the first to develop this application of the electrolytic dissociation theory, but several of his American students quickly became active in the area, filling the need for English-language textbooks and laboratory manuals. A. A. Noyes, O. F. Tower, W. C. Bray, and Miles Sherrill all wrote or contributed to textbooks of qualitative analysis after their return from Germany. The text by Noyes was especially successful, going through ten editions over fifty years.[53] By the beginning of the twentieth century, knowledge of the principles of physical chemistry was fast becoming indispensable to teachers of the humblest introductory courses; those who tried to teach analytical or inorganic chemistry without reference to ions were coming to look old-fashioned indeed.

In addition to seeing physical chemistry as relevant to other branches of chemistry, Ostwald's students also thought that problems in other sciences might be found more tractable when approached from a physico-chemical viewpoint. Medicine and physiology were particularly attractive targets for speculation and investigation. In a paper written in 1898, Theodore Richards sought to relate the taste of acids to their degrees of electrolytic dissociation. This article soon embroiled him in a controversy with Louis Kahlenberg.[54] Kahlenberg himself had spent several years working on applications of electrolytic dissociation to solutions found in nature, investigating methods for testing the toxic action of dissolved salts on bacteria, seedlings, and plants.[55] Knowledge of the toxic effects of ions on bacteria was thought to offer an opportunity to gauge the effectiveness of antiseptics; information on the "death limits" of seedlings might allow for better planning of irrigation projects to lower the concentration of salts in groundwater.[56]

Such work, albeit sketchy, soon attracted attention at the United States Department of Agriculture, which had established itself as the principal public patron of chemical research in America. Work there began when Frank Cameron, a protégé of Wilder D. Bancroft, joined the Bureau of Soils in 1898.[57] Several other physical chemists were hired subsequently, so that by 1908 a substantial colony could be found in Washington.[58] The soil, Cameron argued, should be considered the "stomach of the plant," since before roots could absorb nutrients those nutrients must first be dissolved. Because plant nutrition involves the selective absorption of inorganic salts through the root membrane, Cameron saw it as a problem of applied physical chemistry. The law of mass action and electrolytic dissociation theory might be used to treat the process by which nutrients enter solution in the soil, and the theory of osmotic pressure might be employed to develop a model of the way the root selectively absorbs those nutrients.[59]

These early efforts to extend physico-chemical techniques and principles to biological phenomena proved to be for the most part premature, but work of a more substantial nature soon followed. Jacques Loeb's announcement in 1899 that he had successfully raised sea urchins to the larval stage by altering the salt concentration of solutions containing unfertilized eggs did much to stimulate interest in the relationship between osmotic pressure and life processes. Loeb's discovery, and the physico-chemical methods he employed in making it, received widespread publicity, so that, two years later, when van't Hoff devoted two lectures to the physiological applications of physical chemistry during a visit to Chicago, he was addressing an American audience well prepared for such views.[60] Less sensational, but no less important, was the research of Lawrence J. Henderson, a student—and later brother-in-law—of T. W. Richards. Henderson's work on acid-base equilibria and the action of buffers, which began to appear in 1908, went a long way toward reconstructing the physiologist's understanding of blood by treating it as a solution subject to the laws of chemical equilibrium.[61] Later, in the teens and twenties, physical chemists entered physiological and biochemical research in large numbers, some by way of the study of colloids and others via the study of the structure of biological molecules. The opening of positions for physical chemists in biomedical laboratories, such as at the Rockefeller Institute, and the creation of a department of physical chemistry at the Harvard Medical School reflected the impact of work such as that of Loeb and Henderson, and confirmed the view, already held by many of Ostwald's pupils, that the day was "not far distant when a knowledge of physical chemistry will be an essential part of a physiologist's education."[62]

American physical chemists were no less enthusiastic about applying their knowledge to industry than to problems drawn from other sciences. German firms had demonstrated how important organic chemistry could be in developing profitable new products; synthetic dyes had made fortunes for their manufacturers, destroyed markets for many natural colors, and transformed the chemical industry itself. Physical chemistry, its proponents claimed, might not generate comparable products, but it could lead to new, more efficient, and more profitable processes. And its influence would not be confined to firms making chemicals, but would be felt wherever materials were processed. Problems involving the effects of concentration, pressure, and temperature upon the solubility of substances, the yield of reactions, and the behavior of materials were common to many industries. Physical chemists believed that the solution theory, law of mass action, and phase rule provided the tools whereby such problems could be solved.[63]

The Ionists, both in Europe and America, were quick to recognize the potential their field had for industrial application. In 1894, Ostwald himself had helped to organize the *Deutsche electrochemische Gesellschaft* (later rechristened the *Bunsen Gesellschaft*) in order to foster ties between physical chemists and industrialists. Later, he and a co-worker patented a process for the catalytic production of nitric acid from ammonia that proved critical to German industry during the First World War. Many of Ostwald's colleagues also placed industrial applications high on their lists of priorities. Indeed, the principal leaders of physical chemistry in Germany—Nernst, Haber, Ostwald, and, after he moved to Berlin in 1896, van't Hoff—all were involved in industrial work by the turn of the century.[64]

In America, Ostwald's students echoed this interest. Business firms were not always quick to take advantage of their talents, and it often proved more difficult to convert scientific knowledge into technology than anticipated. Nevertheless, more than half of those who returned from Leipzig engaged in research for industrial firms or took out patents for industrial processes. Several attained renown for such work. Willis R. Whitney's role in organizing the research laboratory of the General Electric Company is well known; E. C. Sullivan performed a similar service for the Corning Glass Company. Colin G. Fink's electrolytic process gave rise to the American chrome plating industry. F. G. Cottrell developed a process for precipitating sulphur trioxide and other substances from smelter smoke that gained general use, and he later adapted the same electrostatic technique to the de-emulsification of crude oil. His firm, Research Cottrell, was a pioneer in the business of environmental controls and resource recovery.[65]

Even among those of Ostwald's pupils who spent their careers in teaching positions, there were many who had more than a passing interest in industrial research. Wilder D. Bancroft, in addition to serving as a consultant for more than fifteen firms, also contributed significantly through his teaching to the development of electrochemical and metallurgical industries in the United States. Bancroft and Louis Kahlenberg both took important roles in organizing the American Electrochemical Society, an organization with the same goals as Ostwald's *Electrochemische Gesellschaft*. Arthur B. Lamb was an industrial consultant and holder of nineteen patents, most of them taken out while he was a faculty member at Harvard, and George A. Hulett, while on the Princeton faculty, was active as a consultant to private firms and the U.S. Bureau of Mines. Even A. A. Noyes, of all Ostwald's students the most prone to draw sharp distinctions between pure and applied science, participated with Willis R. Whitney in developing a profitable process for the recovery of alcohol and ether vapors hitherto lost in the manufacture of photographic supplies. Proceeds from the sale of this process made Noyes a wealthy man and financed much of his subsequent research at MIT.[66]

Although Ostwald's students borrowed much from their master, they did not emulate him in all respects. Perhaps most notable among the differences between them were their respective attitudes toward the juncture of physical chemistry and the philosophy of science. Only a few of Ostwald's American students showed interest in the cause that he championed in the mid-nineties, energetics. J. E. Trevor questioned the explanatory power of the atomic theory in a number of statements during the nineties, but did not go so far as to advocate its abandonment. A. A. Noyes for a time was intrigued by the extent to which natural phenomena might be explored by the concepts of energetics. His textbook, *The General Principles of Physical Science*, used many of the terms and categories of energetics and was dedicated to Ostwald. Nevertheless, Noyes made it plain that he viewed matter and energy to have equal claims as fundamental concepts, thus rejecting Ostwald's contention that matter is merely an aggregate of energies.[67]

Only one of Ostwald's American students, J.L.R. Morgan, adopted his teacher's extreme position on energetics, according to which both the reality and usefulness of atomic conceptions were denied. Morgan came to this position early in his career and held it well after Ostwald himself had recanted.[68] Morgan's viewpoint, however, was considered idiosyncratic, even by fellow Leipzig alumni, as is clear from Willis R. Whitney's review of Morgan's book, *Physical Chemistry for Electrical Engineers*:

> The first chapter deals with fundamental principles, among which are included the author's decision to use in his book the term "combining weight, meaning by it that combining weight which is usually designated as the atomic weight." This is to free the work from any conception of an hypothesis or inaccuracy. This laudable ambition has, in general, cost as much or more than it is worth. The common terms of the average physical chemist are the ones which the electrical engineer ought naturally to wish to learn.[69]

CHARTING THE EXPANSION OF THE DISCIPLINE

We have seen then that in addition to sharing a common instructor, Ostwald's American students were also unified through the possession of certain common views, namely, that chemists and especially physical chemists should take the understanding of chemical processes as their primary goal, and that the work already accomplished toward that end offered valuable information and techniques to scientists working in other fields of study and in industry; they were also unified by their common propensity to avoid the more philosophical and emphasize the more empirical side of Ostwald's program.

We have also seen several lines of evidence indicating that Ostwald's former students enjoyed considerable success in propagating their research interests following their returns to America. Those who returned in the 1890s, such as A. A. Noyes and T. W. Richards, were able to institute undergraduate training at American colleges and influenced others to go to Leipzig for the continuation of studies. Chemical and scientific societies, both in America and abroad, bestowed an unusual number of awards and honors on Ostwald's American students. In addition, the proportion of the American chemical literature dealing with physical chemistry grew impressively between the 1890s and 1920s.

Many questions, however, remain to be answered. Was the expansion of physical chemistry in America a smooth and continuous or a jagged and discontinuous process? When did American graduate schools begin to take a significant role in the development of the specialty? Most important, why were Ostwald's students successful in promulgating their science in the United States? What fueled the growth of this field?

To answer these questions, it is necessary to look beyond the circle of Ostwald's own students, and to investigate the larger community of physical chemists active in America. Publication is perhaps the best criterion for membership in such a community, and prosopographical techniques offer the best means for studying training and career patterns. Identifying physical chemists is the most difficult obstacle confronting this kind of analysis. American physical chemists did not publish in any one periodical, nor was their work comprehensively reviewed or abstracted in any one place. The three principal outlets for their research during the late nineteenth and early twentieth centuries, however, were the *Journal of Physical Chemistry*, the *American Chemical Journal*, and the *Journal of the American Chemical Society*. Lacking a better method, I have surveyed the contents of those three journals myself in order to compile a roster of productive American research workers.

Procedures in this survey were kept as simple as possible. Attention was paid those scientists who wrote or coauthored research articles on eleven broad topics that could generally be found as chapter headings in contemporary textbooks of physical chemistry.[70] Authors of notes of less than five pages, reviews, and articles dealing with other branches of chemistry were ignored. The contents of the *Journal of Physical Chemistry* were examined over a 31-year period, beginning with its foundation in 1896 and ending in 1926. The review of articles in the *Journal of the American Chemical Society* also ended in 1926, but began two years earlier—in 1894 instead of 1896—in order to include the years immediately preceding the foundation of the *Journal of Physical Chemistry*. The survey of the contents of the *American Chemical Journal* also began in 1894, but ended in 1913 when that journal merged with the *Journal of the American Chemical Society*. The year 1926 was selected as a terminus for several reasons. The volume of articles in the two journals increased so greatly during the 1920s—the number published by the *Journal of Physical Chemistry* tripling between 1923 and 1926—that a continuation of the survey was impracticable. At the same time, the 33 years that were covered were the period in which physical chemistry became an accepted part of the chemical curriculum at most American universities and a discipline whose value to industry was widely recognized.

The names of 1,244 scientists appeared on articles that were included in this survey. Many of these authors wrote or collaborated in writing only a single paper, often as graduate students; some were interlopers in the field—organic chemists or physicists who wrote an article or two on physico-chemical topics and then returned to research that fit more squarely within their parent disciplines. In order to eliminate the names of those who failed to become independent research scientists and those with only a casual interest in physical chemistry, a threshold of five articles was selected as a standard of productive research. There were 187 investigators who met this standard. Of these 187 scientists, 168 (90%) worked for all or part of their careers in the United States and achieved sufficient notoriety to be listed in such biographical sources as *American Men of Science* and *Who's Who in America*. This group provides a working roster of the productive physical chemists active in America between 1894 and 1926.

The information provided by this survey indicates that the growth of physical chemistry in the United States was a rapid and fairly uniform process, slowed only slightly by American involvement in the First World War. The number of articles dealing with topics in physical chemistry as a percentage of the total literature published by the *Journal of the American Chemical Society* affords a measure of this growth relative to the expansion of American chemical research as a whole (see Table 2.2). This figure increased rapidly during the first twenty years covered by the survey, jumping from an average of 5 percent for the years 1894 to 1898 to an average of 21 percent for the years 1909 to 1913. No further advance was made during the following quinquennium. The pause suggests that physical chemistry was affected more powerfully by the war than other branches of chemistry, perhaps because a larger proportion of physical chemists were involved in war-related work than those in other chemical specialties. This need not mean that physical chemistry was more important to the war effort than other disciplines. Indeed, it is more likely that it merely reflected the youth of the field and its practitioners. The oldest American physical chemists were entering their fifties when the United States began mobilization, and a very large percentage of physical chemists were in their twenties and thirties. Service obligations fell heavily upon these young men, and as a result the normal processes of training and research suffered greater disruption than was the case in fields in which the average age was higher. Immediately after the war, however, physical chemistry resumed its prewar growth; during the 1919–1923 quinquennium 28% of all articles published by the *Journal of the American Chemical Society* dealt with physical chemistry. Data drawn from the subsequent three-year period (1924–1926) suggest that the ratio of physico-chemical to all chemical research may have stabilized at approximately this level in later years.

There is, of course, a strong element of subjectivity in the way in which these statistics were collected. The editors of the *Journal of the American Chemical Society* did not label articles as being physico-chemical; the classification of each article reflects a judgment on the part of the classifier. There is, however, at least one independent standard by which the reliability of this survey may be measured. The *Journal of Chemical Education* published an annual census of graduate students in chemistry during the final three years covered by this survey. The census classified graduate students according to their research specialties. If the present survey offers a reliable measure of the rate of growth of physical chemistry in America, then the proportion of the chemical literature dealing with physico-chemical topics should be of the same order of magnitude as the proportion of graduate students conducting physico-chemical research, as determined by the census taken during the years 1924 to 1926. This in fact is the case. The present survey found that during these three years 27% of all articles published by the *Journal of the American Chemical Society* dealt with physical chemistry. The census takers found just under 29% of all graduate students engaged in physico-chemical research during the same period.[71]

Although the American journal literature in physical chemistry grew rapidly and fairly steadily during these 33 years, the countries in which American physical chemists completed their graduate studies changed significantly. The most striking change was the precipitous decline in the number of American physical chemistry students traveling to Germany (see Table 2.3). The year 1908 may be taken as a convenient mark for dating this decline since it was the year in which the last American to work with Ostwald left Leipzig. Prior to 1908, more than one third of the American chemists writing five or more articles in physical chemistry received their doctoral degrees in Germany. An additional thirty percent (19 of 65) went to Germany just before or just after receiving doctorates at American universities. The number receiving doctorates in Germany in 1908 and after, however, was negligible; nondegree and postdoctoral study in Germany likewise disappeared almost completely. More physical chemists received doctorates in Britain than in Germany between 1908 and 1926, and the number receiving degrees from all foreign universities was itself dwarfed by the number obtaining American doctorates.

Reasons for the declining popularity of German study are plentiful. In 1904, a number of German universities, led by the University of Berlin, began to impose restrictions on American students, which had the effect of prolonging the stay necessary for those working toward doctorates by as much as two years.[72] In addition, Ostwald's retirement in 1906 and van't Hoff's death in

TABLE 2.3
Locations of Institutions Granting Ph.D.s to Productive American Physical Chemists

	Location of Degree-granting Institution			
Date of Ph.D.	U.S.	Canada	Germany	G.B., France, Switzerland
Pre-1908	38	2	23	2
1908–1926	78	—	4	8

1912 deprived Germany of two of its most attractive and influential teachers. After Ostwald's retirement, only one American student included in this survey received a degree at Leipzig. Finally, the outbreak of hostilities in 1914 created formidable if not insurmountable obstacles to study in Germany. Only in the mid-1920s, after lingering hostility toward the "Huns" had dissipated, were contacts with German universities partially restored. During the period between 1926 and 1933 a new wave of Americans traveled to Germany for advanced study. Typically, however, these physical chemists neither sought degrees nor worked in physico-chemical institutes; rather, they were postdoctoral fellows seeking knowledge of the new quantum mechanics in Germany's physical institutes.

AMERICAN GRADUATE CENTERS

American students had many reasons for avoiding study in Germany after 1908, but they also had a potent reason for staying in America. Ostwald's pupils had created graduate programs in physical chemistry which were coming to rival Germany's own. The university movement, which had begun in 1876 with the opening of the Johns Hopkins University and had gathered force in the 1880s and 1890s with the creation of vigorous graduate schools at old institutions like Harvard and Columbia and the foundation of such new schools as Chicago and Stanford, had created conditions in which new research-oriented disciplines could take root. Ostwald himself, in a review of the first issues of the *Journal of Physical Chemistry* in 1896, remarked on the energy with which American universities were cultivating his new field. Three years later, Pierre Duhem, a leading exponent of the new specialty in France, echoed his sentiments, adding that ". . . more than European universities the young American universities regard physical chemistry as an extremely important science." Soon, Americans were congratulating themselves on the progress of their universities as sites for research in physical chemistry. In 1907 C. F. Maberry, then chairman of the chemistry section of the American Association for the Advancement of Science, singled out American research laboratories in physical chemistry as being especially deserving of notice for their standing *vis-à-vis* German research schools.[73]

Although physical chemists received training at many American universities during the years between 1894 and 1926, several schools stood out as centers for graduate training and research: Johns Hopkins, Harvard, Cornell, Chicago, MIT, California, and Caltech. Together these institutions trained more productive physical chemists than all other American universities combined. Seventy-three productive research workers took their doctorates at these seven schools; the remaining 43 U.S. doctorates were distributed among 19 other universities. In addition, these seven institutions provided employment to a substantial number of physical chemists. In 1899, six of America's most productive physical chemists were on the Cornell faculty; in 1910, ten were at MIT; in 1920, ten were on the staff at the University of California. Even in national terms their patronage of physical chemistry was impressive: in 1900, nearly one half (16/34) of America's productive physical chemists were employed by these seven institutions; in 1910 this figure remained above one third (34/97), and in 1920 it stood at nearly 25 percent (30/131). When contemporaries spoke of American centers of training and research in physical chemistry, they consistently chose their examples from this group.

A number of circumstances conspired to give these institutions such prominence, as will become clear in later chapters, in which the histories of several will be treated in detail. It is important to note here, however, that the leadership of a single individual was critical to the development of all but one of these programs. With the exception of Chicago, a former student of Ostwald was at the heart of each graduate program, and, again with the exception of Chicago, the research schools that developed at these centers reflected their leaders' personal research interests and goals. When a discipline is mature, when it is intensively cultivated by many well-trained scientists, the presence or absence of a single scientist, no matter how exceptional, is not likely to matter greatly. But when a discipline is just beginning to grow, when it is struggling to find resources and recruits, the prescience and energy of a few individuals often are critical both in building institutions and in determining the paths along which research develops. The history of science is not simply the story of a few great heroes, but neither is it a story in which personalities matter for naught.

The importance of personalities becomes evident when one considers fluctuations in the effectiveness of these seven nascent graduate programs. Not all were productive throughout the period under consideration. In the years prior to 1908, Johns Hopkins and Harvard were the only American universities to rival Leipzig as sites for advanced study in physical chemistry. Just before World War I, however, Harvard was eclipsed by Cornell and MIT, and after the war new centers on the west coast—California and Caltech—rose to prominence, while such established leaders as Johns Hopkins, Chicago, and MIT went into decline (see Table 2.4). Changes in the fortunes of these graduate programs were closely related to the vigor of the individuals around whom they were built.

At Johns Hopkins the standing of the program in physical chemistry was tied directly to the fate of Harry Clary Jones.[74] Jones was a Hopkins Ph.D. who spent two postdoctoral years, 1893 and 1894, touring the leading European physico-chemical laboratories, including those of Ostwald, Arrhenius, and van't Hoff. Arrhenius described him to Ostwald as

a very energetic worker. He has already spoken with you about his work. But he was like other Americans and Englishmen are for the most part. He took the whole thing

TABLE 2.4
Institutions Granting Ph.D.s to Productive American Physical Chemists

Institution	Number of American Ph.D. Recipients
Ph.D.s Granted prior to 1908	
Leipzig	11
Johns Hopkins	11
Harvard	7
Chicago	5
Göttingen	4
Cornell	3
Pennsylvania	3
Wisconsin	3
Others (14 institutions)	18
Ph.D.s Granted from 1908 to 1919	
Cornell	8
Chicago	6
Johns Hopkins	6
California	5
Columbia	5
MIT	5
Others (20 institutions)	23
Ph.D.s Granted from 1920 to 1926	
Caltech	5
California	4
Princeton	4
Cornell	3
Others (15 institutions)	16

as "business," almost like a competition, where one uses physical strength, but he was completely lacking in imagination and time for reflection.[75]

Six years later, John Ulric Nef, professor of chemistry at the University of Chicago, echoed this assessment. Jones, he remarked succinctly, "seems a hard worker with no originality."[76]

Despite these critical assessments by his seniors, Jones achieved some impressive successes after returning to a position at Johns Hopkins. His laboratory was extremely productive and became a magnet for research students. His many textbooks of physical chemistry, among the first in the English language, sold well. And, in recognition of his accomplishments, Jones was promoted rapidly. By 1903 he was a full professor.

Jones's research, and that of his students, soon focused on the problem of the deviation of concentrated solutions from theoretical predictions. According to Arrhenius's theory, dissociation should diminish with increasing concentration. In fact, it was found that dissociation values, as measured by a number of techniques, actually increased in many instances. By 1905 Jones had become a leading proponent of a "hydrate theory" to account for these deviations. According to Jones, anomalous dissociation values occurred because dissociated ions of solute combine with water molecules to form complexes, or hydrates. The formation of such hydrates would have subtle but important effects on the properties of solutions. Methods for determining dissociation depended approximately upon the ratio of the number of particles of the dissolved substance to the number of particles of solvent. If the solute combined with the solvent, then the number of particles of the solute would not change, but only their size. There would be, however, a reduction in the number of molecules of solvent; the hydrates would, in effect, eliminate part of the solvent from the system. In dilute solutions this effect would be minute, but in concentrated solutions it would become appreciable. As a result, the freezing points of concentrated solutions would be depressed, the boiling points elevated, and the conductivities increased beyond predictions. The solutions, in other words, would give deceptively high measures of dissociation. Not stopping there, Jones went on to argue that hydrates might be treated as compounds of definite proportions and conducted an extensive series of experiments to determine the composition of hydrates in a large number of salt solutions at various temperatures and degrees of concentration.[77]

Jones's views, which resembled those of some of Arrhenius's English critics, received a frosty reception from his colleagues. While it was generally admitted that solvation might play a secondary role in explaining the behavior of concentrated solutions, few were willing to admit the possibility that solvation was the only factor at work. As one reviewer pointed out, Jones's theory disregarded the law of mass action. Whereas that law would require the greatest hydration in the most dilute solutions, in which the concentration of water molecules would be at a maximum, Jones was using the theory to explain the behavior of concentrated solutions.[78] Another concluded that Jones's attempts to determine the composition of hydrates was valueless, since the very existence of hydrates was an unproven assumption.[79] Jones, however, never abandoned his position and, much to annoyance of his critics, insisted on defending his hydrate theory at great length in his numerous and widely used textbooks.

Jones's adherence to this position had a disastrous effect upon his reputation, his research school, and his psyche. Earlier doubts about his judgment seem to have been borne out, and even his erstwhile admirers began to have second thoughts. Whereas T. W. Richards had commended him in 1899 as being one of America's most promising physical chemists, by 1914 he had come to believe that Jones's "enthusiasm leads him to do things too hastily." Others were less charitable. A.W.C. Menzies, in a scathing review of Jones's

Introduction to Physical Chemistry, criticized Jones for cannibalizing his earlier texts and for making numerous factual errors and omissions. E. C. Franklin a few years later explained his lengthy critique of Jones's *A New Era in Chemistry* by saying that "no one who permits so many inaccurate, careless and exaggerated statements to creep into his work should go unrebuked."[80]

Jones had long been sensitive to criticism. When Ostwald had written an unfavorable review of one of his earlier textbooks, Jones wrote what Noyes described as "a rather pitiful letter over his disappointment. . . . Unfortunately," Noyes added, "I agree with Ostwald and can't give him much consolation." As criticism mounted, so, apparently, did Jones's anguish and frustration. He quarrelled with his colleagues at Johns Hopkins and engaged in public polemics with his adversaries elsewhere. As Jones's reputation waned, so too did the standing of the program in physical chemistry at Johns Hopkins. E. Emmet Reid later recalled that when he joined the department in 1914 Jones was "doing his best to hold a few students." Two years later, Jones committed suicide. Joseph Howard Mathews, discussing the vacant position in physical chemistry at Hopkins a few months later, remarked: "As I understand the situation there, a general reorganization is what the department needs, and they certainly need a new building and adequate equipment."[81]

The history of the program in physical chemistry at Harvard was not so dramatic, but there too much depended on the energy and skill of a single chemist, T. W. Richards. Whereas Jones always seemed gripped by a fever of activity and ambition, Richards displayed the cool rationality and reserve that Americans often associate with the English gentry. He loathed administrative responsibilities and more than once declined to run for office in professional societies on the grounds that he had to husband his energies for research.[82] Judiciousness and prudence were qualities he much admired.

> There have always been two parties as regards any question brought forward by mankind for discussion—the conservative party and the radical party. The former has a tendency to cling to old ideas simply because they are old, and the latter has a tendency to adopt new ideas simply because they are new. It seems to me that neither of these tendencies is legitimate. One should seek new points of view continually, but he should hold to that which is good until something is proposed which seems to him better. In every case he should weigh the respective arguments for and against the new point of view with a mind as free as possible from prejudice, and with a single eye to the truth. In short, the ideal investigator is the scientific independent, the chemical "mugwump."[83]

Richards' research reflected these values. A consummate experimental scientist, Richards achieved success through extraordinary patience and attention to detail. His accomplishments fit squarely within the tradition of American research on physical constants linked with the names of the physicists Rowland, Michelson, and Millikan. Much of his life was dedicated to determining atomic weights, a task that had been done many times before, but never with the accuracy Richards demanded. In recognition of the precision of his work and of its importance at a time when the study of radioactive decay was raising new questions about the elements, Richards was awarded a Nobel Prize in 1914.

Had Richards confined his work to the study of atomic weights, traditionally a part of inorganic chemistry, there would be no reason to call him a physical chemist. But in 1895 Richards, recently promoted from instructor to assistant professor at Harvard and already engaged in his research on atomic weights, traveled to Germany to study physical chemistry. The trip was sponsored by Harvard, which was interested in grooming Richards to take over the courses taught by the lately deceased Josiah Parsons Cooke, Richards' mentor. Richards, after working with both Ostwald at Leipzig and Nernst at Göttingen, returned to Harvard with new interests in thermochemistry and electrochemistry. While his interests were expanding, however, his style remained unchanged; painstaking attempts to perfect the measurement of heat were added to his repertoire of techniques for measuring weight.[84]

In 1901 the University of Göttingen invited Richards to accept its chair in physical chemistry. This was an unprecedented distinction for an American chemist and it elevated Richards to the pinnacle of his profession. In order to retain him, Harvard not only promoted him and reduced his teaching obligation (to three lectures a week), but also promised to raise three quarters of a million dollars to build and endow a new chemical laboratory.[85] When completed in 1912, the Wolcott Gibbs Memorial Laboratory, although small, was among the best equipped in the world. In it Richards enjoyed many of the prerogatives of a chair-holder at a German institute. The wife of one of his graduate students marveled at the working conditions:

> He [Farrington Daniels, Richards' graduate student] had two rooms of his own and a half share in balance and dark rooms. Everything was modern. There were many windows, good ventilation, curved surfaces as in a hospital so dust could not collect in corners, a telephone in each room, constant temperature, fire-proof construction. . . . There were only seven men to work in the building, including Richards, who had the whole second floor.[86]

A call to Göttingen, a princely new laboratory, medals from the Royal Society, the Chemical Society of London, and the American Chemical Society, honorary degrees from Oxford and Cambridge, and a Nobel Prize—awards fell on Richards like rain during the years from 1901 to 1914. But, as is often the case in science, the Nobel Prize came near the end of the recipient's creative years. By 1914, Richards was leaving nearly all of the work of his laboratory to students and junior colleagues, and he was growing more and more detached from the concerns of his discipline.[87] His attention increasingly dwelt on what he called his "hypothesis of compressible atoms."

Richards' starting point was van der Waals's equation, which connected the three factors (pressure, p; volume, v; and temperature, T) that determine the physical state of a substance:

$$(p + a/v^2)(v - b) = RT$$

The equation contained two constants dependent on the nature of the substance: a, which expressed the mutual attractions of atoms or molecules, and b, which represented their volume. In fact, the second of these was not truly constant; atomic and molecular volumes were known to vary with pressure. It was this fact that stimulated Richards' curiosity. It suggested to him that the prevailing picture of atoms and molecules, that they were hard and incompressible, was invalid. It also suggested to him that the volumes of atoms might change during chemical reactions as a result of changes in internal pressure, that is, in response to mutual attractions. Richards began to collect data and to conduct experimental studies of atomic volume; he quickly was led to suggest that a host of the chemical and physical properties of the elements could be correlated with atomic volume. He further speculated that atomic volume was the key to understanding the mechanical basis of such thermodynamic values as enthalpy and free energy. Richards suggested, for example, that the heat of reaction represented the work done in the compression of atoms by the force of chemical affinity.[88]

When Richards developed these ideas around the turn of the century, he was hardly alone in seeking to develop a new understanding of the atom; this was a quest on which many physicists were engaged. Richards, however, was a physical chemist, a former student of Ostwald, who still, at this time, condemned the atomic theory. Nevertheless, Richards had been measuring atomic weights long before he met Ostwald, and it was only natural for him to ponder the nature of the entities whose weights he had measured so accurately. Moreover, he developed these notions in the company of an unusually imaginative student named G. N. Lewis, who contributed suggestions and who may have, through his presence, provided the appropriate mix of competition and sympathy necessary to embolden the cautious Richards to publish his ideas.

Richards' work on the theory of compressible atoms proved a *cul-de-sac*. Sympathetic biographers have noted that it led him to anticipate certain later developments in atomic theory, but such anticipations appeared amidst a mélange of speculative musings that exercised little influence on those who were actually responsible for reshaping atomic theory in the early twentieth century—Thomson, Rutherford, Bohr. More important was the experimental work that it generated in Richards' laboratory, work that produced new and accurate values for the heats of many common reactions. Nevertheless, it was not this experimental work that preoccupied Richards during the last fifteen years of his career, but rather the problem of formalizing his theory of compressible atoms. Here he faced formidable obstacles, not least of them his own uncertain grasp of mathematics.

Throughout his career, Richards was of two minds regarding the value of mathematics in the physical sciences. On the one hand, he recognized that it was sometimes a necessary tool. On the other, he never believed it of sufficient value to merit an important place in the training of chemists, and in his private correspondence often expressed a suspicion of the work of those scientists who made significant use of it. In 1923, for example, he expressed his views forcefully in a letter to his friend and colleague Svante Arrhenius. The Nobel Prize Committee, he suggested, had erred in awarding prizes to Einstein and Bohr in 1921 and 1922. "There is no question that both of these men are very brilliant," he wrote,

> and that their hypotheses are highly ingenious. As mental feats they are certainly most remarkable, and from this point of view the prizes seem to me certainly to be deserved. I can not help thinking, however, that it remains to be proved whether or not the hypothesis of either is consistent with reality. But you know that I am a rather conservative person, who, while believing most ardently in the usefulness of mathematics, is exceedingly cautious about accepting the premises on which the mathematical superstructure may be built. It seems to me that these premises are the most important part of the matter. Any good mathematician can put on the mathematical frills according to the most recent mathematical fashion, but the result is unsatisfactory if the figure inside is a doll stuffed by human hands, and not a real being of flesh and blood.[89]

How much mathematics did Richards know? Not very much, at least by the standards of twentieth-century physics. His papers on chemical thermodynamics and the compressible atom depict a chemist struggling to handle the elementary calculus. Nor do they reveal a truly expert grasp of physical concepts. In a famous paper in 1902, Richards came tantalizingly close to stating what would later become known as the third law of thermodynamics. Looking at the temperature dependence of free-energy change and entropy change for several reactions, he constructed a graph that showed how these values approached each other asymptotically with contrary slopes as temperature approached absolute zero. This insight would later be critical in Nernst's successful development of the third law. But Richards never recognized the significance of his data.[90] When in 1925 an undergraduate by the name of J. Robert Oppenheimer took Richards' course in physical chemistry, he came away feeling that it had been "a great disappointment . . . a very meager hick course. . . . It was formal and tentative and timid; Richards was afraid of even rudimentary mathematics. . . ."[91]

Richards' stubborn labors on behalf of his theory of compressible atoms and his neglect of physical achievements of the early twentieth century raised the eyebrows of some of his colleagues and eroded his leadership among American physical chemists. His isolation was reflected in Wilder D. Bancroft's assessment: "it is only in the most formal way that one can consider his elastic atoms as standing in any close relation to the modern atom, either of the physicist or the chemist." Richards himself understood that his theoretical contributions were of a lesser order than his experimental. The theory of compressible atoms was his hobby, the determination of constants, his vocation. "[T]he working out of some of the consequences of atomic compressibility," he wrote near the end of his life, "[has] given me more intense and vivid intellectual pleasure than anything else—a pleasure which was very exciting but perhaps not as durable as the satisfaction which comes from careful experimental work."[92]

While Johns Hopkins and Harvard were declining as centers of training for physical chemists by or shortly after 1908, MIT and Cornell were in the ascendant. The chemistry department at Cornell began offering graduate courses in physical chemistry in the mid-1890s when two of Ostwald's former pupils joined its staff: Joseph E. Trevor and Wilder D. Bancroft. Prior to 1908, their success in training productive researchers compared poorly with that enjoyed by Jones at Johns Hopkins; but during the subsequent eighteen years Cornell's was the most effective American program, turning out eleven productive scientists between 1908 and 1926.

Central to Cornell's success were Bancroft's talents as a teacher and research director. Bancroft followed his own injunction that physical chemists should look past the problems of aqueous solutions and should focus instead upon relatively unexplored topics of equal or greater practical importance. Here he made a name for himself, first through his studies on applications of the phase rule to alloy systems and work on electrochemistry, and later through his promotion of the study of catalysis, photochemistry, and colloid chemistry.[93] Fond of dividing scientists into two groups, the diligent "accumulators" and the inspired "guessers," Bancroft stood proudly among the guessers; he insisted that the industrious compilation of exact measurements could serve as no substitute for clear thinking, a firm grasp of the issues, and an ability to devise crucial experiments to choose between competing hypotheses. Bancroft, Willis R. Whitney wrote in 1917,

has prepared, at Cornell, a great many good chemists for the industries. He is greatly loved and admired by those men. He has been criticized by others for superficiality. On the other hand, he knows chemical literature as well as anyone in the country. He reads and writes a very great deal, and takes unusual interest in what might be called the novelties of the science. He is not at all a grind or a disciplinarian. One might almost say that he touches only the high spots, but he is far above the average in intelligence and has contagious enthusiasm. . . . He creates enthusiastic students, and his scientific papers are usually full of suggestions for others. He is nowhere near as sound as A. A. Noyes, but he is continually doing something. He is as full of ideas as anyone I know, and nothing discourages him.[94]

Bancroft had a penchant for playing the iconoclast that more than once led him to champion lost causes—as when, for instance, in the 1930s he stoutly defended the notion that proteins are colloidal agglomerates rather than macromolecules. Nevertheless, it also allowed him to perform the genuine service of alerting his colleagues to the potential inherent in a neglected idea, as he did in the 1890s when he publicized the uses of Gibbs's phase rule. Through his editorship of the *Journal of Physical Chemistry*, and through his students, Bancroft exerted considerable influence on the direction and pace of development of his field.

The chemist that Whitney used as a standard for judging Bancroft, A. A. Noyes, was primarily responsible for creating a vibrant research center at MIT. As we have seen, MIT began to offer undergraduate courses on physical chemistry in the 1890s, but graduate work was slower to develop. MIT had been chartered in 1861 to train engineers, it had grown and prospered doing exactly that, and its administrators felt little desire to tinker in matters that were best left to such rich institutions as Harvard. Consequently, MIT was rather slow to join the parade of schools that began to march to the tune of basic research and graduate education during the 1890s.

In 1903, however, MIT's administrators took steps that led directly to the graduation of its first Ph.D.s. The initiative was not their own; rather it came from Noyes. Indeed, he made the administration of MIT an offer it could not refuse. In return for providing a building to house a research laboratory, Noyes agreed to supply half of the operating expenses of the laboratory, including payroll, out of his own pocket. The Research Laboratory of Physical Chemistry, as Noyes christened it, gave rise to MIT's successful graduate program in physical chemistry and served more generally as the nucleus of basic research and graduate study at the Institute.[95]

Noyes's research, like that of Richards at Harvard, focused on the determination of fundamental constants. Beginning with the behavior of solutions of electrolytes at high temperatures, and later shifting to the measurement of the free-energy changes accompanying chemical reactions, Noyes and his coworkers developed data that served as a basis for subsequent advances in the understanding of chemical equilibria. Perhaps even more important was the role of Noyes's laboratory in training the leaders of the next generation of American physical chemists. Between 1903 and his departure from MIT in 1919, Noyes surrounded himself with a group that included William D. Coolidge, G. N. Lewis, Charles A. Kraus, E. W. Washburn, Richard Chace Tolman, W. C. Bray, Claude S. Hudson, William D. Harkins, Frederick G. Keyes, and Duncan MacInnes—the strongest ensemble of chemical talent America had yet seen.[96]

Not all of these chemists took Ph.D.s at MIT, and this is a significant point, for MIT and, to a lesser extent, Cornell were both more important institutions than Ph.D. statistics suggest. In the first place, prior to World War I they gave employment to more physical chemists than any other American institutions. Secondly, most of these positions were not professorships locked up by a few senior chemists, but rather were research positions given to recent Ph.D. recipients. Promising young chemists typically held these positions for a year or two and then moved on to more permanent jobs at other universities, in industry, or in government. As a result of this turnover, many productive physical chemists worked at one or the other of these schools for brief periods early in their careers; of the 168 Americans who published five or more articles on physico-chemical topics, almost one third (54) worked under the direction of Noyes or Bancroft. The availability of full-time junior-grade research positions gave many talented young physical chemists opportunity to mature under the guidance of experienced research workers, and made the MIT and Cornell programs far more influential than less flexible departments elsewhere.

The influence of Noyes's laboratory was felt most directly at the California Institute of Technology, where Noyes himself organized graduate work after leaving MIT, and at the University of California, where the program in physical chemistry began as an MIT transplant. At Caltech, Noyes, to a remarkable degree, managed to duplicate his earlier success at MIT. Noyes increasingly played an executive role in the affairs of the laboratory, but he continued to show sensitive insight into promising paths for research and a sure touch in selecting talented co-workers. Although Noyes himself remained interested in the study of chemical equilibria until his death in 1936, he encouraged his younger associates to use X-ray and electron diffraction techniques and infrared spectroscopy to study molecular structure. Their research, especially that of Linus Pauling, who began his long association with Caltech as a graduate student in 1922, reshaped chemists' understanding of the chemical bond and molecular structure.[97]

Serious work in physical chemistry had its beginnings at Berkeley in 1912, when G. N. Lewis, then professor of physico-chemical research and deputy director of the Research Laboratory of Physical Chemistry at MIT, was persuaded by the President of the University of California to move to Berkeley and become head of the College of Chemistry. When Lewis went west, he took part of the Research Laboratory of Physical Chemistry with him: William C. Bray and Merle Randall, as well as several graduate students. These investigators, together with the rest of the staff Lewis constructed, transformed Berkeley's chemistry department from a comparative backwater to a major research center within a year. Lewis was unwilling to make distinctions between the branches of chemistry in his new department; all professors were professors of chemistry, not of its subdivisions. Nevertheless, the College's publication record demonstrates that the primary direction of research was along physico-chemical lines. Chemists at California during the late teens and twenties made their reputations largely through studies of chemical equilibria, free-energy and entropy changes in chemical reactions, and chemical bonding. The quality of the research was comparable to that being done at Caltech, and because the program was much larger, the influence it exerted on American chemistry through its graduates was unparalleled during the mid–twentieth century.[98]

Although their fortunes followed different trajectories, the six programs in physical chemistry so far discussed had common features. Each coalesced around one of Ostwald's former students, demonstrating once again the importance of "personal knowledge" in the transmission of research technique. Although related through a common line of descent, each program developed a distinctive research agenda just as a language spoken in a large territory will give birth to many dialects. Each of these agendas or dialects perpetuated the attitudes and special interests of certain dominant individuals: Jones, Richards, Bancroft, Noyes, and Lewis.

The seventh important graduate center in physical chemistry violated these patterns. The chemistry department at the University of Chicago compiled an impressive record in physical chemistry although none of Ostwald's former students was on its permanent staff and its physico-chemical research lacked continuity and coherence.

Indeed, Chicago's was a department of chemistry that succeeded despite itself. When the University, a magnificent monument to John D. Rockefeller's talents for amassing money and giving it away, opened its doors in 1892, little thought had been given to the place of physical chemistry in what seemed destined to be the finest chemistry department in the country. William R. Harper, the university's energetic president, had named John Ulric Nef, a chemist from Clark University, head of the department and had delegated to him responsibility for building a staff. Nef, one of the few American chemists with an international reputation, brought the department prestige but little tact, administrative skill, or talent for recognizing promise in younger colleagues. His ambition was to build at Chicago an institute of chemical research comparable to those he had known as a student in Germany. He claimed to desire a department "in which all sides of chemical work, inorganic, physical as well as organic, should be adequately represented," but in fact fought tenaciously any member of the department who challenged the primacy of his own field, organic chemistry. To seek additional resources for physical or inorganic research or to suggest that the curriculum should include a semester-long course in physical chemistry was, in Nef's eyes, tantamount to mutiny; more than one young chemist was court-martialed during his regime for such crimes.[99]

Constant turmoil within the department soon became a matter of both private and public comment. Letters to Harper from junior members of the department and newspaper accounts of Nef's unhappy ship elicited no action.

As early as 1893, an inorganic chemist had petitioned Harper to divide the chemistry department into two independent parts; in 1895 an assistant professor of analytical chemistry was fired because, according to one newspaper account, he had criticized the curriculum's excessive emphasis on organic chemistry.[100] But in 1903, after Wilder Bancroft in his *Journal of Physical Chemistry* had chided Chicago for ignoring physical chemistry and after Nef had suffered a nervous breakdown, Harper assented to yet another petition to divide the chemistry department in two. General and physical chemistry would be granted a separate budget and the status of a department. New stationery was ordered reflecting the change, but Harper deferred formal action. When Harper died in 1906, the change had not yet been effected. His successor, Harry Pratt Judson, could find no evidence in Harper's papers of an agreement to create a new department, and dismissed the idea.[101]

The intense feelings aroused by these events gradually faded as Nef and other participants left the scene, either by choice or ill health. Nevertheless, the chemistry department's turbulent origins left a legacy: Chicago's chemists were resolved to never again allow one of their number to dominate as had Nef. Nef's successor as head of the department was Julius Stieglitz, of whom Nef had once said: "I do not have any great expectations for his future, but he is very quiet and modest. . . ." Nef was wrong about Stieglitz's ability—he became an outstanding chemist who was among the first to use the techniques of physical chemistry to study organic reactions—but he was correct in calling Stieglitz modest. His diplomacy and ability to understand the needs of both organic and physical chemists made him an ideal peacemaker, a role he played frequently during a twenty-year term as department head.[102]

It is difficult to imagine a setting less congenial for physical chemistry than Chicago during the Nef years. Yet the department did enjoy considerable success in training productive physical chemists. That it could do so was due in part to the appeal Chicago had for serious, independent—if a bit naive—graduate students, and in part to a succession of able docents and professors who endured stints in the Kent Chemical Laboratory. Stieglitz, who joined the department as a docent in 1892 and was made a full professor in 1905, was among those who fostered interest in the new discipline. He possessed no special preparation in physical chemistry, although he did have the advantage of a broad education. Trained entirely in Germany, Stieglitz had completed work in organic chemistry under A. W. Hofmann and Victor Meyer; in inorganic chemistry under C. Rammelsberg and H. W. Vogel; and in physics under Helmholtz. At Chicago, he studied the behavior and sensitivity of indicators, the mechanisms and velocities of organic reactions, and molecular rearrangements of organic compounds containing nitrogen; in all these investigations he made ample use of physico-chemical concepts and techniques.[103] His doctoral student Herbert N. McCoy continued this tradition. Like Stieglitz, McCoy was attracted to topics on the fringes of established disciplines.

On the Chicago faculty from 1901 to 1917, McCoy did work on the ionization constant of phenolphthalein, but he is best known for his work in establishing the relationship between radium and uranium in the radioactive decay series.[104]

Both Stieglitz and McCoy stimulated interest in physical chemistry among Chicago's graduate students; so, too, did Alexander Smith. A Scot who studied in Germany during the late 1880s, Smith joined the department in 1894 and stayed until 1911.[105] Although trained as an organic chemist, Smith had an impressive command of inorganic and physical chemistry as well. His textbooks of inorganic chemistry, the first of which was published in 1906, were among the first introductory texts in English to make extensive use of the ideas of the Ionists. They were also among the most successful of such texts, finding widespread use in America and Britain and being translated into numerous foreign languages, including Russian and Chinese. His skills as a teacher and writer were not matched by his achievements in research, perhaps because he spread his efforts too thinly. Nevertheless, he was an aggressive spokesman for physical chemistry during his tenure in the department and helped to launch Chicago's first systematic instruction in the subject.

After Smith went to Columbia in 1911, a young professor from the University of Montana, William D. Harkins, came to fill his place. Harkins, a Stanford Ph.D. who had done postdoctoral work under Fritz Haber at Karlsruhe and A. A. Noyes at MIT, was the first member of the faculty at Chicago with formal training in physical chemistry. Despite his previous teaching and research experience, Harkins joined the Chicago faculty as an instructor, but through intense work he managed to advance to the rank of full professor in only six years. Like Stieglitz, McCoy, and Smith, Harkins was a man of catholic interests. His work at Chicago ranged over solubility phenomena, surface chemistry, nuclear structure, and isotope separation. Possessed of undeniable talent and energy, he was also a scientist with enormous pride and ambition. Always ready to assert rights to discoveries, even when title was not entirely clear, he was christened "Priority Harkins" by those with whom he clashed.[106]

None of these chemists established a true research school in physical chemistry at Chicago in the manner of Noyes at MIT or Lewis at California. Nef denied them necessary resources; perhaps more important, none possessed a single-minded and enduring dedication to any single research topic. Nevertheless, collectively they gave imaginative students the opportunity and stimulation necessary to embark on productive careers in the new discipline.

WHY PHYSICAL CHEMISTRY PROSPERED IN AMERICA

Having discussed the rapid and sustained growth of physical chemistry in America, it is time to ask why that growth occurred. Specialization is a common phenomenon, but it should not be considered uninteresting for that reason. There are powerful intellectual and social forces that foster the growth of new disciplines, but there are also powerful countervailing forces. New disciplines focus human and material resources tightly on specific problems of interest to their founders, but this may occur at the expense of existing disciplines. They offer the promise of rapid development but may threaten the integrity and unity of science. They promise to affect concepts and practices in other specialties, but this promise, while stirring to the partisans of the new discipline, may seem a curse to those in affected fields; intellectual retooling is a time-consuming and difficult process. Scientists, and intellectuals more generally, value innovation, but innovation is not the only quality they value. Conservative biases are built into both the intellectual enterprise and its institutions, and these can effectively stifle specialization. Of disciplines no less than saints it may be said that many are called but few are chosen. The question therefore is why certain disciplines succeed while others fail. How does a mental map that is first drawn by a handful of visionaries become the blueprint for institutions?

Let us translate these abstract queries into terms more concrete and manageable: Why did the discipline of physical chemistry take root and grow in the United States? Why did it expand rapidly in America while growing at more moderate rates in Britain and Germany and slowly in France? An answer to these questions must consider both the motives of those young Americans who entered the field when its future was uncertain and the market their services commanded in the United States. Since students of Ostwald constituted the core around which the discipline coalesced in America, we may begin by briefly re-examining their reasons for becoming physical chemists.

It is difficult to reconstruct the motives of individual scientists even when armed with ample source materials, so the prospect of analyzing the motives of many young and generally inarticulate workers—scientific commoners, in Daniel Kevles' apt phrase—is daunting.[107] Each of Ostwald's students was no doubt drawn to Leipzig by a peculiar combination of circumstances. Nevertheless, common factors do exist in those combinations. There is the disillusionment, voiced during and after their studies in Leipzig, with traditional paths of chemical study, and especially with organic chemistry: recall Noyes's despair over his work under Wislicenus and Loeb's disparagement of organic chemists. These early-comers to Ostwald's laboratory were rather naive young pupils from a country that was still scientifically backward. They were easily impressed by the scientific acumen of their hosts, and their vision of German scientific and academic life was distorted by an astigmatism common among émigré students. But, as economic historians have observed, backwardness has its advantages.[108] Although their vision suffered from some distortions, they discerned quite clearly the asymmetrical character of German research and study in chemistry: the crowded organic laboratories, the competition for position and choice research problem, and the disappointment of those whose research did not strike a rich vein. Coming from a country where there was neither a well-developed industry built upon organic chemistry nor a set of powerful research schools committed to older specialties, American students were able to recognize flaws in German chemistry that those enmeshed in the system tended to overlook, minimize, or dismiss as inevitable.

At the same time, recruits to physical chemistry felt the excitement attached to exploring a new field of study. The magnetic appeal of Ostwald certainly helped to generate that excitement, but Ostwald was hardly the only galvanizing stimulus. Opportunities for intellectual achievement seemed bountiful and this was a powerful lure for young chemists eager to make a mark in science. F. G. Cottrell, reflecting upon his reasons for taking up physical chemistry, stressed this sense of openness and opportunity:

> Very possibly the fact that I grew up in the still pioneer atmosphere of the far west, unconsciously predisposed my natural interest and impulses into the borderline and less formally recognized fields of study and research. This soon forced upon my attention the relative lack of organization and support for just such work, for the larger and stronger an educational or research institution grows, the more conservative and rigidly classified its activities are pretty sure to become. The most fertile fields for really new fundamental developments are almost always to be found between the edges of the orthodox ologies and professorships of the colleges (subjects recognized in our standard classifications of human knowledge). No one has had more striking proof of this in a single generation than the chemist in his own field. . . . Just as the prospector finds his best hunting grounds along the fissures and contact planes between different massive formations, so in the lines of human thought and invention . . .[109]

Cottrell's suggestion that his upbringing in the far west predisposed him toward the study of borderland fields may be discounted, since most of his compatriots at Leipzig came from the east. Nonetheless, his characterization of the impulse to explore the gaps between established subjects of study is revealing of that sense of where fruitful research topics lie that is at the foundation of the process of specialization. Such insight does not know national boundaries and was no doubt shared by German and American scientists, but young German chemists, familiar with the structure of their educational system, aware of the power of entrenched organic chemists, and conscious of the industrial demand for yet more organic chemists, must have had perceptions of career choices that were very different from those of their American counterparts. The decision to forsake organic chemistry for a new field was fraught with greater risk for German students in 1890 than for Americans like Loeb, Noyes, and Bancroft.

Ostwald's American students returned to the United States full of an enthusiasm for their specialty that had several bases: a conviction that physical chemistry was fundamental in ways that other branches of chemistry were not; a confidence that rapid progress in their field was likely; and a belief that advances in their specialty promised enormous benefits for industry and other sciences. This enthusiasm is essential to understanding their success in promulgating physical chemistry in America. But however far enthusiasm can go toward accounting for the prosperity of this specialty in the United States, it cannot constitute a full explanation, since it does not account for how and why opportunities for physical chemists arose in American institutions. Who employed the first generation or two of American physical chemists and why did they do so?

Table 2.5 presents information relevant to the first of these questions. Data are drawn from the survey of productive physical chemists. As the table makes evident, universities were far and away the largest employers of productive physical chemists throughout the years from 1897 to 1926. During the years prior to 1908, when, as we have seen, many were obtaining their training in Germany, employment outside the university context was very low. Many of the nonacademic jobs that did exist were concentrated in the Bureau of Soils of the Department of Agriculture.

American businessmen evidently did not heed the paeans to the usefulness of physical chemistry sung by Ostwald's disciples. To be sure, it would be surprising if they had. American industry was just beginning to make systematic use of science at the time when Ostwald's first students returned to America. Numerous positions existed for analysts in those industries in which the purity of materials or quality controls on products were considerations, and the country's expanding transportation, power, and communications networks

TABLE 2.5
Employment of Productive American Physical Chemists by Sector, 1897–1926 (in percent)

Years	Sectors of Employment			
	Universities	Government	Industry	Other[a]
1897–1900	94	3	—	3
1901–1904	90	6	2	2
1905–1908	84	10	3	3
1909–1912	79	9	6	6
1913–1916	74	10	8	8
1919–1922	68	8	17	7
1923–1926	66	9	19	6

Note: Statistics for 1917–1918 are omitted because of the temporary dislocations caused by World War I.

[a] Includes self-employed chemists and those working for the Geophysical Laboratory of the Carnegie Institution of Washington, the Rockefeller Institute, and several private hospitals.

were hungry for engineers. Nevertheless, American industry saw little need for scientists with sophisticated research training, and managers in most industries felt unable to justify basic research to stockholders. And well they might think so: research is an inherently risky undertaking; the risks cannot be quantified; and there were still but few examples of profit being derived from systematic scientific research. The research laboratory of the General Electric Company, arguably America's first true industrial research laboratory, was created in 1900. Similar laboratories at such firms as Eastman Kodak, AT&T, and Du Pont were organized during the following fifteen years, and these establishments did seek scientists capable of original research. Nevertheless, prior to World War I, the openings were few by comparison with those available in academe.[110]

During and after the war, however, American industry went on a research binge that was begun out of necessity and sustained by handsome new earnings. The Allied blockade begun in 1914 forced American industry to set about finding ways to replace or find substitutes for thousands of products and chemicals for which Germany had been the only or chief supplier: dyes, pharmaceuticals, optical glass, insecticides, fungicides, fertilizers, and tanning chemicals, among others. Urgent domestic and foreign demand and worldwide shortages caused prices of these, the high-technology products of the early twentieth century, to soar. To meet the demand and to realize some of the exceptional opportunities for profit, American industry frantically expanded its research capacity. Shortcuts were sometimes taken, as when German patents were seized in 1917. But patents were sometimes intentionally deceptive, and even when accurate they hardly could substitute for the multifarious skills acquired through long experience in the large-scale production of these critical products. Well-appointed laboratories and Ph.D. scientists became essential to scores of American firms that, prior to the war, had made do with small test stations and a few analysts and engineers. When the crisis was over, the new facilities and scientific staffs were retained. Indeed, after the war, expansion in industrial research continued as American firms fought to retain markets that had been won from German producers during the war and now were jeopardized by the reappearance of German competition.[111] During the 1920s both the number of companies with research laboratories and the number of persons employed in industrial research tripled.[112]

American chemists were among the major beneficiaries of this rapid expansion in industrial research. In 1921, approximately one out of every three persons engaged in industrial research was a chemist.[113] Although there are no comparable statistics on physical chemists in industry, there is no reason to think that physical chemists were forgotten during this expansion. Major industries based on new processes for the fixation of atmospheric nitrogen and the cracking of petroleum depended on physico-chemical knowledge. Physical chemistry benefited both directly and indirectly from this. An increase in the percentage of productive physical chemists with jobs in industry commenced in 1908 and became pronounced during and after the war.

Even as late as the mid-twenties, however, the number of productive physical chemists in industry remained small by comparison with the number in academic life. The significance of the data presented in Table 2.5 would seem to be clear. The growth of physical chemistry, especially during its first two decades, was largely a function of the availability of teaching positions. American universities were expanding rapidly during this period, and physical chemists capitalized on the situation. Only comparatively late, after 1914, did business become a major patron. R.G.A. Dolby has adopted this position and also credits the prosperity of such other new disciplines as experimental psychology to the rapid growth of American institutions of higher education.[114]

The evidence for this position seems irresistible, but it begs the additional question, why were American universities hiring physical chemists? Were they responding to a demand? But what possible source was there for such a demand prior to World War I? If businessmen were not hiring productive physical chemists, could they have been hiring unproductive (by the standard of publication) ones? It is a possibility. Talented young graduate students had many reasons to prefer an academic to an industrial career; the former carried greater prestige, more flexible working hours, and, perhaps most important, greater freedom in the choice of research topics. Moreover, industry sometimes discouraged employees from publishing the results of their research work. We should expect, therefore, that industrial chemists would be underrepresented in a survey based on publication. Nevertheless, as George Wise has observed, not all chemists found industrial research devoid of appeal. Salaries were generally higher in industry than at universities; equipment and laboratory assistants were often more plentiful; and there was no need to spend long and unproductive hours teaching novices in the laboratory or grading papers. Such talented chemists as Willis R. Whitney, William D. Coolidge, Irving Langmuir, and Saul Dushman preferred industrial research to academic, and there is little doubt but that a scientist was better able to conduct research in an industrial laboratory of the first rank than in a college or university of the second or third.[115] If industrial employment is understated by the figures in Table 2.5, the bias is not likely to be large.

Another possibility is that demand for courses, and hence teachers, in physical chemistry originated from below rather than above. Universities respond not only to the needs of businesses for employees with certain skills, but also to the desires of undergraduates. From time to time fads grip undergraduate education; one need only recall the strong interest in psychology in the 1960s and early 1970s and economics in the 1980s. Enrollments fluctuate, not only in response to the economic forces of the marketplace, but also in response to social circumstances and academic fashions. And when changes occur in the preferences of undergraduates, universities must hire teachers in response.

There is, however, little evidence to suggest that physical chemistry was ever a fashionable course of study among undergraduates. Henry Adams tried to mold human history to Gibbs's phase rule, or Adams's version of it, but here, for once, Adams's self-deprecating estimate of his own influence was probably accurate. Sinclair Lewis made physical chemistry a symbol of pure science in his popular novel *Arrowsmith*, but that novel was perhaps the only one ever written in which the protagonist spends evenings thinking feverishly about the law of mass action.[116] The writings of Adams and Lewis may have stimulated curiosity about the discipline in a few students, but these were of the sort who would abandon the subject after the first laboratory exercise.

A third explanation of the expansion of this discipline within universities might stress the competitive nature of American universities. Joseph Ben-David has put great emphasis on this feature of American education in explaining the rise to leadership of American science in the twentieth century.[117] Flush with money from wealthy benefactors, state governments, and the tuition fees of a rapidly expanding undergraduate population, universities could afford to compete with one another in specialties for which there was limited industrial or undergraduate demand. Once a few major universities established programs in a new field, it was inevitable that others do so. Perhaps universities hired physical chemists in a simple effort to keep up with one another in a competition for honor and prestige.

Competition between universities no doubt existed and played some part in the explosive growth of physical chemistry. But persuasive as Ben-David's argument is, it has some significant flaws when imposed on the history of physical chemistry in America. It is, for instance, difficult to find evidence that university presidents and department heads were eager to initiate instruction in physical chemistry, and it is easy to find evidence of indifference. This evidence is most abundant in those years when the discipline grew most rapidly, the period from 1890 to 1910. Ira Remsen, professor and later president of the Johns Hopkins University and dean of American chemists, gave the first volume of Ostwald's *Lehrbuch der allgemeinen Chemie* a glowing review when it appeared in 1885, but his interest turned to skepticism as subsequent work on the theory of solutions unfolded. While admitting the usefulness of the theory of electrolytic dissociation, he nevertheless thought that "[i]ons can easily be overdone. It will hardly be claimed that the ion is entirely free from speculation."[118] Later he praised the dissociation theory for having stimulated research but at the same time referred to ions as a chemist's plague. Remsen gave scant attention to electrolytic dissociation and omitted all discussion of chemical thermodynamics in a late edition of his popular textbook of theoretical chemistry; for Remsen chemistry was still the descriptive study of the composition and properties of the elements and compounds.[119] When his former student Harry Clary Jones returned to Baltimore from studies in Europe, he won a place on the Johns Hopkins faculty more through his persistence than the efforts of Remsen.

Remsen's long-time friend Edgar Fahs Smith shared many of his doubts about the new physical chemistry. Smith, like Remsen, was a member of that small circle which had created the first American graduate programs in chemistry during the 1870s and 1880s. A professor and later provost at the University of Pennsylvania, Smith and his students had a special interest in the use of electrical currents as analytical tools. Their research in electrochemistry was closely related to the topics studied by Ostwald's research school. Nevertheless, Smith was no theoretician, and, like Remsen, he gave only grudging recognition to the Ionists and their ideas.[120]

Such established leaders in the chemical profession as Remsen and Smith met the theories and claims of the new discipline with indifference and skepticism and so too did many college and university presidents. Although some major universities began continuous course offerings in physical chemistry during the 1890s, others ignored the field entirely until after the turn of the century. The University of Chicago, as we have seen, possessed scientists with considerable interest in physical chemistry, but offered no courses in the field until 1902–1903.[121] The chemistry department at Wesleyan University listed courses in the specialty from 1896 to 1898, but when their instructor left they neglected to replace him until 1908.[122] As late as 1903 Henry S. Pritchett, a German-trained astronomer who was then president of the Massachusetts Institute of Technology, could ask A. A. Noyes to inform him about "what part of chemistry is embraced in the term physical chemistry." Pritchett expressed this interest when Noyes asked for an appropriation for his research laboratory.[123]

Nor were academic administrators generous in providing the wherewithal for original investigations in the discipline despite frequent genuflections before the altar of scientific research. A fortunate few, including T. W. Richards at Harvard, enjoyed good working conditions during the 1890s, but more common was the experience of Noyes, who financed research with personal funds, of Kahlenberg, whose first laboratory at Wisconsin was a cellar with a dirty concrete floor, and of Hulett, whose laboratory at Michigan was at the top of three flights of stairs in an abandoned water tower.[124] Physical chemists by and large found themselves occupying out-of-the-way chinks and niches in American universities during the 1890s, and this is hardly in keeping with a model that explains the rapid growth of the specialty by reference to a competition for services.

The meager attention given the discipline by academic administrators was commensurate with undergraduate enrollments in physico-chemical courses during the 1890s and the first years of the twentieth century. Undergraduate education was still the principal concern at nearly all American universities, and the great majority of undergraduates who took chemistry courses did so as parts of programs of general education or to meet the requirements of engineering and medical schools. At MIT, for example, 764 students were matriculated in chemistry courses in 1910, but only 28 undergraduate degrees were awarded in chemistry.[125] Physical chemistry was not generally required for admission to professional school and hence the undergraduate audience for courses in the field was confined to the rather small number of chemistry majors, and very few schools demanded that these students take a course in the subject. Consequently, there was a great disparity between the number of undergraduates enrolled in introductory chemistry courses and the number in courses on physical chemistry.

Table 2.6 illustrates this point with statistics drawn from a national survey of enrollments in chemistry conducted in 1901. Only at Harvard did the enrollment in physical chemistry exceed 10 percent of the enrollment in the introductory inorganic course. Despite the thin undergraduate interest, however, all of these institutions save Yale boasted two or more teachers of physical chemistry. Two of Ostwald's students were teaching at Cornell (Bancroft and Trevor), in addition to a young instructor, Hector Carveth. Three veterans of Leipzig were on the faculty at both Michigan (Bigelow, Hulett, and Sullivan) and MIT (Noyes, Whitney, and Goodwin [physics]); at Harvard, Richards had one junior colleague in physical chemistry (G. W. Heimrod).

In trying to solve the problem of why physical chemistry prospered in America, we encounter serious paradoxes. American colleges and universities were hiring young physical chemists before there was any clear demand for their knowledge, either on the part of undergraduates or nonacademic employers. And although academic administrators showed little scholarly interest in the new discipline and generally offered its practitioners scant research support, they often were liberal in appointing physical chemists to faculties.

A resolution of these paradoxes may be found, however, by considering the nature of the demands being placed on American colleges and universities

TABLE 2.6
Enrollments in Chemistry in 1901, Selected Schools

Institution	Area of Chemistry			
	Inorganic	Analytical	Organic	Physical
Sheffield (Yale)	199	80	60	15
Harvard	410	195	104	73
MIT	385	145	35	30
Michigan	800	600	460	40
Cornell	417	679	157	23

Source: Charles Baskerville, *et al.*, "Report of the Census Committee," in American Chemical Society, *Twenty-Fifth Anniversary of the American Chemical Society* (Easton, Penn., 1902), Table I.

around the turn of the century and the character of physical chemistry itself. As we have seen, university administrators by and large had little knowledge of physical chemistry. Although most possessed some training in the physical sciences, few were current with recent developments. Whatever their degree of scientific sophistication, however, they were aware of one pressing problem: the need for competent and versatile teachers to handle bloated enrollments in introductory science courses, and perhaps especially in chemistry.

There are few firm statistics on chemistry enrollments during the late nineteenth and early twentieth centuries, but there is little doubt but that their growth was impressive. Complaints regarding overcrowded classes and teaching laboratories were common in presidents' reports from the years surrounding the turn of the century. A multitude of new chemistry buildings constructed during the period 1890–1916 further testifies to the increasing pressure on teaching facilities. New chemistry laboratories were built at Cornell (1890, expanded by 90% of its capacity in 1898), Wisconsin (1905), and Harvard (1912), to cite a few examples.[126] More persuasive evidence comes from statistics on the number of chemists in the work force, as determined by the U.S. Census. The number of chemists per ten thousand workers came close to doubling every ten years from 1870 to 1900, and although the rate of growth decreased somewhat in later decades, it remained impressive until 1930. A similar trend is visible in statistics on the number of chemists per ten thousand professional, technical, and kindred workers. Growth was most rapid during the late nineteenth century, but remained strong well into the twentieth (see Table 2.7).

The growth of enrollments in chemistry courses was itself part of a rapid expansion in the number of students in American colleges and universities. Although the number of Americans in college as a percentage of total population grew only slowly during the 25 years following the end of the Civil War, a period of rapid expansion began around 1890. Between 1890 and 1900, the number of students in college grew from 65,274 to 104,098, an increase of almost 60 percent. During the next ten years this figure increased an additional 67 percent, and between 1910 and 1920 matriculations went up 96 percent.[127] It is likely that attendance in chemistry courses grew at least as quickly as overall enrollments: in every year between 1890 and 1920, between 7 and 10 percent of all college graduates majored in chemistry.[128] In addition, it was during this era that chemistry became the backbone of preprofessional undergraduate programs. Aspiring engineers or doctors might not choose to major in chemistry, but by 1920 few could escape the necessity of taking at least a year or two of work in the field in preparation for admission to professional school. This was hardly the case in 1890, when requirements for professional schools were much less demanding, especially in medicine.[129]

The picture that emerges from this welter of statistics is that of a science that enjoyed an unprecedented boom during the period 1890–1920; the places

TABLE 2.7
Chemists per Ten Thousand Working Population, 1870–1970

Year	Chemists	Chemists per Ten Thousand Workers	Chemists per Ten Thousand Professional, Technical, and Kindred Workers
1870	774	0.6	22.6
1880	1,969	1.1	35.8
1890	4,503	1.9	51.4
1900	9,000	3.1	72.9
1910	16,000	4.3	91.0
1920	28,000	6.6	122.6
1930	45,000	9.2	135.9
1940	57,000	11.0	146.9
1950	77,000	13.1	151.5
1960	84,000	12.4	114.5
1970	110,000	13.8	95.1

Source: Arnold Thackray, Jeffrey L. Sturchio, P. Thomas Carroll, and Robert F. Bud, *Chemistry in America, 1876–1976: Historical Indicators* (Dordrecht, 1985), pp. 248 and 256.

TABLE 2.8
U.S. College Chemistry Enrollments by Specialty, 1901

Specialty	Enrollment	% of Total Enrollment
Inorganic	13,296	59.2
Analytical	5,404	24.1
Organic	2,693	12.0
Physical	554	2.5
Agricultural	493	2.2
Total	22,440	100

Source: Charles Baskerville, *et al.*, ''Report of the Census Committee,'' in American Chemical Society, *Twenty-Fifth Anniversary of the American Chemical Society* (Easton, Penn., 1902), Table I.

where the boom had its greatest impact were the classrooms and laboratories in which undergraduates took the traditional introductory courses: inorganic and analytical chemistry. The 1901 census of U.S. college enrollments in chemistry courses cited earlier offers ample evidence of this; 59 percent of all students in 185 colleges and universities covered by this survey were registered in inorganic courses (see Table 2.8). An additional 24 percent were in analytical courses. Registration in all other categories, fields usually studied by those planning to major in chemistry, was limited to approximately 17 percent of total enrollments. Clearly the greatest need felt by college presidents and by chairmen of chemistry departments was for instructors who could handle laboratory sections and classroom duties in inorganic and analytical chemistry.

Physical chemists believed that they met this need and were successful in convincing others of this as well. Their confidence was founded on their teacher's emphasis on the study of general principles. Ostwald, despite the initial skepticism of some colleagues, had written influential textbooks of analytical and inorganic chemistry and, in fact, had effected major changes in those fields through the systematic use of the principles of mass action and electrolytic dissociation. As we have seen, many of his American students followed his lead by writing texts on these introductory subjects, especially analytical chemistry. Indeed, by the turn of the century the didactic value of casting these traditional subjects in physico-chemical terms had been amply demonstrated. Physical chemists, with their broad training and aggressive claims to being practitioners of an *allgemeinen Chemie*, appeared far more versatile than narrowly trained organic chemists, and more up-to-date than traditionally trained specialists in chemical analysis. They offered chemistry departments some glitter, that of possessing a specialist in an esoteric new research specialty; more important, they were able, and often eager, to teach the bread-and-butter courses of every chemistry department. This was a winning combination.

Physical chemists, by and large, found their first positions in America by dint of their versatility as teachers. A careful examination of the careers of Ostwald's American students provides evidence bearing on this point. Most of the twenty physical chemists who returned to America from Leipzig prior to 1900 obtained positions having little or no apparent relationship to their backgrounds in physical chemistry. Many were employed by institutions with large enrollments in chemistry and with a need for assistants to supervise laboratory work or to provide instruction in analytical, inorganic, and even organic chemistry. A. A. Noyes joined the MIT faculty as an instructor in analytical chemistry and moved up the academic ranks during the 1890s as a teacher of analytical and organic chemistry. The great majority of his research papers dealt with physical chemistry, but the majority of his classroom hours were devoted to other branches of the science. Despite his efforts to develop courses in physical chemistry, it was not until 1899 that he was appointed professor of theoretical chemistry and relieved of most of his other duties. When Louis Kahlenberg returned from Germany it was to a teaching position at the School of Pharmacy of the University of Wisconsin. It no doubt enhanced his case that he had passed his doctoral exams at Leipzig *summa cum laude*, but his expertise in the arcana of physical chemistry mattered less than his ability to teach the rudiments of analysis to future apothecaries. Wilder Bancroft taught inorganic chemistry during his first years at Harvard, as did George Hulett at Michigan. E. C. Sullivan, also at Michigan, was responsible for instruction in qualitative analysis until he joined the Geological Survey in 1903. Others among Ostwald's students, such as C. W. Coggeshall, J.L.R.

Morgan, C. S. Palmer, E.H.S. Bailey, and J. White, found their first positions at small colleges, where elementary chemistry was their major responsibility. Still others, such as H. M. Goodwin and Edgar Buckingham, were hired to teach physics.

Once in American universities, physical chemists seized opportunities to develop upper-level courses, to initiate research, and to develop graduate programs in their specialties. Given the essentially egalitarian structure of academic departments, these efforts to develop their field could not long be denied. Equally important, large new sums of money began to flow into American institutions of scientific research just as the new discipline was starting to find its feet. Academic science, having done little to create the vast fortunes of such moguls as Andrew Carnegie, John D. Rockefeller, and George Eastman, became a major beneficiary of their wealth. Physical chemistry, a science that brandished both an impressive record of recent conceptual advances and the promise of manifold applications, benefited handsomely. As we shall see, the Carnegie Institution of Washington, founded in 1901 and a stimulant to American research in many sciences, provided physical chemists with especially critical support.

Through intensive effort, physical chemists eventually did deliver on the promise that their discipline would become a donor science, one capable of contributing both to the development of other sciences and to industry. But the discipline was not called forth in America by any demands for the services of physical chemists *qua* physical chemists; its explosive growth during the 1890s and the first years of the twentieth century was an unintended consequence of the country's insatiable need for teachers of introductory chemistry, a need generated by the expansion of student populations and changes in patterns of professional education. Physical chemistry encountered a more congenial reception in America than in other nations neither because American universities had a stronger commitment to abstract research nor because American business had a stronger commitment to industrial research, but rather because American colleges and universities were growing at a rate unsurpassed in the western world and physical chemists were able and willing to adapt themselves to a role as generalists, jacks of all trades as it were, in those growing institutions.

King Arthur's Court: Arthur A. Noyes and the Research Laboratory of Physical Chemistry

"How LONG will it be before we build laboratories especially for physical chemistry?" This was the question T. W. Richards asked his colleagues in the chemistry section of the American Association for the Advancement of Science in August 1898. Richards had just described Ostwald's impressive new laboratory at Leipzig, and it rankled him that Americans should "allow Germany to outstrip us so far."[1] There was an element of self-interest in this preaching. If an American laboratory for research in physical chemistry were to be built, Harvard would be a likely site for it and Richards an obvious choice as laboratory director. Richards would have been surprised to learn on that occasion that five years hence a laboratory would be built, not by Harvard, but by its poor neighbor, the Massachusetts Institute of Technology.

This chapter will treat the early history of that laboratory. Not only was it the first American institution dedicated to research in physical chemistry, it was by far the most influential. Although physically unimpressive, MIT's Research Laboratory of Physical Chemistry was among the two or three institutions throughout the world that played crucial roles in redefining the contours of physical chemistry during the early twentieth century—by extending and reworking the chemical thermodynamics of van't Hoff and Gibbs, by exploring more fully the dissociation theory of Arrhenius, and, perhaps most important, by developing new ideas on the structure of atoms and molecules. It was a place where new patterns of thought and activity gradually emerged from old. Van't Hoff and Arrhenius had made a revolution; Ostwald had announced it; Noyes, Lewis, and company followed up on it and reshaped its outcome.

Like all institutions, this research laboratory was a place where private and public worlds met and merged. Significant to us for its roles in the development of physical chemistry and of science in America, it was important to its founder and to those who worked in it for personal reasons as well. In order to explain its genesis, then, it is also necessary to enter into the world of its makers, especially that of its founder, Arthur Amos Noyes, and this we shall also try to do in the pages that follow.

POVERTY AND PROMISE: BOSTON TECH AT CENTURY'S END

Boston Tech, as MIT was often called at the turn of the century, still occupied the original site granted it by the Commonwealth of Massachusetts in 1861, a cramped patch of land near the tracks of the Boston and Albany Railroad in Back Bay. The Tech differed from its aristocratic neighbor across the Charles in almost every way. Harvard, richly endowed by generations of alumni, was wealthy; MIT, whose first graduates were only now approaching the apogees of their careers, was poor. In 1884, five sixths of MIT's operating expenses were paid by tuition fees, and as late as 1913–1914 MIT's yearly income was still substantially less than that of Smith, Vassar, or Wellesley and a mere one tenth that of Cornell.[2] Whereas Harvard possessed a long tradition of leadership in education and a growing reputation for scientific research, MIT was an engineering school of good—but primarily local—reputation, with modest resources and few pretensions. Of research there was little; although not explicitly discouraged, it received little official encouragement. Boston Tech lacked the resources necessary to afford faculty and students the time, space, and equipment for original research. Though chartered to grant master's and doctor's degrees in 1872, prior to 1900 it had conferred only 22 of the former and none of the latter.[3] MIT was a school of applied science offering undergraduates a broadly based technical education that would equip them for careers as engineers and industrial managers. Departments in the basic sciences existed to assist in the preparation of technologists. Intellectually no less than geographically, Boston Tech seemed to exist within the matrix in which it had been conceived.

Yet despite what one observer called the "pinching poverty which keeps its President and his colleagues always anxious," MIT had taken at least one step toward the development of research by the turn of the century: the recruitment of a well-trained and keen-minded group of junior faculty.[4] The chemistry department had been particularly fortunate in this respect: it acquired eight young German-educated chemists during the 1890s, who were joined by still others during the subsequent decade. By 1903, 11 of the 15 members of the chemistry department with doctorates were "made in Germany."[5]

With hindsight it is easy to imagine that genius was behind this recruiting effort, for among these young chemists were two who would become presidents of the American Chemical Society (Arthur A. Noyes and Willis R. Whitney), two who today are celebrated as the fathers of chemical engineering (William H. Walker and Warren K. Lewis), two who would organize and make successful General Electric's research laboratory (Whitney and William D. Coolidge), two who would become internationally renowned for contributions to physical chemistry (Noyes and G. N. Lewis), and such distinguished educators and chemists as Samuel P. Mulliken, Henry P. Talbot, Miles Sherrill, Frank H. Thorpe, and F. Jewett Moore.[6]

In fact, the Institute was doing what came naturally—hiring the best and brightest of its own. The great majority of these young chemists were, like Noyes and Talbot, graduates of Boston Tech. More bookish than most of their fellow students, but possessed of the discipline that MIT sought to foster among its students, they had gone to Germany to obtain the postgraduate training unavailable at the Tech. Returning to America, they made desirable candidates for appointment to the faculty: not only did they hold prestigious German degrees, they were known to their would-be colleagues and knowledgeable about the Tech and its traditions. The latter qualification must have seemed especially important during a period of such rapid growth as the Tech saw in the late nineteenth century. As enrollment grew from 253 in 1880–1881 to 937 in 1890–1891 to 1,608 in 1902–1903, old-timers must have grasped at familiar faces among the sea of strangers. Those returning from abroad had been exposed to new ideas but could be expected to retain a loyalty to the values of MIT; they promised novelty, but not at the cost of institutional continuity.[7]

These products of Yankee schooling and German scholarship effected no sudden revolution within the Tech but helped bring about gradual changes in the curriculum and staff as they grew into positions of authority. Noyes and Talbot, among the first to return to Boston, proved to be especially effective teachers and rose through the ranks quickly; Talbot became a full professor in 1898 and Noyes a year later. In 1901 Talbot was named head of the chemistry department. Although Talbot had only attended Ostwald's lectures, he was little less enthusiastic about the new physical chemistry than his colleague Noyes, who had worked closely with both Ostwald and Nernst.

The MIT Catalogue reflected both their enthusiasm for the ideas of the Ionists and their growing influence within the department. Talbot introduced the electrolytic dissociation theory to students in the elementary course on inorganic chemistry. It was treated again in courses on analysis by Noyes and Augustus Gill, who had also attended Ostwald's lectures while working toward a Ph.D. at Leipzig. Together, Noyes and Talbot, as we have seen, transformed what had been a catchall course dealing with the theoretical aspects of organic and inorganic chemistry into a course on chemical change that emphasized such topics in the new physical chemistry as chemical energetics and solution theory. Texts by Ostwald and Nernst were used and later supplemented by simpler treatments written by Noyes and Talbot.[8] Laboratory exercises illustrating topics in physical chemistry were made mandatory for all undergraduates concentrating in chemistry. By 1902 electives in electrochemistry and chemical equilibria were available, as well as a graduate course in advanced theoretical chemistry. Efforts were also made to acquaint undergraduates with current chemical literature in English, French, and German.

This litany of curricular developments is apt to leave a reader glassy-eyed; their significance, however, should not be lost sight of. Gradually, almost imperceptibly, chemical education at the Tech was assuming its modern form. The typical nineteenth-century pattern of training with its stress on the elements, their compounds, and their identification was slowly yielding to a new. General principles rather than the properties and characteristics of particular substances were coming to the fore. The authors of this change proceeded cautiously, working largely within existing forms. Not only did they have to contend with colleagues for whom chemistry would always be the science of the unique and specific, but they lacked a precise model of that toward which they were working. Ostwald had given them a sense of direction, but it was now necessary to adapt his lessons to a different sort of institution and a different sort of student: an American institute of technology and an undergraduate auditor.

The Makings of a Physical Chemist: Arthur Amos Noyes at Newburyport and MIT

Arthur Amos Noyes was the principal agent behind the curricular solutions to these problems. He has appeared earlier in this book and will reappear regularly in the coming chapters, and it is worth probing his background, for he was not only an important figure at MIT, but also a central figure in the development of physical chemistry in America.

Descended from Puritans who settled in Newbury, Massachusetts, during the 1630s, Noyes was the son of a lawyer, Amos Noyes, who spent his entire life in Newburyport, a few miles from where his ancestors had landed in the seventeenth century. Possessing neither wealth nor business acumen, Noyes held the respect of the Yankee aristocracy of Newburyport by virtue of his heritage and integrity. An obituary writer described him as "a scholar in legal and historical lore, a man who had proved his ability in important public trust, of an acute and vigorous mind, and a faculty of expression in speech and writing." He was, the writer concluded, a "native to the soil, a peculiar product in character and manner as he was by birth."[9]

Newburyport, from which generations of Noyeses had sprung, is among the most studied of America's small cities because of both its rich history and its proximity to the scholars of Boston and Cambridge. Situated near the mouth of the Merrimack River, it had been a prosperous port during the colonial period, had suffered years of depression and decay following the War of 1812, and had then prospered during the 1840s and 1850s with the coming of the railroad and steam-powered textile factories. By 1880, when Noyes was a high-school student, the population had stabilized at around 13,000. Although a small number of Yankee families dominated the town's institutions, the population was hardly homogeneous. Irish laborers, attracted by the mills in the 1850s and 1860s, were numerous, especially, as the local papers were wont to observe, in the town jail.[10]

Noyes grew to manhood accustomed to living near but not at the summit of the social hierarchy. The family's home was on High Street, the finest street in town, but in one of the more modest houses, not a grand federalist mansion. While in high school, Noyes was able to afford a 14-foot sailboat, but he had to buy it with money earned by delivering papers and making fires for local businesses. Admitted to MIT in 1882, he did not attend until a year later, "largely for lack of means."[11]

Of religion there is little mention in any of Noyes's correspondence or papers, yet the values of his Puritan ancestors were omnipresent in his words and his life. As a boy almost all his time was spent in study, work, and in such recreational pursuits as the planning of new railroad lines and the making of timetables for new and old ones. Years later, Noyes would impress colleagues by rising at 4:00 AM and working till 8:00 PM while on yachting "vacations" off the coast of Maine and by living for months at a time in a room that did triple duty as laboratory, kitchen, and bedroom. As a child Noyes had a strong interest in natural history: he kept a bird journal, collected bird eggs, leaves, and minerals, and worked through the experiments in Eliot and Storer's textbook of college chemistry in an attic laboratory.[12] It is unlikely that Noyes was seeking God through these hobbies, but they serve as evidence that he had accepted, in secularized form, values that had long been associated with Puritan communities: a respect for and desire to control nature, an attentiveness to details and a commitment to accuracy in observation and reporting, a compulsion to keep busy, a concern with improving himself and his society, and a capacity to take delight in intellectually challenging pursuits. For Noyes scholarship and play were not antithetical.[13]

Noyes's decision to attend MIT was consistent with this background. Science and engineering were attracting the sons of many middle-class Yankees in the 1880s; they were vocations that held appeal similar to that which the ministry held for an earlier generation. An education in science and engineering promised to build character, to reinforce the discipline, probity, and independence of judgment fostered by family and church, to harness scholarship to the task of improving society. Although lacking the piety of earlier generations, Noyes and others like him retained their forbearers' reverence for learning and passion to serve the community. "We should choose and follow a career for which we have aptitude and interest," Noyes advised a friend in 1908:

> It should certainly be one that has a high form of service in view; but the activities of humanity are so varied that there are hundreds of such forms; still, we should, of course, consider when we seem nearly equally fitted for different forms, which of them affords the greatest opportunity for service.[14]

Noyes himself was then Acting President of MIT and facing a choice between high academic office and a return to the laboratory. Noyes declined the invitation to remain in office as President of MIT and later opportunities to assume the presidencies of Johns Hopkins and Caltech. For Noyes the work of science represented a higher form of service than administration; indeed it was the highest form of service.

Noyes also retained his ancestors' distaste for the gaudy display of wealth and personal achievements. Here too a career in science and engineering must have seemed congenial. There would be scope for personal achievement, comfortable income, and honor; the overriding aim, however, was not the celebration of individual attainments, but improved understanding of the world and the promotion of social welfare. The sober work ethic of MIT, "a place for men to work, not for boys to play," was drawing serious youths like Noyes from across New England.[15]

When Noyes matriculated at MIT in the fall of 1883, it was as a scholarship student. He had spent the previous year studying on his own the first-year subjects at MIT and had passed examinations in all of them save drawing, which he would make up the following summer. This was no mean feat at a time when it was not unusual for a third of entering freshmen to fall by the wayside during their first year at the Tech. Granted the status of a sophomore, Noyes immediately declared his intention to pursue a degree in Course V, chemistry, a subject he had especially enjoyed in high school.[16]

At MIT Noyes compiled a strong, although not stunning, record. His transcript is peppered with A's (German and organic chemistry) and B's, although there are also many C's, especially in laboratory courses. Described by one of his teachers as "a good student but never very skillful in manipulation," Noyes would have to overcome a certain natural awkwardness to become competent in the laboratory.[17] His program of studies was narrow. Courses on organic, analytical, and industrial chemistry and foreign languages (French, German, and Spanish) left little time to explore other disciplines. He took a year of physics and a semester of botany. Most notable, in retrospect, is the absence of work in mathematics. After passing freshman examinations in algebra, geometry, and trigonometry, Noyes ignored the subject. Like many other chemists of his generation, Noyes was utterly ignorant of higher mathematics when he began his career; unlike many of his contemporaries, he would later acknowledge that this was a deficiency and strive to remedy it both as a graduate student at Leipzig and still later, after his formal schooling was complete.

Upon graduating in 1886, Noyes did not enter an industrial career, as did many of his classmates, but chose to remain at MIT. He had come to know the academic world; the world of business was a foreign place. Chemistry was something he could master; together with his professor of organic chemistry, L. M. Norton, he had just published his first article, a brief paper on the action of heat upon ethylene.[18] The mastery of other men, such as was necessary in an industrial plant, was another matter—especially when the men, like the Irish of Newburyport, were likely to share few of his values. Noyes was never quick to leave the nest. A fear of the outer world may have been as important as a lack of money in his decision to defer admission to MIT. When he did matriculate, he commuted to Boston from his maternal grandmother's home in Salem. Now, with a bachelor's degree in hand, he stayed at the Tech, first to complete a master's degree in organic chemistry and then to work as a laboratory assistant. When he finally left New England, it was only when he was able to transport part of his world with him: his shipmates on the voyage to Europe in the summer of 1888 included two former classmates, one (Samuel P. Mulliken) a boyhood friend from Newburyport.[19]

It should not be supposed that Noyes was weak or timid. He did not enjoy sailing unfamiliar waters, but once committed to a new course, he followed it with tenacity. And, already as an inexperienced instructor, Noyes showed an unusual capacity to influence his students. Like the best educators, Noyes did not try to teach his subject, but sought to teach the student—to convey an enthusiasm for science, a mastery of the principles and not just the data, discipline and independence of mind, and "the spirit" that would lead students to devote their knowledge to "some high form of service."[20] Convinced that those who teach best teach by personal example, Noyes struck up friendships with his pupils. Three became lifelong friends and colleagues: Willis R. Whitney, later the dean of industrial scientists but then a struggling student, who, encouraged by Noyes, persevered in his study of chemistry; Harry M. Goodwin, who, constantly badgered by Noyes to devote more attention to science and less to pretty girls and novels, would follow Noyes to Leipzig and later become a professor of physics at MIT; and George Ellery Hale, whose eventual fame as the world's preeminent builder of telescopes and scientific institutions would cast Noyes, not unwillingly, into the shadows.[21]

Noyes spent the following two years, 1888–1890, in Germany. We have already discussed the decision whereby Noyes surrendered his ambition to win a doctorate in organic chemistry and enrolled among Ostwald's then-small band of disciples. The choice could not have been an easy one. Not only did Noyes retain a strong interest in organic chemistry, the study of physical chemistry entailed much remedial work in physics and mathematics. As Noyes undertook his *Arbeit*—an attempt to verify Nernst's recently announced solubility-product principle—he was still struggling to master the fundamentals of thermodynamics and the calculus.[22]

After returning to the Tech, Noyes entered upon what one of his biographers has called "a quiet although intensely active period of ten years."[23] He taught analytical, organic, and theoretical chemistry and wrote textbooks for all three.[24] Feeling the absence of an abstracting service that covered American publications in chemistry and hoping, no doubt, to increase the visibility of MIT's department, he began abstracting the literature himself in the *Technology Quarterly*. As colleagues joined the enterprise, this "Review of American Chemical Research" outgrew MIT and began to appear in the *Journal of the American Chemical Society*. Later, under the editorship of Noyes's successor and distant cousin, William Albert Noyes, the "Review" became *Chemical Abstracts*. During his summers, Noyes visited laboratories in Europe and Japan, conducted research at the Tech, and read papers at meetings of the American Association for the Advancement of Science and the American Chemical Society. By 1897 he had been elected to the governing council of the American Chemical Society; in 1904 he became that organization's youngest president in its 28-year history and the first to be affiliated with MIT.[25]

In many respects—the textbook writing, editorial work, activity in professional societies, and effort given to reconstituting the curriculum—Noyes appears to have modeled his own career on the successful pattern offered by Ostwald's. But Noyes was working under handicaps that Ostwald did not have during his Leipzig years. When Ostwald went to Leipzig in 1887, he was immediately given authority over a laboratory and access to advanced students. To be sure, the laboratory was the university's second chemical institute and it took two or three years to attract disciples, but Ostwald had started with the raw materials of a research school within his grasp. Noyes, by contrast, lacked the facilities and students to build a research school. The chairman of his department itemized some of the Tech's problems in 1902:

> there is great need of an increase in the instructing force of the department, that each member may be able to devote some time to research and study, and to summon for this work something more than the remnants of energies practically exhausted by long hours of instruction in the laboratory. This increase, however, will be impossible until provision for additional space is made, as nearly all space available for a working table or even a writing desk is at present occupied.[26]

MIT's lack of an endowment made it dependent on tuition income. Increasing enrollments, necessitated by the need for cash, meant longer hours in the classroom and teaching laboratories for its faculty. And since the Tech had little money or room for expansion, larger classes also meant overcrowded laboratories. Even had MIT desired to initiate a doctoral program, which it did not, there would have been little reason for talented graduate students to have appeared. These interlocking problems—lack of money, time, advanced students, and space—must sometimes have seemed insurmountable, even to an energetic young chemist like Noyes.[27]

There is evidence, albeit fragmentary, that as early as 1896 Noyes was looking beyond the Tech for ways to improve his situation. Letters of recommendation from Ostwald and Nernst to T. W. Richards at Harvard hint at an effort to create a position for Noyes in Cambridge. Ostwald's testimonial could not have been stronger. Noyes, he wrote,

> is a man of independent thought for whom I predict an outstanding future in science, if he comes into a reasonably good situation in which he has the time and means for work. He is by *far* the best among those young physical chemists who came to me from America. . . . If you can get him, take him into your charge.[28]

Harvard may or may not have tendered Noyes an offer: the records are not conclusive. But soon after this testimonial was sent to Richards, Noyes won promotion from assistant to associate professor and shortly thereafter MIT expanded and reequipped a small area set aside for work in electrochemistry. Although in the physical laboratory, this area was under the control of Noyes's friend Harry M. Goodwin, recently minted a doctor by Ostwald at Leipzig. Equipped with exactly the types of apparatus Noyes required for his work on electrolytes—precision calorimeters, thermostats, and equipment for measuring the conductivities of solutions and fused salts—it is likely that Noyes conducted much of his research there during the next few years.[29]

Noyes may have explored similar opportunities at Chicago and Johns Hopkins during the 1890s, but he was not anxious to leave MIT, despite its many liabilities.[30] He bore the Tech the loyalty of an alumnus, was attached to New England, and, as we have seen, was always reluctant to strike out on his own across new and unfamiliar terrain. Instead he explored ways to make himself and his work independent of the need of external support. He dipped into the world of commerce.

Noyes was not entirely ignorant of this world. He had taken on small consulting jobs for industrial firms to pay for trips abroad, sailing holidays, and short-term research assistance.[31] But during the mid-nineties he devoted more and more of his time to research projects that might yield more lucrative returns. His first excursion into entrepreneurial chemistry came about almost by chance.

Arthur [Clement, a senior studying chemistry at MIT] doesn't care much for theoretical work. I thought I would try and get up some technical process for him to work on. So I thought for a long time in vain, but finally a "bright idea" occurred to me of trying the electrolysis of organic bodies. I started with $C_6H_5NO_2$, nitrobenzene, & expected to reduce it to aniline; but instead of getting that, a white solid body was obtained. Arthur tried the first experiment, but I soon saw that he wouldn't begin to have time to work the thing out, and it was so interesting that I took it up. Well, I worked on that thing all my spare time for three or four weeks before finding out what the stuff was.[32]

It turned out to be para-amido-phenol-sulphonic acid. Further experiments revealed it to be a photographic developer. For three months Noyes was convinced he had made his fortune. He found his compound to be as effective as any of the developers then in use and much cheaper to produce. As Noyes enthusiastically described the work to his friend and former student Harry Goodwin, then working in Ostwald's laboratory, he expressed some chagrin at his own excitement.

I can imagine the look of scorn which has been gradually deepening on your face, as you have read through these pages—scorn at my turning my attention to industrial work, but I have told you the cause of it—die Liebe—; and besides, think of the assistants I can have to work on ions—one for each ion—if it succeeds.[33]

In the event, the project did not succeed. As Noyes and Clement perfected a method to produce the substance in larger quantities, an article appeared in the *Chemiker Zeitung* that described a process almost identical to the one Noyes and Clement had discovered. "So three months hard work is gone for almost nothing," wrote Noyes to Goodwin.[34]

Prospecting was not as easy as Noyes had imagined, but the dreams awakened by this episode stayed with him. In subsequent years Noyes and his undergraduate assistants undertook other projects with commercial potential.[35] Then, in 1898, Noyes found his main chance; again it was a discovery linked to the photographic industry. Together with his former student, now colleague, at MIT, Willis R. Whitney, Noyes developed a process for recovering alcohol and ether vapors from air for the American Aristotype Company of Jamestown, New York, one of the major producers of light-sensitive photographic paper. The details of the process are unknown, since Noyes and Whitney never filed a patent, but historian George Wise has unearthed copies of the contracts governing use of the process. In return for the chemists' granting the American Aristotype Company exclusive use of the process, the company entered into a partnership with the two MIT chemists to exploit the development, and it guaranteed the inventors one fourth of the earnings. Because the recovery of these volatile solvents represented a major economy in the production of collodion paper, the earnings were substantial. For some years Noyes and Whitney each netted over $1,000 a month—roughly five times the salary of a full professor. Several years later, when the American Aristotype Company merged with George Eastman's firm, Noyes and Whitney exchanged their interest in the solvent recovery plant for shares of Eastman Kodak stock. The enterprise did not make Noyes fabulously wealthy, but it did give him the independence he sought. He never again undertook research with immediate commercial potential, nor did he subsequently publish any papers on organic chemistry. He had found his golden nugget and was satisfied.[36]

Noyes's attitudes toward commercial work, and those of his contemporaries, were complex. He was not of that school, of which few real examples could be found, which would label all industrial work as prostitution. Such a saint could not have survived, never mind prospered, at MIT. But neither did he see applied and basic research as of a piece, merging insensibly into one another. For Noyes, research that could earn him money and research that could earn him honor were separate, just as we often differentiate between the writing of scholarly history and history for popular consumption. Both sorts of effort elicited Noyes's enthusiasm, but he found enduring satisfaction in the esteem of academic chemists, not in the attentions of businessmen. Like his father, who was too much the scholar to build a lucrative legal practice, Noyes did not measure all values in dollars and cents.

Noyes's forays into applied chemistry tell us about more than his attitudes toward industrial research; they also give us a glimpse of some of the uncertainties facing a young American ambitious to make his way in science during the 1890s. Universities were employing many scientists who, like Noyes, had been exposed to Germany's research schools, admired them, and hoped to continue their traditions in America. Yet even in the 1890s it was by no means certain that America's colleges and universities would afford scientists the means to achieve such ends. Nor was it clear whether research was a private or corporate obligation, whether Johns Hopkins and Chicago, with their liberal provisions for research, would be exemplars or exceptions.

Noyes surely hoped that MIT would eventually give genuine support to research, but he could not have expected it. He had little reason to, when many of the nation's most eminent scientists supported themselves and their work with personal resources. At Harvard, Alexander Agassiz underwrote his zoological and oceanographical research with wealth derived from investments in the mining industry. At Yale, Othniel Marsh's paleontological work was bankrolled by the fortune of his uncle, the investment banker George Peabody; Marsh's chief rival, Edward Drinker Cope, had paid his research expenses with inherited money until he dissipated the competence through unbridled spending and poor investments. For these and other eminent scientists, appointment at a university might bring social standing and conveniences, but not money. Nor did Noyes have to look beyond his own circle of acquaintances to find examples of this pattern. The Tech's most eminent chemist was James Mason Crafts, the son of a successful merchant and manufacturer; Crafts achieved international renown for his research in organic chemistry, but only by taking long leaves of absence from the Tech. Noyes's intimate friend and former student George Ellery Hale, after graduating in 1890, built his first observatory in the backyard of his father's home in Chicago. Hale's father, a prosperous maker of elevators, could afford to give his son a telescope that would excite the envy of the builders of the University of Chicago; Noyes, lacking such an angel, resolved to make his own fortune.[37] In hindsight, the 1890s appear as an era when institutional and professional structures for scientific research were being built in America; to contemporaries the trend was not so obvious. Research was no longer a strictly private affair, but it was not yet a public or corporate responsibility; it was no longer an amateur's pastime, but it was not yet a fully professional activity.

A MASSACHUSETTS YANKEE BUILDS A COURT

Emboldened by financial success and a growing reputation among his fellow chemists, Noyes set about to create that laboratory in which ions could receive their due. In a letter to the president of MIT, Henry Pritchett, Noyes proposed that the Executive Committee of the Institute cooperate with him in establishing a department of chemical research. Noyes described the plan as "one which I have had at heart for a long time now," and specified the conditions for the creation of the new department. Noyes agreed to provide $5,000 a year for five years if MIT would appropriate an equal amount, create an independent department of chemical research, and furnish it adequate space and equipment. The primary goal, Noyes wrote, should be "research in pure science," and not instruction. A carefully selected and adequately paid staff should "devote themselves almost exclusively to scientific investigation." The emphasis on research, however, would not preclude the presence of students. Advanced students who demonstrated their competence and wished to devote most of their time to original investigations would be welcome. It would therefore be desirable to organize the department with doctoral training as a secondary goal. A few talented master's-degree candidates might also be admitted. In order to prevent members of the research staff from becoming "too specialized," every member of the department might be required to teach one advanced course each year. Nevertheless, "it should be distinctly understood by them that promotions and increases of salary will be dependent solely on demonstrated ability in research work." Appointments should be made for short periods of time, Noyes urged, since it would then be easier to eliminate inefficient staff members. Noyes recommended that MIT provide somewhat over 5,000 square feet of space to house the new department.[38]

Noyes's plan seems a peculiar one to our sensibilities, although not perhaps to his. In essence, he proposed to enter into a partnership with the Institute; Noyes would carry some of the costs and in return would enjoy an unusual measure of control. The scheme was similar in many respects to the contract he had signed with the American Aristotype Company. In that case, a recovery plant had been built on the site of the American Aristotype factory but the plant was owned by a partnership consisting of Noyes, Whitney, and the American Aristotype Company. Noyes and Whitney were responsible for operation of the plant; all partners shared the costs and profits.[39] The chemists did not work *for* the American Aristotype Company but *with* it. Now Noyes was proposing a similar arrangement to MIT. The new department or laboratory would be on MIT's grounds, and MIT would underwrite a portion of the costs, but Noyes would have a proprietary interest. He would not be an employee, but a partner.

Noyes had contemplated other possibilities, among them, "to go out somewhere into the suburbs where land and building is not too expensive, where gas & an electric current are available, with a golf field near by, and build a small private laboratory to accommodate say 8 or 10"—a plan similar to that adopted 25 years later by a fictional physical chemist in *Arrowsmith*.[40] But building a scientific utopia in the American middle landscape was beyond

Noyes's means: "this is of course an expensive way and at the best the facilities could not be as complete as at a good institution; but it has its advantages—especially the important ones of freedom from interruptions and from those petty annoyances which have consumed so much of my mental energy in the past at the Institute. . . . If I had more money it would probably be the best plan."[41] Instead, he would try to realize at least some of his dream at MIT.

President Pritchett and the Executive Committee of the Institute greeted Noyes's proposal with mixed feelings. Pritchett, in office just a year, faced the problem of maintaining the Tech's leadership among the growing number of institutions, now almost one hundred in number, offering engineering education. Noyes's proposal offered a relatively inexpensive means to do this; research, although beginning to become fashionable among American educators, had not yet found a place in engineering schools. One member of MIT's governing Corporation was so enthusiastic, he volunteered to contribute $500 a year for a period of five years to support the department. Granting Noyes his requests might excite jealousies within the faculty, but other institutions, grander than MIT, had sold fiefdoms to wealthy professors. Nevertheless, Noyes was very young—still in his thirties—and not that wealthy. And he was inviting the Tech to make a sharp departure from earlier policies—to enter into a long-term arrangement for the support of what amounted to a research-oriented graduate school in chemistry. Complicating the situation was uncertainty over the permanency of the Institute's location in Back Bay. Noyes's plan could be executed only through the construction of a new building, but it was becoming increasingly clear that if MIT was to continue expanding it would have to move to a new site. A new building could hardly be justified if a move was to come in the near future. The pros and cons were rather evenly balanced; as it turned out, the Executive Committee voted to postpone action on Noyes's proposal until the spring of 1903, by which time a decision was expected on the Institute's site.[42]

This response may well have proven fatal to Noyes's plan; MIT's Corporation took more than ten years to come to a decision regarding the Institute's location. Noyes, however, did not remain passive. In preparation for the next round of deliberations, Noyes strengthened his hand. Here, external events redounded to his advantage. While Noyes was struggling to initiate research in a new specialty at MIT, others were working to find systematic solutions to the problems facing would-be American research scientists. One such solution was the creation of a philanthropic foundation dedicated to the support of science—an invention devised by Andrew Carnegie and his advisors.

The history of their creation, the Carnegie Institution of Washington, has been related elsewhere.[43] Here it suffices to say that the policies of this institution, especially during its early years, were congruent with the aims of Noyes and his fellow young physical chemists. Impressed by recent successes in physical chemistry and astrophysics, the scientific advisors to the Carnegie Institution of Washington showed special interest in supporting work in these and other new fields of study.[44] Of the $218,000 that the Carnegie Institution gave chemists during its first 15 years of grant-giving (1902–1916), 85% went for research in physical chemistry.[45] Coming as it did just as the oldest of Ostwald's former pupils were moving into professorships, this support often proved crucial to the realization of research ambitions. The money itself was helpful, but perhaps even more valuable was the effect such grants had upon attitudes of academic administrators. Research that carried the blessing of the Carnegie Institution of Washington had to be taken seriously. Many of America's most eminent scientists participated in the review of research proposals; recipients of grants were wrapped in their cloak of honor. Carnegie had charged the trustees of his institution with finding and supporting the work of "the exceptional man"; those winning favors could not fail to gain stature in the eyes of colleagues and employers.

The multiple meanings of a grant from the Carnegie Institution of Washington are well illustrated by Noyes's experience at MIT. Rebuffed in his first effort to create a department of research, Noyes took his plan to Daniel Coit Gilman, the president of the Carnegie's new foundation. Asking for $5,900 to support his own work as well as research by his friends Mulliken and Willis R. Whitney, Noyes told Gilman that the fate of the proposed department of chemical research hinged on his action. "This proposition is now under consideration by the Corporation of this Institute, and will, as President Pritchett informs me, undoubtedly be adopted, if it is probable that additional support from outside sources sufficient to make the proposed department a complete success will be available."[46]

Noyes did not receive all he asked; money for Mulliken and Whitney was denied. But he was granted $2,000 for his own work in physical chemistry. Noyes received formal notification of the grant on 27 January 1903, but it is likely that informal notice reached him earlier, for on 12 January he submitted a revised proposal to Pritchett. Instead of asking for an entire department of chemical research, Noyes proposed the creation of an independent laboratory devoted exclusively to physical chemistry. The laboratory would have a smaller budget and a narrower mission; otherwise the details of this second plan were identical to those of the first.[47]

Confronted with this more modest proposal and a vote of confidence in Noyes from the Carnegie Institution of Washington, the Executive Committee put aside its reservations and granted Noyes his wishes. To house his new laboratory, funds were voted to construct a temporary building, Engineering C. Noyes was named director; Willis R. Whitney and Harry Manly Goodwin, his fellow Ionists, joined him as senior staff members. Construction proceeded quickly; in October 1903 the laboratory was completed.[48]

Engineering C was uncomfortably close to the tracks of the Boston and

Albany Railroad and soon was as overcrowded as the other Technology buildings. To physical chemists of the next generation, the quarters seemed "about
the world's worst."[49] But the laboratory gave Noyes space, equipment, hands,
and freedom he hitherto had lacked. Occupying three rooms in the basement
and nine on the first floor, the laboratory included an instrument-maker's shop,
photographic laboratory, seminar room, and facilities for weighing, pure-wa-
ter distillation, and glassblowing. Each room was connected with a 220-volt
direct current and was supplied with ordinary and distilled water, gas, suction,
and air under pressure. Noyes, who had long been frustrated by the difficulties
in securing even the simplest physical instruments, made sure that sets of standard measuring instruments and apparatus for comparison and calibration
were provided. Although the laboratory would later seem primitive, to
Noyes's contemporaries it was an important precedent. Whitney, who had
recently become a part-time director of research for the General Electric Company, was so impressed with its design that several years later he hired
Noyes's builder to design a new research laboratory for General Electric in
Schenectady.[50]

The laboratories of Europe no doubt served Noyes as models. It was an
economy version of Ostwald's institute that was built at MIT, one lacking
lecture halls and large work areas, a constant-temperature room and vibration-
damping piers. But when Noyes undertook to explain the new facility to the
Tech community, he compared it with an astronomical observatory:

> The maintenance in connection with universities of observatories for original work
> in astronomy has long been a common practice; but it is a rare thing for an institution
> of learning to establish a laboratory of physics or chemistry devoted specifically to
> research work. The Institute, therefore, is to be congratulated upon being a pioneer
> in this important extension of educational work. Not only is a laboratory of this kind
> of direct value because of its contributions to the advancement of science, but it
> reacts in many ways upon the instruction work of the institution with which it is
> connected, in such a manner as to raise it to a higher grade.[51]

Several points may be teased from this flat prose. For one, the passage suggests that Noyes did not expect his readers to understand what a research laboratory was or what it might be doing at MIT. Seeking something in their
experience to which to link his laboratory, he selected the observatory. It was
a natural choice. Scores of American colleges supported observatories, and
although many served only for demonstrations, increasingly they were being
used for research. The comparison came all the more easily since George Ellery Hale had recently opened the magnificent new Yerkes astrophysical observatory in Williams Bay, Wisconsin. Boasting its own physical, chemical,
spectroscopic, and photographic laboratories, a machine shop, and library,
Hale's observatory, like Noyes's laboratory, was practically self-sufficient.
And, like Noyes, Hale enjoyed great independence in running the facility;

although affiliated with the University of Chicago, the Yerkes Observatory was as much a product of Hale's entrepreneurial skill as the Research Laboratory of Physical Chemistry was of Noyes's. Both were research institutions; both were devoted to work in new interdisciplinary fields; both existed in the ill-defined borderlands between private and public realms and personal and institutional control. It was inevitable that Noyes should compare his own accomplishment with that of his slightly younger friend.[52]

The passage also hints at a concern that was to become increasingly important in Noyes's thoughts—that of stimulating the development of MIT. Noyes was a chemist, but he was also an educator. His interest was not simply in creating an oasis in a desert, but in coaxing the entire land to bear fruit. MIT, he believed, had an opportunity to become a science-based university where every student, from undergraduate to doctoral candidate, would be engaged in independent and original investigations. The aim would be not only to produce research scientists, but also to mold better engineers.

The curriculum at MIT, as at other engineering schools at the turn of the century, was under strong pressure to expand. New specialties like chemical and electrical engineering were growing rapidly. Older specialties, like mechanical engineering, were becoming more complex and diversified. Employers demanded engineers who could make immediate contributions to the firm; this seemed to mandate specialized training in narrow areas of expertise. But technology was changing so rapidly that teachers felt a responsibility to give students an education that would equip them to adapt. The list of topics that the well-prepared engineer should know kept expanding, yet the time available to teach them was limited by both tradition and market forces. Engineering students typically wished to get out into the world as quickly as possible; lengthening the course of study beyond four years was fraught with risks for the institution.[53]

In Noyes's view, the solution to these dilemmas was not to teach students the myriad details of engineering practice, but to educate them in the principles of the physical sciences and to give them extensive experience in solving problems, through both their coursework and independent research. Such an education would equip them to face those industrial challenges, as yet unforeseeable in their details, that were sure to lie ahead in the twentieth century. It would prepare MIT's graduates to be "leaders on the scientific side" in the development of America's industries, not merely "skilled artisans."[54]

The Research Laboratory of Physical Chemistry could play an important role in shaping this new education. Although it might afford opportunities for only a handful of undergraduates, it would be a nucleus for the activity of those at the Institute who shared Noyes's beliefs:

> It is to be hoped that this may be the beginning of a new development of advanced work at the Institute, and that it may soon be possible to start similar research labo-

ratories devoted to other branches of science, especially to the other divisions of physics and chemistry. Such laboratories cannot fail to react in an important way on the character of our undergraduate instruction and on the scientific spirit of our students.[55]

If successful, the Research Laboratory of Physical Chemistry would not only contribute to Noyes's discipline, it would also give witness to the efficacy of his educational ideas.

Noyes faced one additional problem once the financial arrangements and plans for the laboratory's construction were settled: finding scientists who would make good use of the facilities and freedom he could offer. His requirements were simple, but exacting. Research associates and assistants could not be rank amateurs and they had to have original ideas. He was not looking for journeymen, chemists who could not hold a job elsewhere or who were between positions. Rather, he sought young men who had shown exceptional promise in their first professional work or who possessed exceptional skills that might complement Noyes's own. Locating and securing such talent was no easy matter. Looking back in 1905, Noyes described it as "this most serious difficulty which during the first two years of its existence seemed to threaten the success of the Research Laboratory."[56] Initially Noyes looked to the Tech faculty for his staff, bringing in Goodwin from physics and Whitney from chemistry as senior associates, and a young instructor named William D. Coolidge and several recent graduates of the Tech as assistants. By 1905, however, all but three of the eleven original staff members had resigned. Whitney left in 1904, when his part-time arrangement with General Electric became a full-time position. Claude S. Hudson, described by Noyes as a "*star* whom I found in van't Hoff's laboratory in Berlin," lasted a year before joining the faculty at Princeton.[57] Goodwin, whose interest in research faded after he won promotion to full professor, gradually drifted away from Noyes's circle.

The most painful loss was that of William D. Coolidge. His temperament and background had made him an ideal collaborator. A taciturn farm boy, he was accustomed to long hours and was scrupulously conscientious; decades after he had become a world-renowned scientist, Coolidge would still suffer occasional nightmares about neglecting to feed or water the farm animals.[58] He also had a native mechanical intelligence that was honed by experience in machine shops, factories, and laboratories. The boy who made a contraption to ring an alarm, close the window, and open the door of his room for heat at a preset time in the morning became a college student with a consuming interest in electrical devices. His tinkering with X-ray tubes made him one of the first Americans to be treated for burns from the penetrating rays.

After earning a degree in electrical engineering at MIT, Coolidge had gone to Leipzig, at Goodwin's suggestion, to study physics under Paul Drude. While earning his doctorate *summa cum laude*, he also attended Ostwald's lectures, so when he returned to MIT in 1900 Coolidge had a superb preparation for work with Noyes. Not only did he have outstanding academic credentials, he also possessed exceptional mechanical dexterity and an intimate knowledge of experimental physics—both of which Noyes lacked. Noyes quickly plucked him out of the physics department, making him his personal research assistant. When the Research Laboratory opened, Noyes relieved Coolidge of teaching responsibilities and set him to work on the construction of a high-temperature conductivity cell—an exquisitely tooled instrument of gold, platinum, and quartz that would, Noyes hoped, permit accurate measurements of the electrical properties of solutions at extreme temperatures and pressures.[59] Coolidge found the work engrossing, but as he neared the end of the project Willis R. Whitney offered him a position with the General Electric Company's growing research laboratory at a salary twice the $1,500 he was paid by MIT. Coolidge, who had contracted heavy debts to underwrite his education, felt obligated to accept the offer.

The losses of Coolidge, Whitney, Hudson, and other collaborators were blows, but Noyes had an uncanny ability to find and secure the services of exceptionally talented scientists. Like a great impresario, businessman, or baseball manager, Noyes could detect obscure potentialities in novices and divine how their talents might complement one another. Just as Coolidge left, Noyes engaged three young chemists as students and research associates: William C. Bray, Gilbert Newton Lewis, and Charles A. Kraus. All would later win election to the National Academy of Sciences. Bray, a well-to-do Canadian who had just completed doctoral studies in Ostwald's institute, was blunt-spoken to the point of being impolitic, but was also an experienced chemist who shared Noyes's interests in solutions, qualitative analysis, and undergraduate education. The ideal understudy, he would serve Noyes faithfully for seven years and then follow Lewis to Berkeley, where he would work for the rest of his career.[60]

Lewis and Kraus were rather awkward and peculiar figures with unusual backgrounds and far-from-certain prospects. Kraus, an on-again, off-again graduate student, was more interested in research than degrees; eventually Noyes would have to talk him into applying for a Ph.D. He had majored in electrical engineering at Kansas, done graduate work in chemistry at Kansas and Johns Hopkins, and taught physics at the University of California, but had found satisfaction neither in formal degree programs nor in teaching. Although deficient in political and social graces, Kraus had training in physics—always a recommendation for Noyes, but especially so since Coolidge was leaving. He also had shown early signs of originality in research; while a senior at the University of Kansas, Kraus and his teacher, E. C. Franklin, had undertaken some of the first systematic investigations of the physical properties of liquid ammonia. Their experimental studies of the solubilities of substances in this solvent and the conductivities and boiling points of such solutions had won international notice and helped win Franklin a professorship at Stanford; they convinced Noyes that in Kraus he had found a physical chemist who had not yet discovered his vocation.[61]

G. N. Lewis, like Kraus, was well-traveled, unsatisfied with a conventional teaching career, and uncomfortable in the academic world and in any other he had yet encountered. Like Noyes, Lewis was both the descendant of a settler who had arrived in New England in the 1630s and the son of a lawyer with a scholarly turn of mind. Although born in New England, he was raised in the middle west, where he was educated largely at home by parents distrustful of schools. Lewis's intellectual obstinacy and originality must have had some of their roots in this education. Childhood isolation from the society of his peers may also have had enduring effects on his personality; it took him years to learn how to develop rapport with students, and, although Lewis would later win the fierce loyalty of co-workers, he retained rough and abrasive edges. Dissatisfied with election procedures in the 1930s, Lewis would become one of the few members of the National Academy of Sciences ever to tender a resignation; although often nominated, Lewis was never awarded a Nobel Prize, making him the most distinguished chemist of the twentieth century to be passed over for that award. An unorthodox chemist and teacher, Lewis neither sought nor won easy acceptance within his profession.[62]

After beginning collegiate studies at the University of Nebraska, Lewis transferred to Harvard, where he took a bachelor's degree in 1896. After teaching at Andover for a year, Lewis returned to Harvard, where he earned a Ph.D. in 1899. There followed a year as an instructor at Harvard, a year at the institutes of Ostwald and Nernst in Germany, two more years of teaching at Harvard, and a year as, of all things, Superintendent of Weights and Measures in the Philippines, where he acquired his legendary fondness for cigars.

The decision to move from Cambridge to Manila would be inexplicable had Lewis been happy at Harvard; there is much evidence that he was not. His relationship with T. W. Richards, who had led Lewis into physical chemistry and had directed his doctoral work, was complex. While teaching Lewis much about experimental technique and not a little about chemical thermodynamics, Richards imposed a discipline on his graduate students that the imaginative and headstrong Lewis surely found confining. "In my experience," Richards once wrote,

> assistants who are not carefully superintended may be worse than none, for one has to discover in their work not only the laws of nature, but also the assistants' insidious if well meant mistakes. The less brilliant ones often fail to understand the force of one's suggestions, and the more brilliant ones often strike out on blind paths of their own if not carefully watched.[63]

Lewis, whose papers on thermodynamics soon displayed a depth of understanding and originality that surpassed that of his teacher, must have resented such surveillance.[64]

No less confining was Richards' empiricism. Of that generation of American scientists who identified good science with careful measurements, Richards' suspicion of theory was reinforced by his exposure to Ostwald's injunctions against confusing hypotheses with facts. Although Richards engaged in the measurement of atomic weights, he refrained from placing full confidence in the atomic theory. In his publications, he repeatedly cautioned readers that the atomic theory and the edifice of molecular-kinetic hypotheses built upon it might, after all, prove ephemeral.[65] In his lectures he consigned talk of chemical bonds to the realm of metaphysics: "Twaddle about bonds: A very crude method of representing certain known facts about chemical reactions. A mode of represent[ation] not an explanation."[66] Lewis, who was an assistant in the course in which Richards made these remarks, was at that very time sketching models of cubical atoms and chemical bonds. Memories of Richards may have been on Lewis's mind when, three decades later, he wrote, "I have no patience with attempts to identify science with measurement, which is but one of its tools, or with any definition of the scientist which would exclude a Darwin, a Pasteur or a Kekulé."[67]

Years later Lewis would complain bitterly about his years at Cambridge. "I went from the Middle-west to study at Harvard," he wrote in 1919,

> believing that at that time it represented the highest scientific ideals. But now I very much doubt whether either the physics or chemistry department at that time furnished real incentive to research. In 1897 I wrote a paper on the thermodynamics of the hohlraum which was read by several members of the chemistry and physics departments. They agreed unanimously that the work was not worth doing, especially as I postulated a pressure of light, of which they all derided the existence. They advised me strongly not to spend time on such fruitless investigations, all being entirely unaware of the similar work that Wien was then doing. A few years later I had very much the same ideas of atomic and molecular structure as I now hold, and I had a much greater desire to expound them, but I could not find a soul sufficiently interested to hear the theory. There was a great deal of research work being done at the university, but, as I see it now, the spirit of research was dead.[68]

When Lewis returned from Manila to an appointment at MIT he found what was lacking at Harvard—a band of talented young physical chemists who exhibited not only the form but also the spirit of research. His new colleagues, in turn, found in Lewis a powerfully original mind equipped with both a thorough and broad knowledge of chemical thermodynamics and a superb command of physico-chemical techniques.

Kraus and Lewis were intellectuals. The word seems inappropriate when applied to American scientists, especially chemists. It conjures up for many of us images of bohemian dreamers, entranced by general and abstract ideas and disdainful of the mundane, practical, and concrete. Philosophers, literary critics, and social theorists are intellectuals; some physicists, especially theoretical physicists, may also merit the title. But of all scientists, chemists have most cultivated an image of worldliness and matter-of-fact simplicity. In chemistry, so the image suggests, only facts matter, not interpretations or opinions; knowledge is good because it yields power; intense specialization brings success.

Kraus and Lewis were not of this breed. They enjoyed exercising their wits, delighted in flouting conventions, cared little for utility, and were excited by unconventional new ideas, whatever their source. Lewis, Kraus, Bray, and other participants in Noyes's research colony found both intellectual fellowship and a discriminating patron able to shield them from the utilitarianism of MIT. Noyes, in turn, found in them the society he had known as a boy in his attic laboratory with Samuel Mulliken; as a young instructor with his prize students, Hale, Goodwin, and Whitney; and as a student at Leipzig with Ostwald and Nernst. Noyes, so fond of Tennyson's Arthurian poems that he could recite long passages from memory, fond of signing letters to close friends with the signature "Arturo", and nicknamed "The King" by students at MIT, had built his own round table.[69]

From Processes to Structures: The Anomaly of Strong Electrolytes and the Problem of the Chemical Bond

The Research Laboratory of Physical Chemistry fulfilled personal dreams, but it was important because of the precedents it set and the work it accomplished. In it private fancies merged with institutional and disciplinary goals. The most important of those goals for Noyes was to extend and refine the science he had studied in Germany. He was an enthusiast for physical chemistry. Through the 1890s, while Noyes was learning to teach, writing textbooks, changing the curriculum, and making his fortune, he was also turning out a steady stream of papers, two or three a year, on physico-chemical topics. Not of the sort that make new epochs in science, his papers fell into the category that Thomas S. Kuhn has called normal science; they were scholarly, careful, and useful investigations of problems that were framed by the work of great masters: Arrhenius, van't Hoff, Ostwald, and Nernst. Noyes was never quick to strike out into virgin territory; he preferred to carefully survey ground already reconnoitered.

Normal science, however, is not necessarily dull science; as Kuhn has observed, it can be productive of novelty.[70] Such was the case at the Research Laboratory of Physical Chemistry. Despite Noyes's habitual caution, his laboratory became one of the world's three or four leading centers of research in physical chemistry, attracting postdoctoral students not just from America but from Japan, China, Britain, and even Germany.[71] This reputation for excellence was based in part on the laboratory's exacting measurements of physicochemical constants, but also in part on its exceptional record of developing seminal new ideas, both about chemical thermodynamics and the chemical bond. The work of the Research Laboratory of Physical Chemistry in the latter field was especially important; between 1907 and 1916 its personnel and alumni moved to the forefront of what would soon become a broad movement of physical chemists toward the study of molecular structure. By retracing the course Noyes and his colleagues charted at MIT, we witness Americans making their first significant contributions to the new science of physical chemistry. Much more important, we also gain a better understanding of how the study of molecular architecture found a place in Ostwald's science of chemical processes and, more generally, of the way in which startling innovations can emerge from what may seem, at first glance, prosaic efforts to tidy up the theory and data of a discipline.

Noyes's most intense commitment as a young scientist was to solving puzzles connected with what has come to be known as the solubility-product principle, and it was out of this research that the program of the Research Laboratory of Physical Chemistry emerged. Framed by Nernst in 1889, the solubility-product principle followed from applying the law of mass action to the equilibrium that exists in solutions of two electrolytes with a common ion. Arrhenius had postulated that in electrolytic solutions an equilibrium exists between two species: the undissociated molecules of the electrolyte and the ions into which that electrolyte dissociates. In a saturated solution, say a solution in contact with a solid binary salt, Nernst supposed that the concentration of undissociated molecules must remain constant. At equilibrium, Nernst suggested, the product of the concentrations of the dissociated ions must also remain constant; this because the law of mass action specifies that at equilibrium the concentration of the species produced by a reaction, in this case the undissociated molecules c, when multiplied by a constant K equals the product of the concentrations of the reactants, in this case, the ions c^+ and c^-

$$cK = c^+ c^-$$

Now, Nernst reasoned, if another electrolyte sharing an ion with the first (say c^+) is added to the solution, the dissociation of the first electrolyte must retreat in such a way that the product $c^+ c^-$ will remain constant. Consequently, the quantity of undissociated molecules c should increase. But since the solution is already saturated with this undissociated species, a part of the dissolved salt must precipitate from the solution.[72]

Although so straightforward as to seem self-evident, Nernst's principle had great potential significance. Chemists had long puzzled over the ability of one salt to depress the solubility of another, even when a chemical reaction between them seemed impossible. Not only did Nernst give a qualitative explanation of this phenomenon, met with frequently in analytical and industrial chemistry, he also developed a simple expression by which quantitative predictions could be made of the effects of the addition of one salt upon the solubility of another with a common ion. The ability to predict with some precision the results of adding, for instance, a certain quantity of potassium chloride to a saturated solution of silver chloride, or of silver nitrate to a saturated solution of silver bromate, promised analytical and industrial chemists a powerful new tool.

Noyes was a student at Leipzig when Nernst proposed the principle, and it was natural that he, fresh from an engineering school and knowing that he would someday be teaching analytical chemistry, should take an interest in it. Indeed, in his doctoral dissertation, Noyes gave Nernst's principle its first extensive tests, comparing experimentally derived solubility values against those calculated by means of the solubility-product principle.[73] After returning to MIT, he continued to explore the applications and limitations of the principle. Whereas Nernst had discussed systems of binary electrolytes, Noyes studied the more complex case of ternary electrolytes, such as the solubility of lead chloride ($PbCl_2$) upon the addition of the chlorides of magnesium, calcium, and zinc. Whereas Nernst had shown that the solubility of a salt should be decreased by the presence of a solute with a common ion, Noyes showed that the solubility of a salt should be enhanced in the presence of another solute lacking a common ion. Confronted with exceptional cases in which the solubility of a salt increased, rather than decreased, with the addition of another salt containing a common ion, Noyes demonstrated that new complex molecules were being formed, thus explaining the anomalies.[74] By the turn of the century, Noyes had made himself an authority, perhaps *the* authority, on the solubility phenomena of electrolytic solutions.

This research earned Noyes a secure place among the leaders in the second generation of Ionists—those who were colonizing and cultivating the territories discovered by Arrhenius and van't Hoff, Ostwald and Nernst. It also gradually led Noyes to recognize and become engaged by a problem that was coming to the attention of others among his contemporaries as well: solutions of inorganic acids, bases, and salts—strong electrolytes—did not behave in accordance with the dictates of ionic theory. Theory suggested that these electrolytes, like others, should undergo partial dissociation in solution and that the degree of dissociation should decrease with increases in concentration in accordance with Ostwald's dilution law:

$$c\alpha^2/(1-\alpha) = K,$$

where α is the degree of dissociation at concentration c, and K is a constant. Since Arrhenius had suggested that the degree of dissociation is proportional to the equivalent conductivity, or the ratio of an electrolyte's conductivity at a finite concentration to its conductivity at infinite dilution (Λ/Λ_0), Ostwald's law could also be written as

$$c\Lambda^2/\Lambda_0(\Lambda_0 - \Lambda) = K.$$

This equation, a consequence of applying the law of mass action to the equilibrium between the ionized and unionized fractions of a solute in solution, had proved spectacularly successful when evaluated against the behavior of weakly dissociating substances like organic acids and bases: values of K remained constant from very dilute up to moderately concentrated aqueous solutions. Although it failed among concentrated solutions of these electrolytes, this was to be expected; even gases showed significant deviations from the gas laws as density became appreciable. It would have been surprising had solutions conformed any better to theory. But among strong electrolytes, values of K increased with increasing concentration, even at concentrations between 0.001 and 0.01 normal, that is, as the concentration changed from one gram-molecule of solute per thousand liters of solution to one gram-molecule per hundred liters of solution. Phrased another way, the dissociation values of strong electrolytes as determined by conductivity measurements exceeded the values predicted by Ostwald's law even for dilute solutions, and the discrepancy between real and predicted behavior typically grew worse the higher the concentration, up to certain limits that depended upon the salt.

Although Arrhenius and Ostwald realized that there were discrepancies involving strong electrolytes even as the dilution law was being proposed, years passed before the full extent and significance of the anomalies became apparent.[75] Not only did it take time to collect data on a significant number of strong electrolytes, but there were good reasons to distrust the dissociation values obtained experimentally for such substances. It was, for example, difficult to evaluate the conductivity ratio Λ/Λ_0 with precision, since that ratio is much closer to unity among strong electrolytes than among weak.[76] Perhaps most important, however, physical chemists had little incentive to publicize their problem. Engaged in a struggle to win recognition for the ionic theory, it made little sense for them to hang dirty linen for all to see. The needs of the discipline dictated a caution; until the mid-1890s discrepancies between the theoretical and actual behavior of strong electrolytes were minimized or ignored.

As physical chemistry grew and prospered, however, and as members of a second generation, less willing to subordinate personal accomplishment to disciplinary interests, began to establish themselves, the anomalous behavior of strong electrolytes began to attract more comment. By 1895 various empirical equations were being proposed to account for changes in the molecular conductivity of strong electrolytes with changes in concentration. None proved entirely adequate. Suggestions regarding the theoretical significance of the anomalies were also proliferating. None proved entirely convincing.[77]

Noyes's work on solubilities brought him into direct contact with the problem of strong electrolytes. The solubility-product principle and Ostwald's dilution law were both applications of the law of mass action to ionic equilibria. The use of both depended upon the accuracy of ionization values derived from conductivity measurements. Since the two generalizations had common foundations, whatever caused the dilution law to break down when applied to strong electrolytes might also cause the solubility-product principle to fail when applied to similar substances. In fact, Noyes, from the outset, observed discrepancies between predicted and real solubility values, even at low concentrations, among strongly dissociating salts.[78] Although not so great as to rob Nernst's principle of all value, they did demand explanation.

Nernst's solubility-product principle and Ostwald's dilution law rested upon three major assumptions: that there was a true equilibrium in solutions of strong electrolytes involving only the solute and its constituent ions; that the degree of ionization could be calculated accurately from conductivity measurements; and that the law of mass action controlled the equilibrium between the ionized and un-ionized fractions of the solute. The anomalous behavior had three possible explanations corresponding to the negation of these assumptions.

The equilibrium, for example, could involve more species than simply the molecules of electrolyte and its ions; in other words, hydrates, complexes, or polymers might form in solution and complicate what Arrhenius and Nernst had supposed to be a simple equilibrium condition between a molecular species and its ions. If, for example, ions become associated with a significant proportion of solvent, as supposed by Harry Clary Jones, among others, the effective ionic concentration would then be the ratio of the amount of the ion present to the amount of the free solvent, instead of to the total amount of solvent, as ordinarily calculated. None of the hydration theories, however, could account for the observed variations of the conductivity ratio with concentration. The formation of complexes, combinations between simple ions and molecules of the undissociated salt, would render the dilution law invalid, but struck Noyes as unlikely in dilute solutions of simple binary electrolytes like potassium chloride, which manifested some of the most extreme deviations from the dilution law.[79]

Noyes, at least initially, found it more plausible to suppose that the conductivity ratio Λ/Λ_0 did not give accurate measures of ionization except when the number of ions present in the solution is very small, as in the case of weak organic acids.[80] Arrhenius and Ostwald had suggested that the equivalent conductivity Λ was a simple function of the number of ions in dilute solutions; in fact, this entailed an assumption—to wit, that the velocity of ions did not change appreciably with concentration. If this assumption were false, then the conductivity ratio Λ/Λ_0 could not be considered a reliable measure of dissociation and the anomalous ionization values of strong electrolytes might simply be illusions. This explanation of the behavior of strong electrolytes was attractive, since it preserved both the law of mass action and the essential features of Arrhenius's dissociation theory, but its central hypothesis was extremely difficult to test. It was not feasible to determine the velocities of ions directly, and measuring their relative velocities was an imprecise art until well after the turn of the century.

In the absence of reliable means for measuring ionic mobilities, Noyes searched for other practicable methods of evaluating ionization constants. If independent methods of determining dissociation values could be developed, then there would be some standard against which to judge the accuracy of the conductivity method. In principle there existed several such techniques. The freezing points, boiling points, vapor pressures, solubilities, and osmotic pressures of solutions all were colligative properties; the value of each was dependent on the sum of the number of molecules of solute and the number of ions present in solution. By measuring the change in any of these properties with concentration, it was possible to calculate ionization constants.[81]

In practice, however, each of these methods of studying ionic equilibria had serious limitations. The freezing points of dilute solutions, for instance, differ so little from the freezing point of pure water that it took years to find ways to measure the differences with precision. When such techniques were developed and applied in the years between 1900 and 1910, the results only compounded the problem of strong electrolytes. Ionization values calculated from these colligative or osmotic properties of solutions conformed neither to the values derived from conductivity measurements nor to those predicted by Ostwald's dilution law. There was no good reason to choose one or another of these values as being more probable than another.

There was a third possible explanation for the anomalous behavior of strong electrolytes—to wit, that the law of mass action did not govern the equilibrium between the neutral molecules and ions of strongly dissociating substances. By 1895 Noyes was leaning toward this conclusion; by 1904 he felt sufficiently confident of his ground to indict the mass-action law in the pages of *Science*.[82] His case rested in large part upon the failure of alternative explanations; adding force to it, however, was a fact that had been largely ignored by his colleagues. The law of mass action required that in the cases of bi-, tri-, and tetra-ionic salts the concentration of the un-ionized molecules be proportional to the square, cube, and fourth power of the concentration of ions. But experimental data showed that the form of the concentration function was independent of the number of ions into which the molecules of the salt dissociate. Whatever the nature of the strong electrolyte, the concentration of undissociated molecules was approximately proportional to the 3/2 power of the concentration of the ions.[83]

In the middle of this article of 1904, prepared as a lecture for the International Congress of Arts and Science in St. Louis that year, Noyes in passing raised one other possible solution to the anomaly of strong electrolytes. What if strong electrolytes are fully ionized in solution? In support of this unorthodox notion Noyes noted the curious fact that many properties of dilute salt solutions could be expressed as the sum of the values assigned to the constituent radicals or ions. Arrhenius had cited the additive nature of many of the physical properties of solutions as evidence for his theory of dissociation. But, Noyes observed, such properties as the optical activity and the color of salts in solution are additive even up to concentrations at which a large proportion of the salt was believed to be in the un-ionized state. "If there were not other evidence to the contrary," wrote Noyes, "the existence of this general principle . . . would almost warrant the conclusion that the salts are completely ionized up to the concentration in question, and that the decrease in conductivity is due merely to a change in migration velocity."[84] As Lewis would later write, "the additivity of the properties of electrolytic solutions, striking as it is, seems to prove too much. If it is an argument for the dissociation of electrolytes, it seems to be an argument for complete dissociation."[85]

Clausius had posited a small dissociation of electrolytes; Arrhenius had proposed the dissociation of a significant proportion of salt molecules; now Noyes raised the possibility of a complete ionization of those substances that were efficient conductors of current in water. Eventually, this idea would serve as the basis of a satisfactory analysis of the behavior of strong electrolytes in dilute solutions.[86] But that treatment would come in 1923 from a physicist. In 1904, Noyes simply raised the possibility without developing or embracing it. There were many reasons for his reticence. To argue that strong electrolytes are completely dissociated while weak electrolytes are partly dissociated (as they had to be since they did obey the dilution law) begged the question of why these categories of substances should exhibit such different behavior in water. And why should the conductivity, freezing points, and related properties of solutions of strong electrolytes all suggest that ionization was incomplete? The ionization values calculated by the study of these properties did not conform to the dilution law, but neither did they denote complete dissociation. To accept complete dissociation meant accepting the corollary that the conductivity ratio and all properties that were thermodynamically related to osmotic pressure were producing misleading dissociation values. The choice was between accepting Arrhenius's view that ionization was incomplete and searching for the physical or chemical factors that distorted or negated the impact of mass action, or accepting the view that ionization was complete and searching for the factors that caused such substances to behave in some respects as if they were partially ionized. At the beginning of the twentieth century, Noyes and other physical chemists were ill-prepared for such a choice. Noyes went as far as he felt the evidence justified: he declared that the law of mass action was not applicable to solutions of strong electrolytes and called for a more critical and searching study of the properties and behavior of such substances.

When the Research Laboratory of Physical Chemistry opened, it was this problem that was foremost in Noyes's mind. The behavior of strong electrolytes, he wrote, "constitutes one of the most serious imperfections of the theory of solutions, and . . . may well conceal a discovery of great importance.'' But before rational decisions could be made regarding the causes of the anomalous behavior, measurements of extreme accuracy would be necessary to define more precisely the boundaries of the problem—measurements of conductivities, of ion transport numbers, of solubilities, and of other properties related to the condition of salts in solution. The empirical discovery of constant relationships and tendencies in the behavior of solutions of strong electrolytes, Noyes believed, joined with the critical correlation and analysis of data from both his own and other laboratories, might yield an explanation. ''The main object toward which I am working,'' he wrote in 1908,

> is to extend the conceptions of the Ionic Theory as to enable it to account adequately for the behavior and properties of neutral salts in solution, which it now fails to do. It is not a question of explaining minor deviations, but of accounting for the radically different behavior of largely ionized substances from that of slightly ionized ones and from that which the principle of chemical mass-action requires. We are, at least, gradually establishing important empirical principles, and it is to be hoped that the fuller knowledge of these will suggest the theoretical explanation of the anomalies.[87]

During his first five years of Carnegie support, Noyes, together with Coolidge and a number of other collaborators, dedicated most of his effort to an expensive and technically demanding study of solutions at high temperatures. Data on the electrical properties of solutions above one hundred degrees Centigrade were utterly lacking; as water moved toward its critical point (the state at which the liquid and gas phases are indistinguishable), enormous pressures were generated and water became a highly corrosive agent, capable of attacking steel and even glass. The technical obstacles and expense of such investigations had deterred other scientists. Noyes, however, believed that the return on invested time and money would be significant. By extending the temperature range over which conductivity values were available, he might better evaluate the various empirical expressions that had been advanced relating concentration and ionization. Once constructed, high-temperature conductivity cells might also be used to study the thermodynamic properties of water and, as a by-product, yield results that ''also would be of importance in supplying the underlying principles involved in industrial chemistry processes that are or might be carried out under pressure, and in the interpretation of many natural phenomena with which chemical geology has to deal.''[88]

With the aid of a grant from the Rumford Fund of the American Academy of Arts and Sciences, the Carnegie subsidy, and Coolidge's technical skill, Noyes completed one such cell and began to collect data. Later, to expedite the work, three other cells were built. By 1907 Noyes and his assistants had amassed information relating to the conductivities of many electrolytes at several temperatures between 18° and 306° and at a number of concentrations. These measurements enabled Noyes to make a number of generalizations: at high temperatures the degree of dissociation, as calculated from the conductivity ratio, always decreases; the migration velocities of different ions approach equality; and the ionization constant of pure water shows an enormous increase. The apparatus was also used by one of his assistants to measure the vapor pressure, density, and compressibility of water up to its critical temperature, information that was used to extend steam tables used by engineers. But these elaborate investigations added little to Noyes's understanding of the behavior of strong electrolytes. The ionization of strong electrolytes at high temperatures, no less than at ordinary temperatures, violated the dilution law. Weak electrolytes obeyed the dilution law at high temperatures, as they did at room temperature. The data suggested no convincing explanation for the difference.[89]

The real importance of this work was not that it led to the solution of the problem of strong electrolytes, which it did not; nor that it prepared chemists to accept the "correct" answer when it came along (although it did furnish data useful for that purpose). Rather, it was important because it led Noyes and his associates to seek new methods, both thermodynamic and molecular-kinetic, of analyzing the issues they confronted.

Let us review Noyes's situation in the years just prior to World War I. Sufficient empirical evidence was available, he believed, to abandon hopes of bringing the behavior of strong electrolytes under the scope of the principle of mass action. A failure of the law of mass action could only be due to the influence of some unanticipated chemical or physical force or forces. Here the possibilities seemed limited only by chemists' imaginations: electrical or chemical interactions between the charged particles in solution; effects of the addition of ions upon the "solvating power" of water; changes in the dissociation of water molecules themselves. Further additions to and refinements of the classical data of physical chemistry, while not without value, did not promise a definitive answer. As Noyes and his associate, K. G. Falk, observed in 1910, evaluation of the factors affecting the chemical and physical properties of salt solutions "is attended with the difficulty that at least two of them are simultaneously involved in any property of the salt solution that may be measured."[90] To determine dissociation values from conductivity measurements entailed assumptions about the mobility of ions, which could not be directly measured; to determine dissociation values from freezing-point measurements entailed assumptions about hydration and complex formation in solution; and so on. What Noyes and his co-workers lacked was precise and reliable knowledge of the molecular-kinetic details of their solutions. The properties and phenomena that could be measured in the laboratory furnished only ambiguous clues as to the condition and motions of molecules and ions; and yet a better understanding of the condition and motions of those molecules and ions was coming to seem indispensable to solving the problem of strong electrolytes.

The significance of this realization should not be underestimated. In the 1890s, Ostwald had promoted physical chemistry as a science of processes, not structures. It was to be a corrective to the organic chemists' speculative musings about molecular architecture. Set aside vain talk and imaginary pictures of atoms and molecules, he insisted, and concentrate on winning reliable knowledge through experiment and through the application of thermodynamics.

Noyes, though not so extreme in his opinions as Ostwald, was skeptical of molecular-kinetic models and reasoning. Reluctant to speculate about the material nature of the species in salt solutions, Noyes preferred simply to speak of the conducting and nonconducting portions of the solute. He shunned use of models and analogies drawn from the macroscopic world to describe the invisible workings in solutions. Yet during the first years of the new century Noyes, thwarted in his efforts to find an empirical solution to the problem of strong electrolytes, found himself increasingly drawn to questions of a structural or molecular-kinetic nature. The concluding passages of his report on his high-temperature studies, for instance, reveal Noyes struggling to visualize the essential differences between strong and weak electrolytes:

> The molecular explanation of these facts and the more general conclusions drawn from them would seem to be that primarily the ions are united somewhat loosely in virtue of their electrical attraction to form molecules, the constituents of which still retain their electric charges and therefore to a great extent their characteristic power of producing optical effects and such other effects as are not dependent on their existence as separate aggregates. Secondly, the ions may unite in a more intimate way to form ordinary uncharged molecules, whose constituents have completely lost their identity and original characteristics. These two kinds of molecules may be designated electrical molecules and chemical molecules, respectively, in correspondence with the character of the forces which are assumed to give rise to them. Now in the case of salts and most of the inorganic acids and bases, the tendency to form chemical molecules is comparatively slight, so that the neutral electrical molecules greatly predominate. On the other hand, in the case of most of the organic acids, the tendency to form chemical molecules is very much greater, so that as a rule these predominate. The facts, moreover, indicate that chemical molecules are formed from the ions in accordance with the principle of mass-action, but that electrical molecules are formed in accordance with an entirely distinct principle, whose theoretical basis is not understood.[91]

The reason strong and weak electrolytes differ, in other words, is to be found in a fundamental difference in the nature of the bonds joining their "ions." Whereas the molecules of strong electrolytes are held together by electrical forces, molecules of weak electrolytes are held together by chemical forces. Noyes is ambiguous about certain crucial issues, especially the nature of the chemical forces, but the specific merits and flaws of his conception are less important than the direction in which his thought was moving. We see here a physical chemist trained in the methods of Ostwald groping to find a new vocabulary and new way of thought. Laboratory measurements and thermodynamics were not adequate to the task of understanding solutions. It would be necessary to acquire a mental picture of the invisible details, a firmer command of the structures and motions of the constituents of solutions. Noyes was not alone in this view; many of his associates at the Research Laboratory of Physical Chemistry shared or would soon come to share it. As Charles A. Kraus put it in 1907, "The study of the conduction process cannot be successfully carried out until the molecular state of the system has been established, even in the case of electrolytes the progress of investigation is governed by this condition."[92]

How to attain this molecular-kinetic picture became the question, and a hard question it was. Learning more about the molecules of substances in solution, learning more about molecules in general, demanded new techniques, more revealing than those to which physical chemists had grown habituated. Once again, as had an earlier generation, physical chemists looked to physics for ideas.

Enthusiasm for the study of physics permeated Noyes's laboratory as it did few laboratories of physical chemists either in America or Europe. At times their ardor for physics led Noyes's co-workers far beyond the realm of concepts that were relevant to classical physical chemistry. The first papers by Americans on Einstein's theory of special relativity, for example, were written by Lewis and Richard Chace Tolman, a graduate student and research associate in Noyes's laboratory.[93] "I remember," Lewis recalled twenty years later,

> one long summer night through which Richard Tolman and I raised one objection after another, but always in vain, until we were convinced of the truth of relativity, and when with this conviction I, a young chemist, had the temerity to present the first paper on the principle of relativity before the American Physical Society it was with a sense of exultation like that which one experiences on hearing for the first time a great symphony or a great poem.[94]

After this meeting, which was held in December 1908, Lewis was invited to give a series of lectures on relativity to the Harvard physics department.[95] One can well imagine the joy with which the iconoclastic Lewis assaulted the assumptions and sensibilities of his former colleagues.

Papers on such exotic topics as special relativity, however, contributed nothing to the clarification of the problems facing Noyes and his associates. Far more promising were the writings of the English physicist J. J. Thomson, who, in the last years of the nineteenth century, had accumulated persuasive evidence of the corpuscular nature of electricity. The existence of carriers of the fundamental unit of negative charge had long been suspected; the name "electron" had even been proposed in advance of Thomson's research. Yet most physicists, Thomson included, had believed that ions were responsible for the conduction of electricity through gases. This is what the analogy between gases and solutions suggested. Thomson's work on electrical discharges in rarified gases, however, revealed the existence of negatively charged corpuscles thousands of times smaller than the smallest ion. Rather than finding that electrical charge was carried through gases by the products of ionic dissociation, he had discovered the first subatomic particle.[96]

Building upon his discovery, Thomson in 1904 developed a theory of atomic structure and chemical bonding. The atom, he suggested, was a sphere of positive electricity in which negative corpuscles were embedded in concentric rings; chemical bonds resulted from the transfer of electrons between the outer rings of adjacent atoms.[97]

Thomson's theory failed to explain the existence of bonds in non-ionizing molecules and shed little light on the differences between strong and weak electrolytes, but it excited considerable interest and discussion at MIT. *The Corpuscular Theory of Matter*, the book in which Thomson developed his electron theory of chemical bonds, was read in seminars. Kraus spent the years from 1907 to 1909 studying solutions of metals in liquid ammonia, "the missing link between the metallic and the electrolytic conductor," to test Thomson's theory that free electrons were the carriers of electrical charges in metals.[98] Tolman, in his doctoral dissertation, developed techniques he would later use to detect and measure the inertial effects of electrons, which were presumed to drift freely in metals.[99] In part these efforts were directed to establishing, on firm foundations, the essential difference between electrolytic and metallic conduction, a difference that some physical chemists were slow to acknowledge.[100] In part they were aimed at winning fuller knowledge of electrons themselves. Electrons seemed to be intimately involved in chemical bonding and a better understanding of chemical bonds seemed essential, among other things, to a fuller comprehension of electrolytic dissociation and the mysterious behavior of strong electrolytes.

The nature of chemical bonds and the role of electrons in bonding, consequently, were vigorously debated within Noyes's laboratory. Prior to 1910, the discussion was almost entirely internal. Indeed, the only clues to its existence come from a few passages scattered through the Laboratory's numerous publications, such as the conclusion from Noyes's report of 1907 cited above; from announcements in the *Technology Review* of such colloquium topics as Werner's coordination theory of valence, tautomerism, and the absolute size of atoms; and from references to an ongoing exchange of ideas about bonds and bonding in later letters and papers.[101] Yet, it is striking how many of the seminal papers of the 1910s on valence and bond theory were by chemists who were at or had recently left the Research Laboratory of Physical Chemistry. K. G. Falk was still on the staff when, in 1910, he collaborated with John M. Nelson of Columbia on an influential essay that marshaled the chemical evidence for Thomson's electron-transfer theory of valence. W. C. Bray, shortly after leaving MIT for the University of California, collaborated with G.E.K. Branch in developing a dualistic theory of valence according to which most inorganic substances formed polar molecules, held together by the transfer of electrons, one per bond, and most organic substances formed nonpolar molecules in which electrons did not migrate. W. C. Arsem, a physical chemist who had gone to General Electric after studying under Bray at MIT, suggested in 1914 that all bonds resulted from the sharing of a single electron. "The main features of the theory," he wrote, "were conceived about ten years ago," that is, while he was working in Noyes's laboratory.[102]

The theory that eventually superseded all of these was, of course, Lewis's. Lewis, who left MIT for the University of California in 1912, published papers on valence and bonding both in 1913 and in 1916. In the former he developed the dualistic theory of Bray and Branch, but in the latter he outlined a novel theory that eventually won widespread acceptance. All chemical bonds, Lewis suggested, should be understood as arising from the sharing of pairs of electrons. In cases in which an electron pair was shared more or less equally by two atoms, charge was evenly distributed and the bond showed no polar characteristics; in cases in which one atom held the shared pair with greater tenacity than the other, charge was unevenly distributed and the bond exhibited polar characteristics. Hence the chemically and electrically active molecules of inorganic compounds and the relatively inert molecules of organic compounds belonged on the same continuum. There is no difference in kind between polar and nonpolar bonds, Lewis emphasized, only differences in degree.[103]

The diversity of these theories indicates that there was no consensus in Noyes's laboratory about the nature of chemical bonds, but their number attests to the intense discussion under way there. The volume and quality of these papers were unmatched during this period by any similar constellation of chemists, in America or abroad.

The historian Robert E. Kohler, who skillfully analyzed these and other relevant papers in his studies of the development of G. N. Lewis's theory, quite rightly stressed the great significance that Lewis's theory would have for organic and quantum chemistry in the 1920s and later.

> Without Lewis's conception of the shared-pair bond, the interpretation of reaction mechanisms already begun by the English school of A. Lapworth (1872–1941), T. M. Lowry (1874–1936), C. K. Ingold (b. 1893), and R. Robinson (b. 1886) would not have gotten very far. Likewise, without the idea of the shared-pair bond, then being used with increasing confidence and success by organic chemists, the application of quantum mechanics to the chemical bond in the late 1920's by H. London, E. Schrödinger, and L. Pauling would have begun on far less certain ground.[104]

He also notes that Lewis did little to develop the numerous applications of his theory in those realms: "The few papers on bond theory that Lewis published after 1916 are now mainly of historical interest: in both organic and quantum chemistry, in which Lewis's theory proved most serviceable, Lewis was never really at home."[105] Much the same could be said of Falk, Bray, and the others from MIT who worked with less success than Lewis on the problem of the chemical bond prior to 1916. Their interest in the subject evaporated quickly; none participated in exploring applications of Lewis's fruitful theory in the 1920s.

Herein lie several puzzles. If organic chemists would derive greater dividends than physical chemists from a theory of the chemical bond, why was it physical chemists who prosecuted work on the theory prior to World War I? Why, among physical chemists, was the group at the Research Laboratory of Physical Chemistry in the forefront of research on this topic? And why did the interest of these physical chemists dwindle once Lewis had outlined, in skeletal form, a plausible and generally satisfying theory?

In resolving these paradoxes it is important to recall that purposive actions often have unintended consequences, and that the consequences of actions sometimes obscure the motives behind them. Workers at the Research Laboratory of Physical Chemistry at MIT, where thought about the problem of molecular structure incubated during the years between 1904 and 1912, were not concerned with organic chemistry; important as a theory of the chemical bond would prove for organic chemists, it was not developed for their sake. If the applications of a theory of bonding to organic chemistry had been the principal interest of Lewis and the others who shared his interest in the chemical bond, we should expect there to have been follow-up studies after 1916 exhibiting some of the applications. Yet such work was very largely left to others. Nor, of course, were Lewis, Bray, Falk, Arsem, and Noyes looking ahead to the as yet unforeseeable problems of a future generation of quantum chemists.

The problem of strong electrolytes was the chief theoretical concern at Noyes's laboratory, and it was work on that problem that prompted Noyes and his co-workers to study the writings of J. J. Thomson and stimulated them to develop their own ideas on molecular structure. To be sure, Lewis, and perhaps others, had entertained private thoughts about the nature of the atom and the molecule before arriving at MIT. Lewis later published a page from his notes, dated 1902, in which he sketched admittedly crude pictures of cubical atoms whose outer electrons might form electrical bonds.[106] Yet such private musings might well have remained just that, private musings, had Lewis's thought not been stimulated and nurtured by his colleagues at MIT and by the research in which they were engaged. The study of solutions furnished incentive to explore theories of bonding and set some of the criteria that a successful theory of bonding would have to meet. An adequate theory, for example, would have to explain why certain substances conducted electricity by ionizing in solution while others did not, and why some electrolytes were far better conductors than others. If it could shed light on why the dissociation of weak electrolytes appeared to obey the law of mass action while the dissociation of strong electrolytes did not, so much the better. In short, a successful theory had to supply at least a qualitative picture that both conformed to and rationalized the findings of those who, like Noyes, had studied dilute solutions.

Bray, Falk, and Lewis were all engaged in study of the problem of strong electrolytes when they wrote their papers on valence and bonding. When Falk and Nelson published their paper on the "Electron Conception of Valence," Falk was collaborating with Noyes on an exhaustive analysis of the properties of salt solutions in relation to the ionic theory. Bray and Branch published their paper shortly after Bray had completed a series of articles with Noyes on the solubility-product principle and just as Bray and Kraus were finishing a long paper on the problem of strong electrolytes. Between 1906 and 1916, Lewis published numerous articles on the anomalous electrical and thermodynamic properties of strong electrolytes.[107] Like Noyes, they were all struggling for a molecular-kinetic understanding of the differences between nonionizing, weakly ionizing, and strongly ionizing substances. In their papers on the chemical bond they all sought to demonstrate that their theories could explain the phenomena of dissociation and electrolysis. Falk and Arsem specifically discuss Noyes's passage of 1907, Falk quoting the long passage verbatim.[108] A letter from Bray to his former colleague at MIT, Charles Kraus, demonstrates how closely linked the problems of bonding and strong electrolytes were in their thinking:

> I am awfully glad that this paper [the paper jointly written by Bray and Kraus on the anomaly of strong electrolytes] is all finished, and I wish to congratulate you on the very satisfactory way in which you have rewritten the last sections. I am still groping for some kinetic-molecular-ionic explanation, but I have not got any further. I often wish that I could talk these things over with you again. We have had a great many arguments among ourselves this year about the old subject of valence, especially Lewis, Rosenstein, Biddle, Branch, and I. Finally, last week, when the other men were away, Branch and I wrote a paper on this subject and expect to send it off for publication in a day or two. You will not find very much new in it, but it may amuse you. I have, of course, given Lewis credit for many of his ideas.[109]

The evidence strongly suggests that there was a continuous and ongoing dialogue regarding chemical bonds that commenced with Noyes's tentative remarks of 1907, was stimulated both by Thomson's public and Lewis's private theorizing about atomic and molecular structure, and burgeoned in the years 1910–1916, when Lewis, Bray, and many of their co-workers were drawn into public discussion of the issue. The spark for this activity was the desire to develop a better physical understanding of the puzzling differences between strong and weak electrolytes. As private conversation turned to public discourse, participants had incentive to incorporate additional data, from organic and inorganic chemistry, into their arguments. Gradually Lewis, Bray, and their colleagues realized that the territory they were reconnoitering was far larger than they had anticipated. Like the early voyagers to America who sought the spice islands and instead discovered a New World, the physical chemists from Noyes's laboratory had sought a new route to reach an old objective and instead discovered a new continent.

And like those voyagers of old, these physical chemists were strangely indifferent to the new world they had found. Eventually their efforts to explain the nature of chemical bonds would lead to a thoroughgoing reinterpretation of the phenomena of both organic and inorganic chemistry, but this would be done by others. Perhaps we should not be surprised. Lewis was little more interested in exploring the applications of his theory of the shared-pair bond in organic chemistry than searchers for spice were interested in the forests of America.

The work of Lewis and his colleagues on molecular structure may have had its greatest impact on organic and quantum chemistry, but it also had meaning for classical physical chemistry. For one thing, it furnished a clearer—albeit qualitative—understanding of the process of ionization. Arrhenius and his successors had given short shrift to the question of why certain substances ionize in solution; as Paul Walden, Ostwald's student and successor at Riga, put it:

> *For what reason does the neutral salt molecule break up into ions at all, as soon as it enters solution?* This fundamental question has, up to the present, not been answered. Strangely enough, we do not even make a serious attempt at its solution.[110]

Lewis suggested a plausible, although sketchy, answer to Walden's question: the capacity to ionize was a property that varied with the polarity of molecular bonds. In strong electrolytes, according to Lewis, the shared electron pair constituting the bond is far closer to one atomic kernel than the other, almost the property of one atom alone. In solution, electrostatic forces between the strongly polar molecules of solute and between the molecules of the solute and those of the solvent are sufficient to sever what are already tenuous bonds. Among non-ionizing substances, Lewis proposed, electron pairs are roughly equidistant between the two atoms they join. In solution, such bonds are relatively immune to attack. And in weak electrolytes, in which atoms or radicals with a strong tendency to take up electrons (like oxygen or a hydroxyl) are present together with other atoms less greedy for electrons (like those of carbon), shared electrons are displaced toward the electrophilic atom or group, introducing a slight polarity and a weak tendency to ionize, especially in a polar medium like water.[111] By proposing plausible physical reasons both for electrolytic dissociation and for variations in the ionizing tendencies of various classes of substances, Lewis was outlining answers to questions that Arrhenius and most of his successors had ignored.

Lewis's theory also diverted the attention of physical chemists from the nature of substances in solution to their nature as solids. By positing that differences between the properties of strong and weak electrolytes in solution were due to differences in the nature of their bonds, Lewis implicitly suggested that study of the molecular structure of solids might yield clues pertinent to understanding the phenomenon of ionization in solution. Seeing the electrons in shared-pair bonds was beyond the capacity of instruments. Nevertheless, techniques were becoming available in 1916 that were beginning to furnish data on bond angles and lengths. In England, W. H. and W. L. Bragg had been making rapid progress in the analysis of crystal structures by X-ray diffraction, a technique invented by the German physicist Laue in 1912. Their early studies of crystals of sodium chloride revealed a network in which each sodium ion was surrounded by six chlorine ions and each chlorine ion by six sodium ions. Measurements of the distances between atomic centers revealed no evidence that sodium chloride existed as a bivalent molecule in the solid state. The links between sodium and chlorine atoms, in other words, were so polar as to hardly merit the name bonds. The implication, at least for Noyes and Lewis, was that there was no discrete molecule in this strong electrolyte to dissociate upon solvation. There was a hint here that strong electrolytes might, as Noyes had tentatively suggested in 1904, be completely dissociated in solution.[112]

Although in an embryonic stage, the Braggs' techniques were sufficiently promising to prompt Noyes to call their field "the most important one in physical chemistry today."[113] In 1916 Noyes invited a former student to go to the Braggs' laboratory in London to master the new art and to bring knowledge of it back to MIT.[114] A physical chemist dedicated to the study of solutions thereby became one of the first proponents in the United States of this powerful new method for the structural analysis of solids. Although it appears to have been a radical departure, X-ray crystallography was an extension of the research program prosecuted by Noyes and his colleagues for 25 years.

Lewis's theory of molecular structure helped clarify the process of ionization and it promoted among classically trained physical chemists a new interest in the architecture of solids. It led to other developments in physical chemistry as well, such as new and broader definitions of acids and bases. But it did not solve the problem of strong electrolytes. Lewis, Noyes, and their associates developed a fuller understanding of the molecular condition of electrolytes; but it was still a long way from their essentially qualitative picture to a full-bodied, quantitative treatment based upon molecular-kinetic considerations. When, in 1923, the European physicists Debye and Hückel developed such a theory, it was by making assumptions that the work of Noyes and Lewis had made seem probable: complete dissociation of strong electrolytes in solution and the existence of powerful interionic electrostatic forces that affected ion mobilities. But the new theory also involved ingenious physical and mathematical reasoning that Noyes and Lewis could follow and admire but lacked the skills to pioneer.

In reviewing the specific consequences of the Noyes school's research on molecular structure it is important, however, not to lose sight of broader themes. There were, of course, many reasons for chemists to be interested in molecular structure, and we need not conclude that the program of the Research Laboratory of Physical Chemistry, with its emphasis on the problem of strong electrolytes, was the only route leading to this topic. Nevertheless, it does seem necessary to explain why physical chemists, rather than organic chemists, moved to the forefront of work on molecular structure in the years before World War I, and it is especially important to understand why the classically schooled physical chemists at MIT in particular were so intrigued by this subject. It is more than a little ironic that Noyes, who was so wary of molecular-kinetic theories as to avoid all reference to electrons in his textbook of physical science of 1902, should have presided over the laboratory in which so many fertile ideas about molecular structure emerged during the next decade, and that Lewis, who felt more secure traveling the broad and safe highway of thermodynamics, should have played a crucial role in the emergence of a chemistry of electrons and electron bonds.

Yet the paradox is resolved if we recognize that the study of molecular structure and chemical bonds emerged in Noyes's laboratory not as an end in itself, but as a means toward an end. Although in subsequent years Lewis's theory of the shared-pair bond would overshadow in importance all the work of Noyes, Lewis and their colleagues on strong electrolytes, the study of structure was initially an ancillary theme in Noyes's laboratory, secondary to the goal of explaining the behavior of strong electrolytes. The need to explain the anomaly of strong electrolytes, urgently felt by all those working at MIT, prompted Noyes and his co-workers to attend to the issue of chemical bonds and bonding; it created an atmosphere in which Lewis, and others less talented, had incentive to develop what otherwise might have remained idle speculations; and it propelled Noyes to sponsor some of the earliest work on X-ray crystallography in America. The work of the Noyes school on molecular structure was, in other words, a logical outgrowth of an established research tradition. Noyes, Lewis, and their co-workers had discovered that in order to understand solutions it would be necessary to understand solids, and that in order to understand the laws governing chemical processes it would be necessary to study chemical structures. Their turn toward the study of structure was not simply a response to a revolutionary event external to their discipline—Thomson's discovery of the electron—but resulted from a gradual evolution of thought about a topic that lay at the core of classical physical chemistry.

CHEMICAL THERMODYNAMICS: FROM ACTIVITIES TO A TABLE OF FREE ENERGIES

The problem of strong electrolytes stimulated more than a discussion of molecular structure at the Research Laboratory of Physical Chemistry; it also prompted a revision and reform of the thermodynamic techniques used to treat solutions. Through years of effort Noyes, his colleagues, and other physical chemists in the United States and abroad had shown that the equations of van't Hoff, Nernst, and their colleagues commanded only a comparatively small realm, governing dilute solutions of weak electrolytes and extremely dilute solutions of strong electrolytes, but failing when applied outside of those rather narrow limits. They had defined many questions: why do strong and weak electrolytes differ? Why does the conductivity method of measuring the ionization of a strong electrolyte give values that conflict with those obtained by the measurement of freezing point and other properties related to osmotic pressure? And why do none of these ionization values correspond to those derived from Ostwald's dilution law? But they had discovered no persuasive solutions to these puzzles. The study of molecular structure and of the molecular-kinetic condition of substances in solution seemed to promise some answers, but this subject was in a preliminary and speculative stage.

Noyes was fond of telling students who were stymied by a recalcitrant problem to follow Napoleon's advice: when one line of attack is unproductive, take another.[115] This is what he and his colleagues did. While reconnoitering the field of molecular structure, workers at MIT also sought methods to make the most of the empirical data and thermodynamic principles they already possessed. They developed techniques for applying the laws of ideal, infinitely dilute solutions to real systems, not by discovering the physical basis of the anomaly of strong electrolytes, but by working around their areas of ignorance. Many of the crucial concepts in the Laboratory's work on this subject, as in the case of its work on molecular structure, came from Lewis. Nevertheless, the direction and expression of his thought and its application are best understood within the context of the research underway around him at MIT.

By the turn of the century, chemical thermodynamics had undergone an extensive but not entirely rigorous development. Like Euclidean geometry, thermodynamics consisted of a body of theorems logically deduced from a small number of axioms: the sum of energy changes in an isolated system equals zero; the entropy of an isolated system tends to increase. These two laws of thermodynamics were among the most general and certain principles known to science. But as with any body of law, applications were by no means easy or straightforward. It is one thing to know the laws, it is another to use them to analyze specific chemical systems—a solution or Voltaic cell, for example. The challenge was in finding ways to connect measurable laboratory quantities with the abstract entities of thermodynamics: energy, entropy, heat, and work. Van't Hoff and Gibbs had found techniques for doing this, Gibbs by developing analogies between chemistry and mechanics and van't Hoff by following chemical systems through reversible cycles and analyzing the conversion of energy to work. The methods of Gibbs were more rigorous, but those of van't Hoff, which demanded less mathematical cunning and were more readily grasped, won favor among chemists and were extended by later workers, especially Nernst.[116]

By 1900 physical chemists possessed a dense network of interlocking equations describing the effects of variables like temperature and pressure upon various categories of equilibria. While useful, these equations generally were not precise. Just as the carpenter must learn to make adjustments and allowances when translating geometrical forms into wood, so chemists had to make adjustments and allowances when they moved from the realm of thermodynamics to that of real gases and solutions. In neither state did matter behave exactly as the equations predicted; the denser the gas and the more concentrated the solution, the worse the fit between theoretical and actual behavior. The failure of strong electrolytes to obey the law of mass action and the dismaying gap between the dissociation values yielded by measurements of osmotic properties and those yielded by conductivity measurements were but examples, albeit especially egregious ones, of the practically universal tendency of matter to flout the laws that science sought to impose on it.

Yet, the laws of thermodynamics were not at fault; the problem arose because it was necessary to introduce various empirical principles or approximations in order to apply those laws. Thus, for example, in deriving the equation that describes how the vapor pressure of a simple substance varies with temperature, it was assumed that the volume of the condensed phase could be neglected in comparison with the volume of the gas; in the equations describing the osmotic properties of solutions it was assumed that the electrochemical properties of the constituents of a solution have no influence on their vapor pressures. Such assumptions are not strictly true, but they had to be made in order to obtain usable equations.

The young G. N. Lewis saw opportunity here. While teaching at Harvard Lewis conceived the ambition of reforming chemical thermodynamics; "if the present formulae," he wrote, "could in any way be replaced by rigorously exact ones, without sacrificing concreteness or immediate applicability, then these exact equations might be so systematized that one might serve where a number of isolated equations are now in use, with a great gain in simplification." In two papers, published in 1900 and 1901, Lewis made a start on this project.[117]

The papers were unorthodox, brilliant in places, but uneven. To Lewis's mind, the indiscriminate use of such quantities as partial pressure and concentration was a major source of imprecision in the equations of chemical thermodynamics. Partial pressure and concentration appeared in these equations because it was necessary that some links exist between practice and theory, between the measurable quantities of the laboratory and the abstract quantities of thermodynamics. Yet it was well known that values of these quantities diverged from theoretical expectations as gases or solutions diverged from the ideal state. At high pressures, the pressure exerted by a gas in a mixture of gases is not the same as the pressure that gas would exert if the other gases were absent. And in concentrated solutions (and dilute solutions of electrolytes) the vapor pressures of the constituents are not proportional to their concentrations.

Nevertheless, many of the equations of chemical thermodynamics incorporated these assumptions: van't Hoff's equations describing the influence of temperature and pressure on equilibrium, his statement of the law of mass action, and Nernst's equation of the electromotive force of galvanic cells, among others. Lewis suggested that another quantity existed, which was measurable and yet more fundamental than partial pressure or concentration: escaping tendency or fugacity.

> If any phase containing a given molecular species is brought in contact with any other phase not containing that species, a certain quantity will pass from the first phase to the second. Every molecular species may be considered, therefore, to have a tendency to escape from the phase in which it is. In order to express this tendency quantitatively for any particular state, an infinite number of quantities could be used, such, for example, as the thermodynamic potential of the species, its vapor pressure, its solubility in water, etc. The quantity which we shall choose is one which seems at first sight more abstruse than any of these, but is in fact simpler, more general, and easier to manipulate. It will be called the fugacity . . .[118]

When two phases are in equilibrium, Lewis suggested, the fugacity has the same value in both; when not in equilibrium, the fugacity is greater in the less stable phase. As a gas is rarified—that is, as it approaches the ideal state—the value of the fugacity approaches the value of the gas pressure as a limiting value. For nonideal substances, the fugacity is equal to the vapor pressure that the substance would have if its vapor were a perfect gas.[119]

Having defined fugacity, Lewis proceeded to develop equations showing the influence of temperature and pressure on the fugacity of simple substances. Lewis's equations were analogous to those of van't Hoff, but used values for the fugacity of substances in place of vapor pressures. If vapor pressures were systematically replaced by fugacities, Lewis suggested, all of the approximate equations of chemical thermodynamics could be made exact.

Lewis's papers received negligible attention, and not without reason. They were written by a young American who had only two previous publications to his credit. They were extravagantly ambitious; in the article of 1901, Lewis extended his results so far as to propose a single equation that "embraces every possibility of change of state of any simple substance under all conceivable conditions,'' and promised a subsequent paper generalizing the equation so that it would apply to mixtures and to chemical as well as physical processes.[120] But since Lewis did not immediately publish the follow-up paper, it was by no means clear that his elaborate reworking of the equations of thermodynamics would yield anything of interest to chemists. Chemists were concerned with solutions and with chemical reactions, not with behavior of simple substances and physical changes of state. Lewis's papers were also forbiddingly abstract. Although he insisted that fugacity was ''a real physical quantity capable in all cases of experimental determination,'' Lewis discussed only a single method of determining its value, and that entailed the measurement of osmotic pressures over a range of concentrations—a method that would be awkward at best, since it was notoriously difficult to obtain accurate values of the osmotic pressure of dilute solutions.[121]

Perhaps most damaging, Lewis seemed to attain his results through the use of mirrors. The fugacity was simply the vapor pressure a substance would have if its vapor were a perfect gas. Lewis had defined the quantity so that it would have to yield accurate results when used in thermodynamic equations. Wilder Bancroft expressed what must have been the opinion of most of those chemists sufficiently patient to labor through Lewis's papers. ''The author,'' Bancroft wrote in a review of another of Lewis's early articles, ''deduces a single equation which should enable one to predict anything; but which does not lead the author to anything new . . . What we need in physical chemistry is a closer adherence to the facts and less approximation theory.''[122]

Lewis appears to have put aside work on thermodynamic theory after 1901. Whether because of the response of critics like Bancroft, or the advice of his mentor Richards, or some personal crisis, he published nothing between 1901 and 1905, the only such drought in his career. When he resumed writing, while in Manila, it was to report on experimental work. Yet he had not lost all interest in the problem of simplifying and rationalizing chemical thermodynamics. Shortly after joining the staff at Noyes's laboratory, Lewis took up the task again, this time with greater skill and success.

A single paper, published in 1907, ''Outlines of a New System of Thermodynamic Chemistry,'' summarized Lewis's ideas. Although brief, it was a remarkably comprehensive revision of chemical thermodynamics. As in his earlier work, Lewis gave a prominent position to the concept of escaping tendency. But in the paper of 1907 he defined this slippery concept in another and more useful way. In addition to discussing fugacity, which is an expression of escaping tendency using the dimensions of pressure, he made use of another term, activity—an expression of escaping tendency using the dimensions of concentration. The activity of a substance, he suggested, is equal to its fugacity divided by the product of the gas constant and the absolute temperature. Like the fugacity, the activity expressed the tendency of a molecular species to escape from its condition. Just as the fugacity was a measure of the vapor pressure a substance would have if it behaved as a perfect gas, so the activity was a measure of the concentration a substance would have if it behaved according to the laws of perfect solutions. The activity of a species, Lewis wrote, constitutes "a perfect measure of the tendency of the species to take part in any chemical reaction. In other words, the activity is an exact measure of that which has been rather vaguely called the "active mass" of a substance."[123]

Physical chemists had long been accustomed to using the expression "active mass." Even van't Hoff, Arrhenius, and Nernst sanctioned its use when the laws of solution yielded predictions at variance with experimental data. So when Nernst spoke of the anomaly of strong electrolytes in his widely used textbook, he suggested that "the active mass of the ions is not exactly proportional to the concentration."[124] Used in this sense, "active mass" was simply a relatively painless way of saying "the laws don't work and I don't know why."

Lewis's predecessors used the term as a euphemism for ignorance; Lewis shed some of their embarrassment and tried to mold the expression "activity" into a useful quantity. It may not yet be possible to explain why the activities of ions differ from their concentrations, but if activities work in the equations of solution theory and concentrations do not, why not systematically replace concentrations with activities?

As in his earlier papers, in which Lewis had derived new versions of many of the familiar equations of chemical thermodynamics with fugacities substituted for partial pressures, he now proceeded to show how activities could be used in place of concentrations. A typical example is the law of mass action. Given a chemical reaction in solution of the form

$$aA + bB = oO + pP,$$

where the lowercase letters represent mols and the uppercase letters represent chemical substances, van't Hoff had shown that, for ideal solutions at constant temperature,

$$C_O{}^o C_P{}^p / C_A{}^a C_B{}^b = \text{constant}.$$

Having defined the activity ξ so that it equaled the concentration C in an ideal solution, Lewis rewrote this equation so that

$$\xi_O{}^o \xi_P{}^p / \xi_A{}^a \xi_B{}^b = K,$$

where ξ_A, etc. represent activities and K is another constant. In ideal solutions, the values of the two constants are identical. But in nonideal solutions, in which concentrations and activities diverge, van't Hoff's constant will vary while K will remain unchanged. Although identical in form, the two equations yield different results for most real chemical systems. Thus in the case of the dissociation of a strong electrolyte like hydrochloric acid in aqueous solution,

HCl = H$^+$ + Cl$^-$, the mass law would require that $C_H + C_{Cl} - /C_{HCl}$ be constant. Yet no matter how the concentrations of the three species are determined, whether by conductivity measurements or the measurement of such thermodynamic properties as freezing-point depression, the ratio is not constant at all concentrations. Lewis suggested that if concentration values are replaced by activities in the expression, then the ratio will hold constant at any concentration.[125]

But how could activities be measured? Lewis could suggest only a cumbersome means of calculating the absolute activities of gases and undissociated substances in solution. And he acknowledged that it was impossible to determine the absolute activities of ions in solution.[126] Nevertheless, Lewis was able to demonstrate a method (later to be joined by several others) for determining the ratio between the activities of a substance at two different concentrations; for most of the chemist's purposes, a knowledge of such activity ratios was sufficient. The method made use of the proportionality between the change in the osmotic pressure of a substance when it passes from one state at which its osmotic pressure is π to another at which it is π' and the change in free energy that accompanies the process:

$$\Delta F = RT \ln(\pi'/\pi).$$

Van't Hoff had demonstrated that this equation was accurate for perfect solutions at constant temperature; Lewis, noting that values of the osmotic pressure were proportional to activity values, rewrote the equation in such a way as to make it more general:

$$\Delta F = RT \ln(\xi'/\xi).$$

It was well understood that the electromotive force E of a reversible Voltaic cell provides a measure of the free-energy change occurring within the cell:

$$\Delta F = -EmF'$$

where m is the number of faraday equivalents that pass through the cell and F' the faraday equivalent (96,500 Coulombs).

Consequently, when an electromotive force is produced by the reversible transfer of a substance from a solution in which it is at one concentration to another where it is at a second,

$$-EmF' = RT \ln(\xi/\xi'), \text{ and}$$
$$E = RT/mF' \ln(\xi/\xi').$$

This reversible transfer of a substance between two solutions differing only in concentration was precisely the way in which a certain type of galvanic cell, the concentration cell, generates electromotive force. By measuring the electromotive force of concentration cells with a suitably sensitive potentiometer, it was possible to calculate the activity ratio ξ/ξ' of the electrolyte, since all the quantities in the last equation, save the activities, are known or directly measurable.

By similar reasoning, Lewis demonstrated that for a reaction of the form $aA + bB = oO + pP$ the ratio of the mean activities of the ions

$$\xi_O{}^o\xi_P{}^p/\xi_A{}^a\xi_B{}^b$$

could be determined by measuring the electromotive force produced by the reaction in a reversible galvanic cell. In this case,

$$\Delta F = RT/mF' \ln K - RT/mF' \ln(\xi_O{}^o\xi_P{}^p/\xi_A{}^a\xi_B{}^b) - RT \ln K, \text{ and}$$
$$E = RT/mF' \ln K - RT/mF' \ln(\xi_O{}^o\xi_P{}^p/\xi_A{}^a\xi_B{}^b).$$

Here m represents the number of faraday equivalents accompanying the disappearance of a mols of A, b mols of B, etc.[127]

By demonstrating how activities were related both to free-energy changes accompanying reactions and to the electromotive force of galvanic cells, Lewis had now accomplished what his earlier papers had failed to do. He showed both that usable data regarding the activities of substances could be obtained in the laboratory and that such data could be used in equations describing not simply changes in physical state, but also chemical processes. In subsequent years, Lewis would demonstrate that measurements of solubilities, freezing points, and other properties thermodynamically related to osmotic pressure could yield similar information about activities.[128]

Yet Lewis's paper did not have immediate consequences. Prior to World War I few chemists aside from Noyes, Lewis and their co-workers made use of activities. In part this neglect must be ascribed to an allergic response by chemists to the abstract and theoretical. Although Lewis outlined a workable technique for determining certain activity ratios, he did not present a single such value in his paper. The paper was, in this sense, entirely theoretical. Moreover, except for one brief paragraph buried deep in the text, Lewis gave his readers little guidance on where, in the broad domain of physical chemistry, the use of activities might be especially valuable.[129] In fact, most gases at atmospheric pressure behave very much like ideal gases, and many dilute solutions closely approximate perfect solutions. The systematic substitution of activities for concentrations in chemistry, which some readers must have imagined Lewis to be advocating, would have been an arduous and, in many cases, unnecessary exercise.

The real advantages of Lewis's reformulation of chemical thermodynamics were to be found in the treatment of strong electrolytes, substances that, at concentrations as low as 1 gram-molecule of salt per 10,000 grams of solvent, showed marked deviations from the laws of solutions. Here the use of activities instead of concentrations was essential to make the laws of solution predictive of actual behavior. Yet even among chemists working on the problem of strong electrolytes, Lewis won converts only slowly.

The reason is not far to seek. The concept of activity, by itself, explained nothing. Lewis did not furnish a physical explanation of why concentration and activity values differed in nonideal solutions. The use of activities in place of concentrations permitted the equations of chemical thermodynamics to be restated in more accurate and general terms, but this seemed like a rather formal result so long as the calculation of activities would be a purely empirical process. Activities, in other words, had a status much the same as van't Hoff's correction factor i prior to Arrhenius's explication of the theory giving that coefficient physical significance.[130] In a review of Lewis's later work, J. R. Partington, a leading British thermodynamicist, made essentially this point. "It is perfectly obvious," Partington wrote,

> that a set of equations which are made up to retain the form of the ideal gas law, $pv = RT$, although they may serve as a makeshift in the study of solutions, cannot attempt to represent phenomena which can never be covered by equations of the gas law type. It is perhaps not sufficiently emphasized that the use of activity can only be a passing phase in the study of thermodynamics, and that it is desirable to accumulate data which will enable us to apply the fundamental equations without such artificial methods. This is said without the slightest suggestion that fugacities and activities are not useful in the study of solutions, but even here it may be suggested that a more fundamental type of treatment should be kept always in view.[131]

Lewis might protest that activities had genuine physical significance, but, for most physical chemists, they had a distinctly lower ontological status.

Even Lewis's colleagues at MIT, who took Lewis and his ideas seriously, stuttered a bit before using Lewis's vocabulary. But in 1911, in an influential study of the solubility-product principle, Noyes and Bray dropped some of their inhibitions and endorsed the use of activity as a useful tool in the analysis of deviations from the laws of perfect solutions. To make use of the concept more convenient, Noyes and Bray defined a new quantity, the activity coefficient—the ratio of the activity of a substance (or mean activity, in the case of an electrolyte) to its concentration. Typically, as the concentration of a substance in solution decreases, the activity coefficient would be expected to approach a constant value, which could be assumed to be unity at infinite dilution. At higher concentrations, the activity coefficient then would be a relative measure of the effectiveness of a substance in influencing any chemical equilibrium in which it is involved compared with its effectiveness at very low concentrations.[132] The activity coefficient, in other words, was a convenient means of lumping together all the factors, none of them precisely understood, which caused deviation from the ideal state in solution.

Activity coefficients furnished physical chemists with a new means of describing the effects of physical causes that were not understood; they also gave physical chemists a better way of living with uncertainty. No one could say with confidence what percentage of the molecules of a strong electrolyte like potassium chloride was ionized in a given aqueous solution of finite concentration. But with a knowledge of activity coefficients it was possible to state the "thermodynamically effective" degree of dissociation. This in turn made it possible to apply the equations of thermodynamics to nonideal systems with much greater precision. One area in which this had an impact was the determination of the free-energy changes accompanying chemical reactions.

The change in free energy, the maximum work made available for external use by a reversible process, is a measure of the affinity or driving force of a chemical reaction—a point made in various ways by Gibbs, Helmholtz, van't Hoff, and Nernst in the 1880s.[133] If the free-energy change in a reaction is positive, it is thermodynamically feasible for that reaction to occur spontaneously; if it is negative, the reaction can proceed only if energy is supplied to the system from some external source. When the free energy of the products of a reaction equals the free energy of the reactants, equilibrium has been attained. Since the free energy F of a substance, defined thermodynamically, is a function of the enthalpy H and entropy S of a system,

$$F = H - TS,$$

it is, like those quantities, a state function; the change in free energy accompanying a chemical reaction is independent of the path by which the reaction proceeds. Whether a particular substance is synthesized from its elements in one or many steps, the overall change in the free energy of the system will have the same value so long as all the steps are reversible and the temperature is constant. This was of great practical importance: once determined, the free energy associated with the formation of a substance from its elements might be used in all of the reactions in which that substance was a participant. A table containing values of the free energy of formation of a relatively small number of substances could, with a little ingenuity, serve as an affinity table covering a vast number of chemical reactions. The change in free energy, in short, was that quantity which had been sought since the eighteenth century, when the notion of quantifying chemical affinities first came to exercise the imaginations of chemists.

All of these ideas about free energy were in circulation by the time Noyes and Lewis were beginning their careers. Yet prior to 1905, little progress was made toward actually constructing such a free-energy table. The obstacles were at least as much experimental as theoretical.

In principle there were many ways to determine the free energies of formation. Van't Hoff had shown that the change in free energy accompanying a chemical reaction was related to the equilibrium constant of the reaction:

$$\Delta F = RT \ln K.$$

And Nernst had shown that the free-energy change accompanying a reaction was proportional to the electromotive force generated by the reaction in a Voltaic cell:

$$\Delta F = EmF'.$$

Hence, it was theoretically possible to determine the free-energy change of a reaction either by measuring equilibrium constants or electromotive forces. In practice, however, neither method could be readily applied to study large numbers of reactions. The direct measurement of equilibrium constants entailed the analytical determination of the concentrations of the reactants and products in the reaction vessel at equilibrium. This was feasible only if the equilibrium constant was near unity. Although many of the reactions of organic chemistry satisfied this requirement, the equilibrium position in most of the reactions of inorganic chemistry was displaced very far toward one or the other side of the chemical equation, making an equilibrium analysis of all constituents impossible.

Many of the reactions of inorganic chemistry, of course, could be used to generate electrical action in galvanic cells. Here, however, the experimenter encountered another set of obstacles. To measure the electromotive force generated by a reaction, it was necessary to find or design a Voltaic cell in which the reaction occurred in the absence of secondary reactions. It was also necessary that the process taking place at the electrodes be reversible and that the electrodes give a constant and reproducible potential difference. Very small errors in the measurement of potential differences could give rise to very large errors in the free-energy values. Taken together, these technical considerations placed very stringent constraints on the experimenter.[134]

By the early twentieth century, advances in the design of Voltaic cells and techniques for measuring potential differences had greatly enlarged both the number of reactions that could be studied electrochemically and the accuracy of the resulting data. But another obstacle had to be overcome before free-energy measurements could realize their promise: that of finding a means whereby the free-energy values obtained for a particular reaction under one set of experimental conditions could be made useful in calculating the free-energy change associated with the same reaction under other conditions. To measure the free energy of formation of a substance using a Voltaic cell, it is necessary that the reaction take place in solution, typically a fairly concentrated solution. Yet the chemist may wish to know the free-energy change of a reaction when one or more of the reactants is in a gaseous state, or in a dilute solution.

Here is where Lewis's notion of activity assumed importance. Because the use of activities rendered the equations of thermodynamics relevant to nonideal systems, it furnished chemists a means whereby they could freely translate free-energy values obtained when one or more of the constituents of a reaction are in a nonideal state into corresponding values when all the constituents are in standard states. If the standard state of a substance is defined as that in which the activity coefficient of the substance equals unity (hence the standard state of a solute in solution is an infinitely dilute solution), then it is necessary only to measure the activities of substances in their nonstandard states in order to perform conversions. So, for example, if ΔF is the change in free energy experimentally determined for the reaction $aA + bB = oO + pP$ when the substances A, B, etc. are in nonstandard states, and if $\Delta F'$ is the unknown change of free energy accompanying the same reaction when each substance is in its standard state, then

$$\Delta F - \Delta F' = RT \ln(\xi_O{}^o \xi_P{}^p / \xi_A{}^a \xi_B{}^b),$$

where ξ_A, ξ_B, etc. are the activities in the nonstandard states. This equation furnished a general method for moving back and forth between the free-energy changes associated with reactions as they were typically performed in the laboratory and a general table describing free-energy changes when all substances were in standard states. "The problem of converting free energies in various states into free energies in standard states," Lewis wrote, "is therefore the problem of determining the activities of the various substances concerned."[135]

Armed with a technique for converting the free-energy values, the task of preparing a comprehensive table of free-energy changes could be begun in confidence that the results could find general application. Both Noyes and Lewis were conscious of the enormous value such a table would have for chemistry, and by 1905 both had begun a preliminary reconnaissance of the subject. While in Manila, Lewis experimentally determined the free energy of formation of water from its elements and became intrigued by the experimental challenges posed by such determinations. At the same time, Noyes was discussing with friends and colleagues the possibility of mounting a cooperative attack on the problem through the National Academy of Sciences. At Noyes's behest, the National Academy created a committee to consider the subject in 1906 and commissioned Lewis, by then at MIT, to prepare a critical summary of the existing data relating to the free-energy changes accompanying chemical reactions. Eventually, Noyes hoped, the complex and demanding tasks involved in measuring free energies of formation might be apportioned among several laboratories.[136] But, as with most cooperative schemes, little progress was made and Noyes soon developed doubts about the wisdom of his proposal. The project, he later wrote,

was postponed partly through the pressure of other obligations, but also partly by a deep-seated feeling that after all this cooperative plan was a laborious and uncertain method of obtaining the desired result, which could be much better secured by concentrated effort in a single laboratory provided with a competent staff and complete experimental facilities.[137]

Heartened perhaps by his colleague's enthusiasm for the project, Noyes gave Lewis free rein to undertake the job in the Research Laboratory of Physical Chemistry. Noyes himself, burdened with the presidency of MIT between 1907 and 1909 and preoccupied thereafter with his work on strong electrolytes, took a secondary role in these studies, but encouraged graduate students and research associates to collaborate with Lewis.[138] By 1912, when Lewis left MIT, substantial progress had been made. The free energies of formation of such common substances as water, nitric acid, and urea had been measured to new standards of accuracy, and techniques had been perfected that would permit subsequent determination of many additional values. Most important, Lewis and his associates had determined standard values for the potentials of thallium, sodium, chlorine, and potassium electrodes, from whence it was possible to obtain the free energy of formation of many electrolytes in solution. These studies, especially of the highly reactive sodium and potassium electrodes, demanded extraordinary care and ingenuity in experimental design. Years later Lewis cited this as the work that gave him the greatest satisfaction. "It is one thing," he wrote, "to learn an experimental method and apply it with great exactness to all the problems which come to hand; it is another thing to have a definite problem which requires the use and often the invention of many different methods."[139]

Lewis's concept of activity, invented as part of personal quest to reform chemical thermodynamics, ended up being a useful tool both in the study of the anomaly of strong electrolytes and in the calculation of free-energy values. Lewis's original idea, that the concepts of fugacity and activity might play a central role in the formal structure of thermodynamics, never caught on, but by 1920 physical chemists everywhere were beginning to use activities and activity coefficients in their calculations. That activities and activity coefficients became part of the vocabulary of physical chemistry was due in large part to the support given Lewis and his ideas by Noyes and his associates at MIT.

There is an instructive analogy here between the course of Lewis's work on the chemical bond and that of his work on thermodynamics. In both cases, Lewis developed unorthodox lines of thought that were rejected or ignored while he was at Harvard; in both cases, these ideas were resuscitated during his years at MIT. In part this surely was due to the presence of a generally more sympathetic audience at the Research Laboratory of Physical Chemistry, where Noyes had assembled a young, clever, and ambitious band of young chemists and fostered among them an *esprit* missing at Harvard. But more important than this happy fit of personalities and styles was the conceptual fit between Lewis's musings and the research program of Noyes's laboratory. Noyes, Bray, Kraus, and their associates recognized in Lewis's work interests and ideas that were related to their effort to understand the behavior of strong electrolytes. Lewis's interest in molecular structure corresponded to the conviction, reached independently by Noyes and his colleagues, that the key to the puzzle of strong electrolytes might lie in a better understanding of chemical bonds. Lewis's concern with finding ways to make chemical thermodynamics pertinent to nonideal systems corresponded to their interest in finding better ways to treat strong electrolytes, that class of substances which, in solution, showed the most extreme departures from ideal behavior.

Lewis and the chemists of the Research Laboratory of Physical Chemistry had independently begun to move in the same direction; when they met, each party found its resolve strengthened and its self-confidence renewed. Like two generals facing a common adversary or two explorers surveying a single territory, Noyes and Lewis were able to accomplish more by working together than by working separately; at the Research Laboratory of Physical Chemistry, the whole was greater than the sum of the parts.

THE RESEARCH LABORATORY OF PHYSICAL CHEMISTRY IN RETROSPECT

Noyes's laboratory, conceived at the turn of the century as an Ionist's utopia, came much closer than most daydreams to realization. It was one of a handful of scientific institutions in early-twentieth-century America where theory and experiment were both cultivated, where undergraduate education was subordinated to research, and where Europeans could come not just to teach but to learn. It played a role in the development of physical chemistry in America comparable in many respects to that of T. H. Morgan's "fly room" at Columbia in the development of genetics.

But utopian communities do not endure. In 1912 Lewis left MIT for Berkeley, where he became dean of the College of Chemistry. Surviving records give no hint that he was dissatisfied with his position at the Research Laboratory of Physical Chemistry. Rather, they suggest that he was disappointed by MIT's uncertain commitment to basic research and flattered by the assiduous courting of Benjamin Ide Wheeler, the newly appointed president of the University of California, who promised Lewis a salary double that he was earning at MIT and complete freedom of action in making Berkeley into a research center of the first rank.[140] When Lewis left MIT, he took Bray and several graduate students with him, thereby giving his new department an immediate infusion of talent but at significant cost to his old. During the subsequent two years, Noyes lost three additional members of the team that had made the Research Laboratory of Physical Chemistry so successful: Charles Kraus, who was called to the directorship of the chemical laboratory at Clark University, and F. G. Keyes and Roy D. Mailey, who both went to the Cooper Hewitt Electric Company to direct research on mercury-vapor lamps. The success of the Research Laboratory of Physical Chemistry had earned for its members reputations—local, in the case of Mailey, and international, in the case of Lewis; but in a country that was becoming ever more conscious of the value of scientific research, even modest accomplishment could have generous rewards. Noyes's collaborators, accustomed to low salaries and meager facilities, found it hard to refuse such lucrative opportunities.

Noyes sought to rebuild as he had following the earlier losses of Coolidge and Whitney, but on the eve of World War I it was far more difficult than it had been in 1905. "I see no immediate solution for my own laboratory situation," Noyes wrote to Hale in 1914,

> I have always had (in Lewis, Kraus, Keyes, Mailey) men who were first-rate on the experimental physical side—men therefore who *supplement* my ignorance on that side in an important way; but since Dr. Bray left me, I have had no one who, as a kind of understudy, could look after in detail my own research men, help in the serious task of writing up their work for publication, give a part of the two year advanced course in theoretical chemistry, and conduct independently the three or four undergraduate theses in physical chemistry, which we have each year. It is extremely difficult to find a man who combines the *ability* to do these things with the *cooperative, non-individualistic* spirit.[141]

The task of rebuilding was hard in large part because Noyes was no longer alone in offering young physical chemists the opportunity to do research and be compensated for it. It was impossible for Noyes, with his limited resources, to match the salaries and facilities offered by the likes of the University of California. Nor could Noyes expect additional support from MIT. Intent on raising funds for its new campus in Cambridge, MIT could spare Noyes little money. Even had funds been available, it would have been politically difficult to invest them in Noyes's laboratory. Noyes's independence from the chemistry department, his freedom to pick and choose his teaching assignments and departmental responsibilities, his sometimes self-righteous glorification of pure research, the privileged position of members of his staff, and the international reputation he and his lieutenants had won—all this gave rise to envy and resentment among his colleagues. Noyes, it was suggested, had sought to become the tail that wagged the dog. Not a few greeted his misfortunes, his "generally unsettled state of mind and apparent lack of enthusiasm for the work here," with some secret satisfaction.[142]

Noyes eventually found able new associates: three talented graduates of MIT—Roscoe Dickinson, Duncan MacRae, and James A. Beattie; C. Lalor Burdick, who had taken a Ph.D. at Basel and done postgraduate work with the Braggs in London; and Duncan MacInnes, an instructor from the University of Illinois who had earned his doctorate under Noyes's former students R. C. Tolman and E. W. Washburn; in addition, F. G. Keyes, after two years at Cooper-Hewitt, returned to Noyes's circle. All of these chemists went on to win distinction in their fields, but they could not compensate for the losses of Kraus, Bray, and, especially, Lewis.[143] Noyes himself, approaching fifty, spent less and less time in the laboratory and more and more traveling—to Pasadena, where he was helping George Ellery Hale build a technological institute in the shadow of the Mt. Wilson Observatory, and, after 1916, to Washington, where Noyes became active in the campaign to mobilize American science for war. The publication record of the Research Laboratory of

Physical Chemistry testifies to the decline: between 1908 and 1912 the laboratory issued an average of more than 14 papers a year; between 1913 and 1917 that figure fell to a little over 4 per year.[144]

The decline of the Research Laboratory of Physical Chemistry after 1912 raises the question of how much its earlier success was due to Noyes, how much to Lewis, and how much to circumstances or forces beyond the range of individual initiative or plan. The expansion of graduate education in America, the growth of interest in research among educators and philanthropists, the opening up of career opportunities for research scientists—all of these trends contributed to making a Research Laboratory of Physical Chemistry possible. Without a subsidy from the Carnegie Institution of Washington or the availability of young chemists eager to participate in Noyes's dream, the Laboratory would not have opened. At the outset, Noyes's laboratory benefited from a temporary imbalance between the number of American physical chemists with advanced training who wished to do research and the supply of suitable research positions. When, on the eve of World War I, demand for able physical chemists came to equal or perhaps exceed supply, Noyes had to struggle to meet earlier standards.

Surely social trends and market forces affected the Research Laboratory of Physical Chemistry, yet they do not fully explain its success. It should be remembered that when the Research Laboratory of Physical Chemistry was founded, MIT was a reluctant partner. Noyes's personal resources were essential to the creation of this laboratory and remained essential to its maintenance until after World War I. The Research Laboratory of Physical Chemistry did not appear and flourish because MIT was seeking to become a science-based university; indeed, MIT began to take cautious progress in that direction only after Noyes and his associates proved that such a course could be navigated.[145] Nor did market forces dictate Noyes's selection of co-workers. Both the quality of the material available and the relative value of Noyes's resources in the academic marketplace might fluctuate, but Noyes never lost his exceptional talent for assessing character and intellectual potential. At Pasadena, where greater resources were at his disposal, Noyes would duplicate his achievement at MIT, building a small engineering school into a world center of physico-chemical research by finding bright new stars, like Linus Pauling, in unlikely places.

Individual initiative and skill played essential roles in the history of the Research Laboratory of Physical Chemistry, but the initiative and skill did not belong entirely to Noyes. There is no denying the powerful intellectual and personal stimulus given the Research Laboratory of Physical Chemistry by Lewis, nor is there any question but that, as a chemist, Lewis outshone Noyes. Whereas Noyes, in Linus Pauling's words, ''was a good chemist,'' possessing a clear head, a sure grasp of thermodynamics, and considerable sensitivity to trends in physical science, Lewis was a great chemist whose every paper gave evidence of imagination and ingenuity. By the time Lewis left for California in 1912, he had also acquired the managerial skills necessary to construct and direct a research school. Yet it was in Noyes's laboratory that the socially awkward and intellectually immature Lewis first flourished, and it was under Noyes's tutelage that he acquired many of the managerial skills that would win him fame as the dean of Berkeley's chemists. Lewis took as much as he contributed to the Research Laboratory of Physical Chemistry.[146]

Lewis was not the only chemist to benefit from an association with Noyes's laboratory. It was a portal to careers in research for dozens of talented young chemists. Statistics presented in a previous chapter offer testimony, albeit rather colorless, to this point. Both as an institution that trained productive physical chemists and as an employer of such scientists, Noyes's laboratory compiled a remarkable record between 1903 and 1919. Yet statistics do not convey a sense of the quality of individuals with whom Noyes surrounded himself nor a sense of their importance in the subsequent development of physical chemistry in America. When one of Noyes's former associates remarked in 1936 that "the old roster of the research laboratory now reads almost like a membership list of the National Academy of Sciences," he was hardly exaggerating. By then eleven of Noyes's former students and associates at the Research Laboratory of Physical Chemistry had been elected.[147] Former co-workers were sprinkled across the country, holding positions of influence in universities and research institutions (Lewis, Bray, and Merle Randall at Berkeley; Kraus at Brown; Tolman, William N. Lacey, and Roscoe G. Dickinson at Caltech; J. A. Beattie and F. G. Keyes at MIT; William D. Harkins at Chicago; E. B. Spear at Yale; Robert Sosman at the Geophysical Laboratory of the Carnegie Institution of Washington; and Duncan MacInnes at the Rockefeller Institute) and in government (E. W. Washburn as chief chemist at the National Bureau of Standards, C. S. Hudson as director of the Chemistry Division of the National Institute of Health, and Duncan MacRae as chief chemist of the Chemical Warfare Service).

Noyes never cultivated industrial patronage for his laboratory, nor selected research topics on the basis of their economic significance; nonetheless, a large number of his former students and co-workers advanced to senior positions in industry. Between them, Whitney and Coolidge directed the fortunes of the General Electric Research Laboratory through its first forty-four years; John Johnston and R. E. Zimmerman both served stints as directors of research at U.S. Steel, Guy Buchanan became director of research and R. P. Rose vice president in charge of research at American Cyanamid, and Roger Williams rose to vice president in charge of research at Du Pont. Others worked for Bell Laboratories, Westinghouse, Technicolor, Inc., and a number of smaller firms.

Given that the Research Laboratory of Physical Chemistry was a small institution, having space for no more than twenty researchers at a time and a budget that never exceeded fifteen thousand dollars a year, this is a remarkable record.[148] It testifies both to Noyes's extraordinary talent for discerning ability, ambition, and imagination in young chemists and to his skill in designing and directing an institution in which those gifts could be developed and refined.

But the Research Laboratory of Physical Chemistry was not simply a way station in the careers of many eminent chemists; it also helped to shape their attitudes and values. Noyes built not just a laboratory, but a research school. Members of his school, or, perhaps better, the Noyes-Lewis school, were distinguished by their common conviction that science was more than precise measurement, by their eclecticism—their willingness to use both thermodynamic and molecular-kinetic lines of reasoning—and by their confidence in the relevance of modern physics to their enterprise. While some physical chemists were particularists, content to cultivate their discipline in blissful ignorance of twentieth-century physics, the Noyes-Lewis school constantly looked to physics for new ideas and techniques. For Noyes, Lewis, and their associates, the future of physical chemistry lay not in insular independence but in intimate interdependence.[149]

The three lines of research that had evolved at the Research Laboratory of Physical Chemistry—work on the anomaly of strong electrolytes, on the construction of a table of free energies, and on the nature of the chemical bond—came to fruition in 1923. In that year, efforts to account for the deviation between the activity coefficients and dissociation values of strong electrolytes culminated in the development of the theory of interionic attraction by two European physicists, Peter Debye and Erich Hückel. Based in part upon a molecular-kinetic analysis of solutions and in part on thermodynamic reasoning, the Debye-Hückel theory solved the puzzle that had obsessed Noyes and many members of his generation of physical chemists. The effort to measure the free energies of formation of chemical substances was crowned in that same year with the publication by Lewis and his junior associate, Merle Randall, of their comprehensive textbook on chemical thermodynamics, a work that ended with a table of standard free energies of formation of 139 common substances. Through use of this table, unsurpassed for its range and accuracy, chemists could calculate the equilibrium constants and electromotive forces associated with thousands of chemical reactions. And in that same year, 1923, Lewis published his exceptionally rich *Valence and the Structure of Atoms and Molecules*, a book that presented the theory of the shared-pair bond in clear and vivid prose and forcefully brought the fruit of Lewis's twenty years of thought about molecular structure to the attention of all chemists.

The solution of the problem of strong electrolytes, the construction of what amounted to a table of chemical affinities, and the development of the theory of the shared-electron-pair bond were symptoms of a broad and deep transformation in physical chemistry that had been in the making for two decades.

When the determination of free energies became routine and when the theory of solutions was extended to embrace strong electrolytes, two of the major goals of classical physical chemistry were attained. Ample opportunities for research on these topics still existed, but such work would no longer command the general interest and urgent concern of chemists like Noyes and Lewis. By contrast, the theory of the electron-pair bond generated a plethora of new questions, not the least of which was that of integrating this view of the chemical bond with physicists' rapidly evolving ideas about atomic and molecular structure. Solutions and the forces governing chemical change, the central concerns of physical chemists trained under Ostwald, had lost their undisputed position at the core of physical chemistry, displaced by that molecule whose very existence Ostwald had questioned.

By 1923, MIT and the Research Laboratory of Physical Chemistry had moved across the Charles River to Cambridge, and Lewis and Noyes had both moved to California. Yet Noyes, Lewis, and their associates, while in the cramped laboratory on Trinity Place, had made essential contributions to the emergence of the new physical chemistry of the 1920s. Their work had helped define the parameters of the problem of strong electrolytes and had afforded chemists new techniques for describing and analyzing the behavior of solutions. They had developed many of the theoretical and experimental tools necessary to measure free-energy values. And, convinced that a knowledge of molecular structure would illuminate processes such as electrolytic conduction, they had helped point physical chemists toward the study of chemical bonds.

CHAPTER 4

The Phase Ruler: Wilder D. Bancroft and His Agenda for Physical Chemistry

ARTHUR AMOS NOYES was a physical chemist who put great emphasis on the modifier, "physical." From the time he enrolled under Ostwald's banner, Noyes labored to remedy his deficiencies in mathematics and physics; throughout his career he was receptive to new techniques borrowed from physics, even when they were as foreign to his prior experience as X-ray diffraction. Both chemistry and physics, according to Noyes, rested upon a common body of principles relating to matter and energy; it made little sense to introduce formal distinctions where nature recognized none. Consequently, Noyes believed it important that physical chemists have a broad knowledge of both chemistry and physics, and the mathematical skill to understand and creatively use the latest theory and techniques of both sciences. Few could master all this; Noyes himself fell short. But by constructing a laboratory staff properly and fostering cooperative effort, the group could achieve what lay beyond the reach of most individuals. Hence Noyes's concern with finding and retaining "men who are first-rate on the experimental physical side." Hence too his frequent homilies on the importance of cooperation in research and his insistence that co-workers be selfless, nonindividualists. Noyes's science, no less than the religion of his ancestors, seemed to demand such qualities.[1]

Through personal example, textbooks, and, most important, through the powerful research schools that Noyes founded at MIT and Caltech and that Lewis founded at Berkeley, these ideas and attitudes exerted enormous influence on physical chemistry in America. By the mid-1930s, Noyes's views on the relations between physical chemistry and physics and on the role of mathematics in physical chemistry were widely accepted, as was his conviction that thermodynamic and molecular-kinetic modes of analysis are complementary. The decisions that he, Lewis, and their associates had made—to emphasize the quantitative study of equilibrium relations in solution, of free-energy changes in chemical reactions, and of the nature and properties of chemical bonds—had been sanctified by success. Their choices seemed so right and natural as to make it appear as though there had never been options. Their mental map of physical chemistry and its relations to other sciences, itself a revised version of their European teachers', had become the point of departure for future generations.

But there were and always had been other options. Disciplines, and especially new disciplines, are constantly confronted with the task of defining the research frontier and identifying those sectors on the frontier where investigations are most likely to result in advance, that is, the solution of old problems, the discovery of fruitful new questions, and the development of new methods, instruments, and applications. Rarely is there unanimity of opinion regarding the route of march. Indeed, clashes over the direction of research and priorities in problem-choice may be as frequent in science and as important in the development of disciplines as conflicts over the merits of particular theories. Recent debates over the allocation of resources among the various branches of physics illustrate this point. Clashes between physicists over the construction of particle accelerators are, in large part, battles over priorities—are further large investments in particle physics likely to be as fruitful as the distribution of similar sums along other sectors of the research frontier? Whether physicists accept the existence of quarks matters less to the funding of particle physics than whether physicists and their patrons acknowledge the claims of subatomic structure to intense and sustained attention.

Tensions between scientists over research agenda are more clearly visible today than they were prior to World War II. It is easier to perceive intellectual uncertainties in contemporary debates over the funding of particle accelerators than in past debates whose outcome is long settled. Moreover, the allocation of resources among specialists has become a matter of public interest only in recent years, when science has become so expensive and so closely linked to military and economic power. Prior to World War II, decisions regarding the funding of science were decentralized, often informal, and usually private. Officials at philanthropic foundations might sometimes aspire to shape science, but their power was limited and their role unscrutinized by the public. Scientists themselves, and administrators of the universities and industrial research laboratories in which they worked, more often than not determined the direction of research through hosts of small decisions about hiring, the purchase of equipment, the design of the curriculum, the publication of journals, the acceptance or rejection of papers, and other matters seemingly too inconsequential to note. The sum of these small judgments meant prosperity for some fields and research programs and poverty for others; some paths were worn smooth by constant traffic and others allowed to remain so overgrown as to seem impassable.

In the last chapter we saw how a powerful research school emerged, grew, and prospered in early-twentieth-century America. Through its products—pupils and publications—this Noyes-Lewis school shaped the agenda of physical chemists for decades to come. But in a world of limited resources, the success of one group comes at the expense of others. Noyes and Lewis were not the only physical chemists to formulate a program for their discipline. Others, even those with backgrounds similar to those of Noyes and Lewis, could and did differ with them over issues as fundamental as the role of mathematics in physical chemistry, relations between chemistry and physics, the importance of dilute solutions, and the significance of the study of molecular structure.

Among those who competed for leadership of American physical chemists, Wilder Dwight Bancroft offers the greatest contrasts with Noyes and Lewis. Today his name is unfamiliar to physical chemists and historians of science; his work is no longer cited, and his vision of physical chemistry no longer finds adherents. But in his own day, Bancroft was a figure to be reckoned with. The founder and editor of the *Journal of Physical Chemistry*, Bancroft spent much of his career jousting with Noyes and Lewis—not so often over issues of theory as over matters of emphasis and priority. They held different agendas for their science. Whereas Noyes and Lewis valued mathematics and looked to physics for ideas and techniques, Bancroft was interested neither in mathematical chemistry nor in cultivating relations between chemistry and physics. For Bancroft, chemistry was the central science, relevant to the practical concerns of technologists and scientists in other fields in ways that an increasingly esoteric physics was not. Whereas Noyes and Lewis set high value on achieving a quantitatively exact understanding of dilute aqueous solutions, Bancroft was far more interested in having a qualitatively valid understanding of the more complex systems so important to industry, geology, and medicine: alloys, magmas, concentrated aqueous solutions, colloidal suspensions, and gels. And whereas Noyes and Lewis came to view the study of molecular structure as a critically important part of physical chemistry, Bancroft long remained skeptical of molecular-kinetic reasoning and indifferent to the study of chemical bonds.

Bancroft's agenda for physical chemistry was not nearly as influential as that of Noyes and Lewis. Bancroft trained many students, published some significant research, and attained many high honors, but he failed to mold physical chemistry to his vision. While his program won sympathy among some journeymen in research, it swayed few leaders in the field, either of his own generation or of the next. By the end of his career, Bancroft had come to be viewed by many of his ablest colleagues as an interesting but wrongheaded gadfly, a victim in equal parts of his intellectual limitations and his obstinacy in the face of a changing science.

By studying the work of Noyes and Lewis, we follow a thoroughfare along which many of their contemporaries and successors moved. By studying the work of Bancroft, we inspect a path fewer trod. But such an excursion has rewards. In taking it, we may come to appreciate better the uncertainties facing the practitioners of a young discipline as they seek to forecast future trends in their science and to invest their efforts as productively as possible. More important, the history of a program that failed may tell us much about the conditions that made for the success of others. Contrast is necessary to sight.

WILDER BANCROFT: GENTLEMAN CHEMIST

Bancroft, like Noyes, was the son of an old New England family. His fore-bears had settled in Lynn, Massachusetts, in 1632, a year before and a few miles distant from Noyes's. The Bancroft family, however, was far more distinguished than the Noyes clan; along with deacons and prosperous farmers, it included two who gained fame as authors, Aaron and George Bancroft. It was in their shadows that Wilder Bancroft grew, and it was by their standards that he measured himself. When at the height of his career, Bancroft told former classmates at Harvard that he "had worked hard," and had "much less to show for it than I expected"—as if a professorship at Cornell and the presidency of the American Chemical Society were gentlemen's C's on the Bancroft scale.[2] But while ambitious, he took a not-so-secret delight in championing unpopular ideas. After retiring, he described himself as a prophet without honor:

> Owing to my lifelong habit of being a minority of one on all occasion, my research work does not look convincing to most people. Since I have become avowedly a specialist in unorthodox ideas in the last decade the situation is getting worse, because now I irritate more people.[3]

Both his ambition and his pride in being an iconoclast were traits long present among Bancrofts.

Aaron Bancroft, Wilder Bancroft's great-grandfather, exhibited both characteristics. A graduate of Harvard, a minister, and the author of a popular life of George Washington, Aaron broke with the rigid Calvinism of his youth and embraced the idea that all religions should bow before the tribunal of reason. Ostracized by the orthodox, he nevertheless used his skill with words to build a congregation in Worcester and, eventually, to win a position of leadership in the American Unitarian Association and acceptance in all but the most conservative clerical circles. For Aaron Bancroft, unorthodox belief proved compatible with achievement and recognition.[4]

The most prominent of Aaron's thirteen children was Wilder Bancroft's grandfather, George Bancroft: historian, cabinet member, and diplomat. Raised to habits of "plain living and high thinking," George Bancroft took a degree at Harvard at the age of seventeen and then became one of the first Americans to seek and earn a doctorate in Germany.[5] There, at Göttingen, his theological interests and clerical ambitions gradually were obscured by an enthusiasm for languages, literature, and history. There was no crisis—a life ordered according to the ideals of German scholarship differed little from one placed in the service of God—but upon returning to New England, secular concerns had largely displaced religious. Despite the ease of this transition, George was no less intellectually restless than his father. But whereas Aaron had rejected New England's established theology, George renounced New England's established political traditions. Surrounded by Whigs, he joined the Democratic party of Jackson and Van Buren and rose to prominence in its national councils. While holding a succession of responsible positions in government, he made a tidy fortune through an advantageous marriage and shrewd investments and pursued his passion for history.

Wilder Bancroft's personality and career are best understood in the light of these forebears. Like his great-grandfather and grandfather, he was adept with the pen; like them, he believed learning was linked to action—that a scholar could and should mix in the world of practical affairs; and like them, he took pride in having the courage of his convictions. Indeed, he gloried in challenging the assumptions prevalent among his colleagues and took satisfaction in viewing himself as a dissenter in an established community. The dissenting creed he espoused was not religious, as in the case of his great-grandfather, nor political, as in the case of his grandfather, but scientific.

Of Wilder Bancroft's father, John Chandler Bancroft, much less can be said. Lacking the discipline and talents of Aaron or George Bancroft, he compiled a poor record at Harvard and failed in a succession of pursuits, legal, artistic, and commercial. Unable to succeed on his own, John spent much of his adult life in the orbit of his famous father, depending upon him for money and eventually for shelter. John Chandler Bancroft's first wife died when their son, Wilder, was four; subsequently Wilder Bancroft spent much time in his grandfather's households in Newport and Washington. Although aging, George Bancroft was still vigorous—addicted to writing and horseback riding—and welcome in the homes of the nation's rich and powerful. Chester Arthur is said to have remarked that the President is "permitted to accept the invitations of members of his cabinet, Supreme Court judges and—Mr. George Bancroft."[6] The bright youngster could not have ignored the contrast between his grandfather's power, wealth, and fame and his father's failures and dependence. Nor could he have missed the lesson that vigorous effort and independence of mind would reap rewards.

Following what was by then family custom, Wilder Bancroft attended Harvard after completing preparatory studies at the Roxbury Latin School and the Milton Academy. Entering in 1884, he compiled a record that was better than mediocre, although not brilliant. Football, a sport in which Bancroft was sufficiently adept to win a place on the Harvard eleven, competed with books for his attention. The few A's on his transcript (in English and Political Economy) were balanced by a few D's (in natural history and chemical analysis) and several C's (in mathematics, physics, and German); but for the most part he did B work. Harvard's controversial elective system permitted Bancroft to concentrate heavily in the sciences, and this he did. His program was consistent with a major in either physics or chemistry until his senior year, when, by taking three electives, he met the requirements for a degree in chemistry. After his graduation in 1888, Bancroft stayed at Cambridge for an extra year, having been invited to serve as a laboratory assistant.[7]

Little evidence exists to explain Bancroft's interest in science or his choice of chemistry as a career, although his grandfather may have encouraged such thoughts. George Bancroft had written his undergraduate thesis on astronomy and had done his graduate work in oriental languages and philology, fields that were considered no less scientific than chemistry. During his years in Europe he had relished the company of Alexander von Humboldt, Charles Babbage, Charles Lyell, August Wilhelm von Hofmann, and many other luminaries of European science. As a private citizen in Washington, his dinner companions included the neurologist S. Weir Mitchell and others prominent in American scientific circles.

The models available to Wilder Bancroft at Harvard must have been no less influential. Prominent among these was a chemist under whom Bancroft did much of his coursework: Josiah Parsons Cooke. A wealthy gentleman of distinguished family who spent summers near the Bancrofts in Newport, Cooke's fame was sufficient to draw talented young men like T. W. Richards to Harvard. Joining the faculty in the 1850s under the worst possible conditions— his incompetent predecessor had been hanged for murdering a colleague— Cooke had built a place for chemistry in the Harvard curriculum.

Cooke's success was due in part to his introduction of laboratory practice into his courses, but in part as well to his skill in describing the cultural rewards of chemical study. "Success in the observation of phenomena implies . . . quickness and sharpness of perception, accuracy in details, and truthfulness; and on its power to cultivate these qualities a large part of the value of science, as a means of education, depends. . . . Slovenly work means slovenly results, and habits of carefulness, neatness, and order produce as excellent fruits in the laboratory as in the home."[8] Nor did the study of chemistry simply foster the development of socially desirable traits; it also afforded striking evidence of the existence of a beneficent God. "Illustrations of the Divine attributes," Cooke wrote, "lie all around us, in the air we breathe, in the water we drink, and in the coal we burn." Like his colleague, Louis Agassiz, Cooke believed that "the laws of nature are the thoughts of God . . . the most direct evidence possible of Infinite wisdom."[9]

Colleagues acquainted with his unfortunate predecessor must have found this new chemist's sincere expressions of piety reassuring. Students like Wilder Bancroft discovered in Cooke an example of how secular scholarship might be integrated comfortably into what was still largely a religious culture. Cooke, like George Bancroft, had one foot planted in the patrician world of old New England, where learning was linked with religion, where scholarship was undifferentiated, and where educators were dedicated to molding character; another, in a new world where the traditional virtues were valued for their contributions to secular success, where knowledge was specialized, and where teachers transmitted expertise.

Cooke's special interest in chemistry was a field he called chemical physics, meaning by that simply those portions of physics "which are more closely connected with Chemistry than the rest."[10] Similar in many respects to the old physical chemistry of Regnault, Kopp, Bunsen, and Landolt, Cooke's chemical physics dealt with the relation between the physical properties and chemical constitution of substances, the action of heat on matter, and methods of measuring the weight and volume of bodies. Cooke bequeathed a strong interest in these topics to his student and successor, Theodore William Richards, who took his Ph.D. at Harvard in the same year that Bancroft completed his undergraduate studies. He also may have planted seeds in Bancroft; after studying organic chemistry for two years under C. Loring Jackson at Harvard and under Rudolph Fittig at Strassburg, Bancroft entered Ostwald's institute at Leipzig.

Bancroft arrived in Leipzig in 1890, the year Noyes left. Though Noyes and Bancroft were very nearly contemporaries, their educations differed in subtly significant ways. While Noyes had been at Leipzig, Ostwald and his assistants had been preoccupied with following up on the work of Arrhenius and van't Hoff. Ostwald expressed skepticism about the atomic theory during those years, vigorously warned his students about the dangers of such hypotheses, and stressed the fundamental importance of energy in physical science. But he had not yet made these themes central to his thought. During Bancroft's tenure at Leipzig, however, energetics became a far more important motif in Ostwald's work. It was during these years, 1890–1892, that Ostwald translated the works of J. Willard Gibbs, finding in them sophisticated tools for treating equilibria without the use of any hypotheses regarding atoms or molecules. During this period too, Ostwald began to assert publicly and vigorously that matter is simply a bundle of energies coexisting in one place. The skirmishes between Ostwald and defenders of atomism of the late 1880s became war in the early 1890s.[11] Although both Noyes and Bancroft were touched by Ostwald's energetics, the impact was greater on Bancroft, who saw Ostwald in his full martial colors, than it was on Noyes, who was back in Boston when intense combat commenced.

Different too were the paths taken by Noyes and Bancroft after winning their doctorates. Whereas Noyes lacked the money to prolong his stay in Europe, Bancroft could afford to make a leisurely pilgrimage to scientific shrines. His first stop was Berlin, where he spent the autumn of 1892 attending the lectures of the feeble but legendary Helmholtz. Then he moved on to Amsterdam, where he worked in van't Hoff's laboratory and developed a strong and enduring affection for the Dutchman.

After completing this tour of Europe, Bancroft returned to Cambridge, where he served for two years as a laboratory assistant and instructor. It must have been an uncomfortable position. Josiah Parsons Cooke was in failing health, and it was clear that the College would soon need someone to assume responsibility for Cooke's courses in inorganic and theoretical chemistry and chemical physics. Bancroft had fine credentials, but so did T. W. Richards, who had stayed one rung above Bancroft since both had arrived at Harvard in the mid-1880s. When Bancroft had taken his A.B., Richards had taken a Ph.D.; while Bancroft worked as a laboratory assistant, Richards had done postdoctoral work in European laboratories; while Bancroft was in Europe, Richards was teaching at Harvard; and now Richards was an assistant professor and Bancroft an instructor. Shortly after Cooke died in 1894, the Corporation of the University voted Richards a leave of absence to spend a year with Ostwald and Nernst, thereby anointing him as Cooke's successor.[12] A few months later Bancroft accepted an offer of an assistant professorship at Cornell.

BANCROFT AT CORNELL

The chemistry department that Bancroft joined at Cornell was hardly yet a research center to compare with Leipzig or even Harvard, but it was beginning to show promise. A new laboratory had been built in 1890, and under the direction of an aging but competent graduate of Göttingen, C. G. Caldwell, a corps of well-educated, research-oriented junior faculty was being assembled. L. M. Dennis, a German-trained analytical chemist who made rare earths his specialty, had been hired in 1887; in that same year a recent graduate of Johns Hopkins, W. R. Orndorff, was appointed to teach organic chemistry. Nor was Bancroft the first physical chemist on the faculty; another of Ostwald's students, Joseph E. Trevor, had been hired in 1892 to assist in teaching elementary chemistry and to start elective courses in physical chemistry for upperclassmen and graduate students.[13]

As at MIT and other American universities, growth in enrollments prompted expansion of the faculty. Even though the country had been in the grip of a depression through much of the 1890s, the fraction of the American population attending college increased. At Cornell, the fourteen hundred full-time students in the 1890–1891 academic year increased to over two thousand by 1899–1900.[14] This growth created rich opportunities for ambitious young academics to advance their careers and their disciplines. But it did not dictate which specialists would be seated at the banquet. Physical chemistry was but one among many fields, new and old, competing for the attention and resources. Why did Cornell make room in its growing, but still small, department of chemistry for two Leipzig graduates?

Comments made by Jacob Gould Schurman, Cornell's vigorous president, provide some evidence. "[I]t has been necessary," he wrote in 1893, owing to the increase of undergraduates and graduates, to make provision for increasing the corps of instruction in the department of Chemistry. A professorship of applied chemistry, with a special laboratory, and a professorship of physiological and sanitary chemistry are called for both in the interests of humanity and science, and these subjects should be embraced in the curriculum of the University; but though compelled to await the receipt of gifts for this new development, the Board have provided advanced students with instruction by an assistant professor in physical chemistry, which is the most advanced field of chemical theory, and, by increasing the number of instructors and assistants they have enabled the head of the department to make the laboratory work more efficient in the elementary course.[15]

Although cluttered with commas, this passage exposes Schurman's concerns and the complexity of judgments about staffing. He wished to initiate programs in industrial, physiological, and sanitary chemistry. Work in these fields fit easily within his conception of Cornell as a university dedicated to public service. Utility had been a cornerstone of Cornell's educational philosophy since its establishment, and Schurman's administration was studded with new projects that conformed to this goal: colleges of agricultural science, forestry, veterinary medicine, and a medical school. The university, however, lacked the resources to launch all of these initiatives simultaneously; expensive new undertakings in industrial and sanitary chemistry, for example, were postponed until after the turn of the century.[16]

Nevertheless, the chemistry department needed help with its oversubscribed introductory courses, and Schurman wished to make appointments that would improve Cornell's standing as a center of research and advanced study in chemistry. His description of physical chemistry as "the most advanced field of chemical theory" suggests that Schurman had only a fuzzy idea of what the specialty was about. But he must have known, or been told, that physical chemists had stirred excitement and controversy in the previous few years. He also seems to have believed that physical chemistry had the virtue of being inexpensive. Since it was a new field, there was no question of making a senior appointment; there were no senior professors of physical chemistry in America. And because Schurman considered the field theoretical, he felt no obligation to invest in equipment; Cornell would make no separate provision for laboratory instruction in physical chemistry until 1898. The appointment of one or two assistant professors of physical chemistry entailed minimal cost and no long-term risk; at the same time these physical chemists would reinforce the university's reputation for curricular innovation and lend some assistance in supervising the hordes of students taking elementary inorganic and analytical chemistry.[17]

If Schurman was concerned with keeping expenses down, Joseph E. Trevor must have pleased him. A physical chemist whose favorite instrument was a piece of chalk, Trevor's great passion was thermodynamic theory. He did little laboratory research, but was indefatigable in his work in the introductory course in inorganic chemistry. The president's annual report for 1895–1896 showed Trevor in attendance at laboratory practice for twenty-three hours a week. This, of course, was in addition to his duties in recitation sections and his own elective courses. In recognition of these contributions, Trevor soon was named to fill a new professorship in physical chemistry.[18] Enrollments in Trevor's own courses on physical chemistry, however, must have been a disappointment. In 1894–1895, six students attended his advanced course in physical chemistry, five attended his seminar on the physico-chemical literature, and only three enrolled in his seminar on chemical equilibrium—a course in which the German edition of Gibbs's papers was used as a text. Trevor was simply too demanding to be popular. One of his few pupils later advised others: "If you want to bust, take Trevor's course."[19]

Trevor had been at Cornell for two years when his former classmate at Leipzig, Wilder Bancroft, joined the faculty. Although Bancroft was also a German-trained specialist, he promised to be a far more popular teacher. In addition to his impeccable credentials as a chemist, Bancroft also embodied the ideals of the gentleman-scholar. Comfortably supported by money his grandfather had earned, Bancroft refrained from drawing a salary until well after the turn of the century.[20] Well-bred, physically imposing, possessed of a lively mind and a keen wit, Bancroft seemed the perfect choice to add vitality to Cornell's program in physical chemistry. This he did.

Bancroft's addition to the faculty had the desired effect on course enrollments. Dividing the teaching responsibilities with Trevor, Bancroft took charge of all courses that were nonmathematical in character: a course entitled "qualitative physical chemistry," devoted to uses of the phase rule; another consisting of a nonmathematical exposition of the principles of mass action, chemical kinetics, and solution theory; still others on electrochemistry and physical chemistry for engineers. Trevor taught all courses requiring calculus. Bancroft's lectures soon outstripped Trevor's in attendance, and eventually Trevor moved from the chemistry to the physics department. Nor did Bancroft appeal only to undergraduates; he soon began to attract graduate students to Ithaca: Hector R. Carveth from the University of Toronto and D. McIntosh from Dalhousie in 1896, Frank K. Cameron from Johns Hopkins and O. W. Brown from Indiana in 1897. By 1900 Cornell had become one of the two or three leading American centers for graduate training in physical chemistry, at least in terms of the number of doctorates conferred.[21]

Bancroft's presence, however, was not without drawbacks. He was highly opinionated and somewhat eccentric. A former student later described him as

a wild man on committees. One time . . . , on a Ph.D. examination, he asked some student what there was in water that put out fire? . . . the poor guy was completely flabbergasted, and Bancroft said, "It's easy, there are fireboats."

Bancroft's wit, which some found engaging, was detested by others; he soon became a center of controversy within the department.[22]

Dennis, the heir apparent to Caldwell as department chairman, became his special nemesis. Their differences were both personal and intellectual. Bancroft, according to one colleague, refused to show Dennis the deference others in the department granted him; Dennis, for his part, thought Bancroft a disruptive "non-conformist." Dennis believed a thorough knowledge of analytical chemistry was fundamental in an undergraduate's training and was skeptical of theoretical chemistry and theoretical chemists.[23] Bancroft, like Ostwald, maintained that the significance of analytical procedures could be appreciated only when set in the context provided by physical chemistry.[24] A system of instruction like that at Cornell, in which elective courses in physical chemistry could be taken only after extensive training in analytical practices, was, to Bancroft, placing the cart before the horse. After Dennis succeeded Caldwell as department chairman in 1903, the faculty divided between Dennis and Bancroft men. As one faculty member put it, "[a]ny man who could be considered an adherent or protégé of Professor Bancroft lost all of the regard and respect of Professor Dennis."[25]

Despite his differences with Dennis, Bancroft was promoted to full professor in 1903. The reason is not hard to find. Bancroft, although careless of the feelings of his colleagues, had cultivated Jacob Gould Schurman. He had, moreover, made contributions to the chemistry program far in excess of his teaching duties and was coming to be seen in some quarters as one of Cornell's brighter ornaments.

The key both to Bancroft's position at Cornell and to his growing reputation outside Ithaca was the *Journal of Physical Chemistry*. Following its first appearance in October 1896, the *Journal* was published monthly at Ithaca throughout the academic year. Although the names of both Bancroft and Trevor graced the title page, Bancroft was the senior editor from the outset. It was Bancroft who promoted the *Journal* by means of letters to physical chemists around the country, who worked hardest to fill its early issues with articles and reviews, and who paid the difference between the cost of publication and income from subscriptions.[26] What the Research Laboratory of Physical Chemistry would become to Noyes, the *Journal of Physical Chemistry* was to Bancroft.

Just as Noyes's laboratory answered both personal and professional needs, so too did Bancroft's *Journal*. Close to home, Bancroft's enterprise strengthened his standing with the Cornell administration, since it meshed well with Schurman's conception of what progressive faculty members ought to be doing. Schurman was of that new generation of college presidents for whom research was as important as teaching. Ambitious to see Cornell become a center for scholarship and graduate study, he strongly encouraged professors to be active in their disciplines. One method to assert leadership in a discipline, a method much favored at Cornell, was to establish and edit scholarly journals. Schurman had himself founded the *Philosophical Review* in 1892, the year he was inaugurated president of Cornell; among his first acts as president was to encourage members of the engineering faculty to set up the *Sibley Journal of Engineering* and members of the physics department to organize the *Physical Review*.[27] Journals had brought prestige to Liebig and Giessen and to Remsen and Johns Hopkins. Why not Cornell? Bancroft, who had greater facility with the pen than most chemists and rather less skill in the laboratory than many, took to the idea of editing a journal with enthusiasm.

Nationalistic sentiment also played a part in the creation of the *Journal of Physical Chemistry*. At the turn of the century, chemists, like other American scientists, were beginning to feel restless under the scientific hegemony of Germany. The grievances were many: the need for American students to learn German; the patronizing tone of some German scientists; the delays in obtaining imported glassware, fine chemicals, and equipment; the slights Americans felt when their work was ignored or needlessly duplicated abroad. ''[M]any of our German friends are apparently of the opinion that unless work has been done in Germany it has not been done,'' lamented one prominent electrochemist.[28]

The grandson of a historian who celebrated America's independence from Europe and the great-grandson of a biographer of George Washington, Bancroft was eager to see America attain scientific parity with Germany. The success of German science, he suggested, was due in large part to the existence of specialized, German-language journals that helped to develop the fields they represented. If American scientists were to improve their international standing, Bancroft believed, they would have to emulate the German example by concentrating their best work in a few good disciplinary journals. Organs that attempted to cover the whole field of chemistry, such as Remsen's *American Journal of Chemistry* and the *Journal of the American Chemical Society*, were simply too diffuse to compete for the best articles with the likes of Ostwald's *Zeitschrift für physikalische Chemie*. ''I see no reason myself,'' wrote Bancroft, ''why we should not take what is good from the Germans and modify it to suit our needs. After studying four years in Germany, two of them with Ostwald, I became convinced of the desirability of special journals in special languages. . . .''[29] By bringing together the best that Americans had to offer in physical chemistry, Bancroft hoped to compel Europeans to take notice of American work and, at the same time, to quicken the pace of research in the United States.

The new journal reflected favorably on Bancroft and was born amid hopes that it would become the alternative to publishing in Germany for American physical chemists, but its most important function was as a medium for a viewpoint. From its inception the *Journal of Physical Chemistry* gave expression to the independent, even idiosyncratic, conception of physical chemistry that Bancroft was developing in his courses and writings.

BANCROFT'S CONCEPTION OF PHYSICAL CHEMISTRY

For Bancroft, chemistry was a "coordinate science," the aim of which was the "study of all properties and changes of matter depending on the nature of the substance concerned." Following from this definition, Bancroft saw large areas of overlap between chemistry and other sciences. Thus, geology was fundamentally the study of the chemistry of the earth, and engineering the art of making the structural properties of substances useful to man. The growth and functioning of organisms resulted from chemical changes, and even physics, apart from "the law of gravitation, the laws of motion and a few other abstract formulations," was, according to Bancroft, "chemistry pure and simple."[30]

Physical chemists investigated and ordered the laws governing chemical change; they were, in Bancroft's view, those chemists who aimed "to present the science of chemistry as a clear and complete whole."[31] Much taken with Ostwald's rhetoric about founding a new science of *allgemeinen Chemie* that would unify the various branches of chemistry, Bancroft saw his lineage as a chemist stretching back through Ostwald to the great generalists of the early nineteenth century. In a review of Mitscherlich's collected works, he asked where the spirit of Mitscherlich still dwelled: "Not among the inorganic chemists for they know little of organic chemistry; not among the organic chemists for they care less for inorganic chemistry. It is only the physical chemist who is of necessity interested in the whole field and he is the legitimate successor of Berzelius, Mitscherlich, Gay-Lussac, Dumas, Liebig, Davy, and Faraday."[32]

Organic chemists may have scoffed at Bancroft's profession of concern; physical chemists were notorious for their indifference to the charms of organic chemistry. But Bancroft did not mean that physical chemists had a genuine interest in the details of the synthesis and structure of organic molecules. Rather, he was suggesting that by making the laws governing the direction and yield of chemical reactions their concern, physical chemists transcended the traditional division between organic and inorganic chemistry. Physical chemists did not define their subject according to the nature of the substances under consideration. They made a specialty of generalizations applicable to all branches of chemistry.

As Bancroft saw it, physical chemistry's central role in chemistry and chemistry's central position among the sciences gave practitioners of his young specialty special opportunities and responsibilities. Many of the problems and phenomena of geology, biology, and industry were known to be related, although the structure of science and scientific institutions often seemed to obscure those relations. Physical chemists, by defining themselves as students of change rather than as masters of some particular form of matter, could appreciate similitudes and affinities lost upon those with narrower training and ambitions. Their knowledge gave them license to act as intellectual brokers—middlemen who might prosper by matching techniques to problems, regardless of traditional patterns of interaction among the sciences. In Bancroft's view, wherever matter underwent change, whether in the interior of the earth or in stars, in human bodies or in industrial vessels, the physical chemist could both learn and teach.

Before physical chemists could meet their duties as generalists, Bancroft believed, they would have to find a rational system for organizing their subject. Entering the field at a time when it was expanding rapidly, Bancroft felt need of a structure into which might be fit that which was known and that which was yet to be learned. "There has been so much work done in physical chemistry during the last ten years," wrote Bancroft in 1897, "that the mass of accumulated material is now too large to be remembered as miscellaneous facts. It becomes comparatively easy to survey the whole field if we consider the phenomena as examples illustrating a few general principles."[33]

The classificatory scheme on which Bancroft settled first divided physical chemistry according to whether the ideas and treatments involved were primarily mathematical or nonmathematical in nature.[34] He never attempted to justify or fully explicate this division, preferring to say simply that it was not artificial but quite natural and widely recognized. To the mathematical side of physical chemistry belonged the formal structure of chemical thermodynamics—the austere work of those who, like Gibbs, Planck, Duhem, and Trevor, could as easily be called physicists as chemists. To the nonmathematical side belonged the work of investigators who, like van't Hoff, Arrhenius, Ostwald, and Nernst, adopted a more empirical approach. These workers might draw upon the mathematical tradition and occasionally contribute to it, but they were experimentalists as well as theoreticians and chemists rather than physicists.

After distinguishing between the mathematical and nonmathematical sides of physical chemistry, Bancroft further subdivided the nonmathematical branch, distinguishing between quantitative and qualitative work. On the quantitative side stood research involving the law of mass action and the principles of thermochemistry, electrochemistry, and reaction kinetics—in other words, the sort of work for which Bancroft's teachers and colleagues were best known. Just as a railroad timetable provides information about the number of trains traveling a route on a given day and their destinations, so these principles provided precise information about the amounts of materials participating in a reaction and the end point of the reaction under specified conditions. By contrast, Bancroft also envisioned a qualitative physical chemistry that would resemble not a timetable but rather a map showing the location of railroad tracks. "If one knows where the railroad tracks are one can predict with absolute accuracy where the trains will run"; likewise, Bancroft suggested, there were means by which the physical chemist could determine the constraints under which a physico-chemical system existed and the directions in which equilibria would be displaced when subject to changes in physical conditions.[35] As Bancroft saw it, the theorem of Le Châtelier and the phase rule of Gibbs constituted the basis for such a qualitative understanding of chemical processes.

The first of these principles was an empirical rule of thumb formulated by Henry Le Châtelier in 1884. In its sparest form, this principle states that any change in the factors of equilibrium from outside a system will result in a reverse change within the system. An increase of the external pressure on a system in equilibrium will result in an increase in the formation of the substance occupying the lesser volume; if heat is added, there is increased formation of the substance whose production is accompanied by the absorption of heat; and so on. Bancroft recognized in Le Châtelier's theorem an exceedingly practical method of quickly and simply determining the direction in which equilibria will shift when subject to changes in condition: whenever a system in equilibrium is disturbed by an external event, there ensues a readjustment in the system so as to relieve the strain.

The principle yielded no exact measure of the degree to which an equilibrium is displaced by specific changes in the factors of equilibrium; for this it would be necessary to apply the law of mass action, the equation of the reaction isochore, or the equation of the reaction isotherm. These, however, entailed laborious measurements and calculations, and were, moreover, accurate only within narrow parameters. Le Châtelier's theorem carried neither liability. Although it yielded qualitative rather than quantitative information, this was frequently sufficient for petrologists, engineers, or industrial chemists. Indeed, Le Châtelier himself had been led to formulate the theorem through his study of the reactions that occur in blast furnaces.[36]

The second of Bancroft's guiding principles was the phase rule. First stated by J. Willard Gibbs in 1876, the phase rule specified the conditions of any system in a state of equilibrium while entailing no assumptions of a molecular-kinetic nature. The rule defined the conditions of equilibrium of a system by stating the relations among the number of coexisting phases, the number of components, and the degree of freedom or variability of a system:

$$F = n + 2 - r,$$

where F is the degree of freedom, or number of conditions such as pressure and temperature that may be varied without causing a change in the number of phases; n is the number of components, or substances of independently variable concentration in the system; and r is the number of phases, or physically distinct parts of the system.

Much as the periodic table permitted a classification of elements on the basis of their physical and chemical properties, the phase rule permitted a classification of physico-chemical systems on the basis of their degrees of freedom, that is, according to the number of variable factors (temperature, pressure, composition) that must be specified in order that the system be fully defined. When, for instance, the number of phases is two greater than the number of components in a system at equilibrium, the temperature, pressure, and composition of the system are all fixed; the system is nonvariant. A system of one component and three phases, such as ice, water, and water vapor, can be at equilibrium only at one pressure and temperature, experimentally determined to be 0.0075 °C and 4.6 mm of mercury. By definition, the phases of a one-component system have the same composition, but if the pressure or temperature is altered, one of the phases will disappear. An equilibrium mixture of two components in four phases, such as ice, solid salt, a solution of the salt in water, and vapor, is likewise nonvariant. If, however, the same two-component system exists in three phases, a degree of freedom is opened up and the system becomes univariant. One factor of equilibrium can be changed without the elimination of a phase if the others assume specified values. The temperature, for example, of a system composed of a solid salt, its aqueous solution, and vapor, can be varied; but at each temperature, the solubility and vapor pressure of the system must assume definite values. If the number of phases is reduced yet again, say by the elimination of the solid, the system becomes bivariant, and two factors, say the pressure and temperature, must be specified before the third (the solubility) is fixed. Using the phase rule as a guide, a simple table covering all cases of heterogeneous equilibria can be constructed:

Nonvariant Systems	1 component	2 components	3 components
	3 phases	4 phases	5 phases
Univariant Systems	1 component	2 components	3 components
	2 phases	3 phases	4 phases
Bivariant Systems	1 component	2 components	3 components
	1 phase	2 phases	3 phases

Whatever their chemical dissimilarities, systems of the same variability share physical characteristics. Given knowledge of the number of components and phases in a system, it is possible to draw conclusions regarding the behavior of the system when the equilibrium is disturbed, conclusions that can be amplified using Le Châtelier's theorem. And given knowledge of the number of components and variability of a system, it becomes possible to draw conclusions regarding the number of phases, to determine, for instance, whether a body is a homogeneous substance or a heterogeneous mixture.

Gibbs, in his characteristically terse style, did little to draw out these and other implications of his rule. Even his derivation of the phase rule occupied only a few sentences. His proof depended on his realization that just as the temperature and pressure in a system at equilibrium must be constant through-out, so too the chemical potential, or derivative of the energy with respect to the mass of each chemical component, must be constant throughout the sys-tem. So, for example, the chemical potential (or, to use the term Lewis later was to introduce, the partial molal free energy) of a salt dissolved in water must be the same as the chemical potential of the same salt dissolved in alcohol if the two solutions are in equilibrium, whatever may be the solubilities or concentrations of the salt in the two solvents. Hence at equilibrium a system of n components will have $(n + 2)$ independent variables: pressure, tempera-ture, and the chemical potentials of the n components. But Gibbs had already established that an equation of state relating the $(n + 2)$ variables could be written for each of the r phases. The degree of freedom or variability of the entire system then follows simply by subtracting the number of equations from the number of variables, or $n + 2 - r$, which is Gibbs's phase rule.[37]

Like much of Gibbs's work, the phase rule was largely ignored following its publication. Tucked away in the middle of a long, abstract article that was published in an obscure American journal, it might have suffered the same fate as Mendel's work and been overlooked for decades had not Gibbs sent reprints of his paper to scores of Europe's leading physicists and chemists. Even so, only one chemist who received such a reprint appears to have read the paper with any understanding, and he needed "a translation of the paper into ordinary language" by James Clerk Maxwell before recognizing its merit.[38] A small circle of French chemists, many of them with strong back-grounds in physics and mathematics, also discovered Gibbs's work in the early 1880s.[39] But the phase rule did not begin to attract serious and widespread attention from chemists until the mid-1880s, when H. W. Bakhuis Rooze-boom and Ostwald independently learned of it.

Roozeboom, a Dutch chemist at the University of Leiden, was studying the hydrates of sulphur dioxide when, in 1886, the professor of physics at Leiden, van der Waals, called his attention to Gibbs's work. About the same time, Ostwald was told of Gibbs by his colleague and former teacher of physics at Dorpat, A. Oetingen. Gibbs had sent both physicists reprints of his work. Within a year, Ostwald had begun to incorporate elements of Gibbs's thinking into his massive textbook of physical chemistry. Roozeboom, even more en-thusiastic, had quickly published a paper arranging all known dissociation equilibria on the basis of the number of components and phases.[40]

By the time Bancroft went to Europe, Ostwald was preparing a German translation of Gibbs's papers, van't Hoff was beginning to study the formation of "solid solutions," or homogeneous mixtures of crystals, and Roozeboom was beginning to use graphical methods to depict temperature-pressure-con-centration relations in heterogeneous systems. Bancroft no doubt learned of the phase rule while studying at Leipzig; he could not have spent the spring of 1893 in Holland, fast becoming a hotbed of interest in the phase rule, without becoming reacquainted with it.[41] Initially the phase rule made no great impression on the young American. His research at Leipzig had dealt with Nernst's theory of electromotive force, a topic that seemed to have little relation to Gibbs's work on heterogeneous equilibrium. As he wrote ten years later, "I can remember the time when I thought that people made a good deal of unnecessary fuss over the phase rule. It seemed to me an interesting mathematical relation but nothing more."[42]

After returning to the United States, however, Bancroft's research interests and his opinion of the phase rule gradually changed. In a series of papers published while he was at Harvard, Bancroft studied solubility relations in ternary systems, those, for example, composed of two immiscible liquids and a salt dissolved in both, or of two immiscible liquids and a liquid miscible in both. Although notable for introducing the word "solute," these papers were, in other respects, dismal failures.[43] Convinced that such heterogeneous physical equilibria could be treated by equations similar to those derived from the law of mass action for homogeneous chemical equilibria, Bancroft fit his data to equations with up to four empirical constants. As A. A. Noyes observed in a review:

> no physical significance whatever can be attached to such results, a fact which will be evident to any one acquainted with the properties of empirical equations; for, if the introduction of so many arbitrary constants be permitted, a great variety of mathematical functions could be found which would represent the experimental results with a practically equal degree of accuracy. The agreement is, in other words, almost a mathematical necessity, and is therefore no evidence of the correctness of the author's application of the mass-action law to the phenomenon in question.[44]

After several such rebukes, Bancroft abandoned his effort to give such systems precise, quantitative treatment and instead began to view them in the context of Gibbs's phase rule. Shortly after moving to Cornell, he began work on a book to inform American scientists of the significance of their countryman's work. When published in 1897, Bancroft's monograph was the first extended treatment of the phase rule and its applications. But this book was just the beginning. For the next twenty years Bancroft continued to publicize the rule and its applications in articles and reviews, making himself the leading American authority on heterogeneous equilibria and prompting his admirers to salute him as the Phase Ruler. Instead of accepting the phase rule as the useful auxiliary in the study of chemical equilibria that most chemists took it to be, Bancroft undertook to restructure much of physical chemistry around the principle, arguing that it should serve as the cornerstone for instruction and research.[45] What had once seemed a curiosity now became the organizing principle of Bancroft's science.

THE APPEAL OF THE PHASE RULE

Bancroft's efforts to reconstruct physical chemistry around the phase rule merit closer analysis. Although there were other physical chemists who disputed central tenets of the Ionists' creed, there were few, either in America or Germany, who challenged the Ionists' assumptions about the structure of their science. Louis Kahlenberg, for example, a graduate of Leipzig and professor at the University of Wisconsin, attracted considerable attention for his extravagant criticisms of both the theory of electrolytic dissociation and the theory of osmotic pressure; but he never stepped outside the framework within which those theories functioned.[46] He continued to believe that the behavior of solutions was the fundamental problem facing physical chemists, rejecting only the treatment given that problem by van't Hoff and Arrhenius. Unable to formulate alternatives that were as comprehensive and effective as the theories he attacked, Kahlenberg eventually was dismissed as a crank.

Bancroft's aims were both more radical and more interesting than those of his colleague from Wisconsin. Rather than concentrating his attack on the theories of his teachers, Bancroft sought to overturn their priorities, their assumptions about what was most important and deserving of investigation. To return to the geographical metaphor, he disputed some of the main outlines of their mental map of physical chemistry.

It is important to understand why the phase rule held such appeal for Bancroft, and not just Bancroft—others too were fascinated by Gibbs's principle. Bancroft found disciples in the United States and colleagues abroad who shared his enthusiasm, although few were so radical as Bancroft in their claims for the centrality of the phase rule in their discipline. The problem therefore is to understand both the sources of Bancroft's passion and the reasons why others could share in it, if only to a limited degree.

Bancroft himself gave several justifications for his enthusiasm. One was founded on the fact that the phase rule was derived solely from the laws of thermodynamics and thus was free from assumptions regarding the atomic, molecular, or ionic nature of the systems under consideration. In treating equilibria with the phase rule, Bancroft wrote,

[t]here is no need of assuming that matter is made up of discrete particles nor that it is continuous; there is even no need of assuming its existence or non-existence. It is immaterial whether there is or is not a distinction between "chemical" and "physical" reactions. It is simply a question of the relative number of independently variable components and phases . . .[47]

Bancroft used this argument on several occasions and took delight in pointing out that investigations based on the phase rule involved fewer assumptions than research based on the supposed existence of hydrates in solution, such as that conducted by Harry Clary Jones at Johns Hopkins, or of research founded upon Arrhenius's dissociation theory, such as Arthur A. Noyes's at MIT.[48]

This justification, however, was most often invoked as a debater's point. Bancroft was not a deep thinker; he did not base his chemistry on epistemology and made fun of those who did or pretended to.[49] Moreover, Bancroft paid surprisingly little attention to the thermodynamics upon which the phase rule was based. He did not attempt to derive the phase rule in his book but rather confined himself to discussing its applications. His one significant foray into thermodynamic theory, a paper relating the equations for electromotive force of Nernst and Planck to Gibbs's conception of electromotive force as a measure of chemical potential, consisted almost entirely of a letter sent by Gibbs to Bancroft and Bancroft's commentary on it.[50]

Bancroft was interested in facts and useful hypotheses; he considered a hypothesis useful if it described relations between known facts in a simple and clear manner. If the hypothesis was truly simple and rooted in the phenomena, it would lead inevitably to the discovery of new facts and broader generalizations. Bancroft preferred such hypotheses to be phrased in words rather than in mathematical equations. His comments on one of Pierre Duhem's works reflected these attitudes; "What we have," he wrote of Duhem's book,

> is an exhaustive study of chemical equilibrium put into mathematical form and expressed in terms of the thermodynamic potential. This application of mathematics to chemistry is unfortunately more ornamental than useful. There are myriads of formulas, but very few can be applied to any concrete case. . . . To the experimental theorist the book is a joy and a sorrow, a joy because it points out so much and a sorrow because it always stops short of becoming practical.[51]

Physical chemists, Bancroft implied, could not derive power from equations; they could not treat the genuine problems of geology, medicine, or industry by scribbling integral signs. For Bancroft, the practical value of an idea tended to be inversely related to its mathematical content. This attitude is a recurring motif in Bancroft's prose. Commenting in 1905 on the development of chemical thermodynamics, Bancroft complained that the enterprise was becoming sterile and "metaphysical": "The mathematical chemistry of the past decade has not done what it should have done [generate new and more useful generalizations], and there is no immediate prospect of any improvement." Writing to T. W. Richards in that same year, Bancroft described the attributes he wanted in a new instructor at Cornell: "A knowledge of chemistry and a distinct manipulative skill counts for more than mathematics. I want a man who will do things."[52]

If physical chemists were to be effective intellectual brokers, suggested Bancroft, they ought not embrace mathematics too closely. Neither could they meet their responsibilities if they became obsessed with quantitative accuracy in matching experimental data to theory. As far as Bancroft was concerned, the time and resources that many physical chemists were investing in the study of dilute solutions were misdirected. "The majority of the papers on physical chemistry published every year," he wrote in 1899,

> deal with so-called dilute solutions, solutions containing less than two per cent. of one of the components. Practically all of our quantitative theory of solutions fails to apply to ninety six per cent. and over of the possible field. We have accomplished a great deal inside the narrow limits we have set ourselves, but it is obvious that we are handicapped seriously in the application of physical chemistry to technical chemistry so long as we discuss quantitatively only such solutions as do not occur in technical work. Quite apart from the technical bearing, we can never obtain for physical chemistry its proper title as the science of chemistry until we can say that we do cover the whole field. . . . the ideal training in physical chemistry cannot be obtained until we have broken away from the shackles of "ideal" solutions.[53]

Chemists like Arthur A. Noyes and G. N. Lewis, according to Bancroft, were constructing an esoteric specialty instead of following Ostwald's bold injunction to develop a new *allgemeine Chemie*.[54]

Bancroft regarded those hypotheses that described the larger number of facts as being the more useful, and, in his estimation, this meant that the phase rule was of much greater significance than the collection of hypotheses that went under the name of solution theory. Whereas the theory of solution was rife with exceptions and adequately described only those solutions approaching infinite dilution, the phase rule was perfectly general. "The beauty of the phase rule is that, though qualitative, it is absolute and applies to every case of equilibrium. . . . It is therefore the framework on which everything must rest."[55]

By comparing the phase rule and the theory of solutions Bancroft was, in a sense, comparing apples and oranges. They were not applicable to the same set of problems, nor did they yield the same type of information. The phase rule described the parameters within which physico-chemical systems could exist; it guided its user in drawing qualitative conclusions about the makeup of such systems and the types of interactions that could exist within them. By contrast, the theory of solution of van't Hoff and Arrhenius yielded quantitative predictions regarding molecular condition and properties, for example, the degree of ionization of salts in solution and the osmotic pressure, vapor pressure, and freezing and boiling points of such solutions.

Bancroft was not so ignorant as to think that the phase rule could be considered a substitute for a comprehensive and quantitatively accurate theory of solution. But he did maintain that the phase rule permitted a classification of physico-chemical systems in which solutions formed a special case. Like the historian who criticizes the political history of the past for being too narrow and who advocates its placement within a broader social and economic context, Bancroft wished to see the study of solutions placed within the broader context of the study of all forms of chemical and physical equilibria. He did not maintain that the phase rule would solve the kinds of problems with which a Noyes or a Lewis dealt; rather he suggested that their questions were not so urgent nor their methods so powerful as to merit the attention lavished upon them. In Bancroft's eyes, the critical difference between his program for physical chemistry and that of Noyes and Lewis was one of agenda.

Just as Bancroft considered the phase rule to be a principle of wider application than the theory of solutions, so too he thought Le Châtelier's theorem to be of more general use than such quantitatively exact expressions as the law of mass action or equation of the reaction isochore. Le Châtelier's theorem, Bancroft observed, could be used to describe any instance of mobile equilibrium, whether between distinct phases of matter in heterogeneous systems or between two components of a homogeneous system, whether the disturbing factor was a change in temperature, concentration, or pressure. All that was necessary for its application was to know if a given process liberated or consumed heat or resulted in an expansion or contraction of volume. The law of mass action and the equation of the reaction isochore were neither so simple nor so general. Their use required precise information regarding such factors as the concentration of the reactants and products at equilibrium and the heat of reaction; there were many systems to which these quantitatively exact principles could not be applied.

Bancroft was again, in a sense, comparing apples and oranges. The theorem of Le Châtelier could be used to make predictions regarding the direction in which equilibria would shift when subjected to external stress. The law of mass action and equation of the reaction isochore could be used to predict both the direction and extent of such shifts. But once again Bancroft could point to the wider applicability of the qualitative principle and the difficulties involved in putting quantitatively exact equations to work in the real world of nonideal substances. Even Nernst admitted that applying the quantitative equations was "none too easy." According to Bancroft, "People would much rather have a first approximation which is easy to apply than an exact theory which is very difficult to apply."[56]

The distaste for mathematics, the suspicion of theoretical physics, the emphasis on the unity of the sciences, the concern with classification, the conviction that scientific knowledge should be of service to industry—all this smacks of that Baconianism for which nineteenth-century America is supposed to have been so fond. But before applying such a broad brush, it is good to remember the words of Bancroft's teacher, Josiah Parsons Cooke: "Practically no great originator in science followed Bacon's rules, or any other rules." Bancroft was not a great originator, but neither was he a Baconian in any philosophical sense. It would have been strange had he been, for none of his teachers fit squarely into a Baconian tradition. In his writings on the history of science, versions of which Bancroft surely heard in lecture, Cooke was kinder to Aristotle's old organon than to Bacon's new.[57] George Bancroft is best described as an idealist, although his idealism was a curious medley of American and German, religious and secular motifs. Nor could Bancroft's masters in Europe, Ostwald and van't Hoff, by any stretch of that elastic term, be labeled Baconians.

Bancroft's habits of thought and work were not clothes purchased off a rack in some bargain basement of American Baconianism. Rather, his thinking and his science were the products of a multitude of influences acting upon a mind with its own native predilections. Bancroft's concern with classification and the unification of knowledge was but natural for one whose heritage included unitarians, a natural theologian, and a prophet of monism. George Bancroft had amalgamated theology and history by folding the history of the United States into the larger story of God's plan for mankind; Josiah Parsons Cooke had dedicated much of his best work to showing that scientific and Christian truths rest upon the same inductive foundations; Ostwald was no less committed to the unification of secular learning. Wilder Bancroft's enthusiasm for the classification of knowledge was, if anything, overdetermined.

So too his commitment to utility. Bancroft had been brought up to believe that the ultimate purpose of learning was action. When his grandfather had first contemplated writing history, he wrote that it "has always interested me, suits well with my theology, and I think I could become useful by it." George Bancroft had made his knowledge useful. He moved easily between archives and chancery offices, between libraries and the smoke-filled rooms of politicos. His history, which celebrated American democracy, and his Jacksonian politics were mutually supporting. Likewise, although in a very different context, Ostwald and van't Hoff had preached the practical value of their science. They did not make utility the principal goal of their learning, but neither did they disdain it. Service to industry was entirely legitimate and desirable so long as it did not infringe upon the freedoms that academics in most times and places have deemed essential to creative work.

At Cornell, Bancroft found the theme of utility emphasized yet again. The years around 1900 were an era of explosive growth for the Empire State's chemical and chemical-process industries, and Ithaca, although not itself an industrial city, was at the hub of a wheel of industrial development. Sixty miles to the northwest was Rochester, the home of George Eastman's photographic enterprise; forty miles to the southwest was Corning, with its glassworks; one hundred and ten miles due east lay Schenectady, site of General Electric's new research laboratory; and at an equal distance to the west lay the huge Niagara Falls power project, near which more than a dozen producers of electrochemicals sprouted during the years around the turn of the century. These firms, the high-tech industries of the day, were taking their first steps toward the development of a capacity for systematic scientific research during the years between 1900 and 1916. Jacob Gould Schurman, Cornell's enterprising president, recognized opportunities in this boom, and so too did Bancroft.[58]

If Wilder Bancroft ever perceived a conflict of interest between science and industry, he did not record it. For him, the significant problems were in making physical chemistry more relevant to industrial practice and in making businessmen more familiar with the benefits of physico-chemical research. There was nothing specifically Baconian about this attitude. That learning should be useful was an idea upon which individuals of diverse philosophical predilections, or of no philosophical sensibility, could agree in the nineteenth century.

Bancroft's aversion to mathematics, and the theoretical physics that took mathematics as its language, are not so easily traced. His European teachers tended to equate mathematization with progress, and, at Harvard, Cooke pronounced his respect for mathematics, calling it "the most important tool in [deductive] processes of thought."[59] But as with his most famous pupil, T. W. Richards, Cooke's use of mathematics did not betray a genuine comfort with the subject; Cooke's chemical physics was a quantitative physics, but not a mathematical physics. While requiring precision, his research required little mathematical sophistication. It was in this tradition that Bancroft was educated; he spent far more time as a student at Harvard in the laboratory than in mathematics classes.[60]

This is the equivalent of saying that he was trained as were most American scientists in the late nineteenth century. The one outstanding difference between the education of physical scientists in America and Europe at the end of the nineteenth century was in their preparation in higher mathematics. In Europe, students of science, especially the physical sciences, typically received a mathematical preparation at early stages in their careers that was both extensive and intensive; although educators might argue about how much mathematics was enough, there was no debate over whether those entering fields like physics and physical chemistry should master the fundamentals of the calculus. In America, secondary schools, colleges, and even graduate programs in physics gave mathematics short shrift. As late as 1914 it was possible to obtain a Ph.D. in physical chemistry at Harvard without the benefit of a single course in the calculus.[61]

The word "Baconianism" can be invoked as an explanation of this disparity only at the risk of fostering myths about the role of philosophical doctrines in the practice of science. American scientists' indifference or hostility to mathematics owed less to the writings of the Lord Chancellor than to cultural and social circumstances quite unrelated to philosophical debates over scientific method. There were, for example, cultural factors that predisposed Americans to view laboratories as the shrines of science. The scientists and educators who imported laboratory instruction to America in the generation before Bancroft viewed it as something more than simply an efficient device for training scientists or producing new knowledge. For many of them, the laboratory was, first and foremost, a place to mold character, to inculcate in youths the virtues of honesty, perseverance, and fidelity in the little things, and to instill respect for painstaking manual labor. "I would that I could convey to you," said Cooke to an audience of ministers,

> the full force of the impression which is left on the mind after repeated experiences [in laboratory research]. . . . The helplessness which one feels while working in the dark gives a reality to the sense of the limitations of our knowledge, of which so much is said and so little appreciated. If anything will lead man to hold his knowledge in humility and reverence it is the consciousness that results so laboriously obtained may be invalidated by circumstances over which he has no control, and of whose existence he is wholly unaware.

"I venture to assert," he continued,

> that there is no class of men in the world among whom is found more unselfish devotion and more personal sacrifice than among the great army of scientific workers. The love of abstract truth may be a much lower motive than the love of man, but it equally calls forth the very noblest qualities of the mind. Moreover, in most cases the constancy and courage of the scientific investigator meet with no reward except the satisfaction which unselfish duty conscientiously discharged always brings; and, as Professor Tyndall has said, "There is a morality brought to bear on such matters which in point of severity is probably without a parallel in any other domain of intellectual action."[62]

Value inhered not just in the product but also in the process of laboratory work; such activity fit easily within the moral universe of Victorian America. By contrast, there was something prideful and arrogant in thinking that man could capture the world on paper, something sinful in believing that he could use mathematical cunning to determine how nature worked, and something dishonest about winning a reputation in science through work at the desk rather than in the laboratory or field.

If the diffuse Protestant values of nineteenth-century America encouraged scientists to dedicate themselves to laboratory work, the abstract and purer-than-pure style of American mathematicians discouraged them from probing far into the realm of mathematics. American mathematics, which had been closely tied to the physical sciences in the first half of the nineteenth century, declared its independence of applications in the second half. By doing so, the nascent profession won recognition in Europe and a place in America's young research universities, but at the cost of a good portion of its natural audience among physical scientists. By the 1890s, American universities offered a multitude of courses in mathematics, but few that treated the subject in a style congenial to science students. When mathematics courses were taken, so much attention was devoted to abstract issues and so little to physical applications that, as one physicist charged, students were left "so scared that they never dare integrate, differentiate, or touch an infinite series, for fear that something awful may happen."[63] But, for the most part, courses in higher mathematics were not taken. Students in the physical sciences preferred to follow their native inclinations and devote themselves to the laboratory.

With experience and maturity, some American scientists came to perceive their ignorance of mathematics as a serious weakness. When the scientist was strong-willed and dedicated to mastering every aspect of his subject, as was a Noyes or Lewis, deficits in early education could be largely, although perhaps never entirely, overcome. But many found it easier to convince themselves that mathematics was not essential to their discipline, or even that it represented a distraction from good science. Bancroft was hardly the most extreme example of this tendency. Louis Kahlenberg at the University of Wisconsin, Bancroft's friend, fellow graduate of Leipzig, and colleague, bluntly asserted that "physical chemists don't need higher mathematics, they only need arithmetic."[64]

Why did Noyes see his inexperience in the uses of mathematics as a deficiency to be remedied, while Bancroft sought to justify his ignorance by celebrating the nonmathematical character of his science? Disparities in native aptitudes surely played a role, but important as well may have been a difference in character. Although that word has less explanatory power today than it did in the age of Noyes and Bancroft, we impoverish history by avoiding it altogether. There was a strong streak of asceticism in Noyes. He appears to have taken satisfaction in denying himself: wife, family, the honors of high office, many of the personal luxuries that his wealth could buy. He was brought up to live simply; it takes little effort to imagine him in the black cloth of his Puritan ancestors. He welcomed the labor of mastering new and unfamiliar skills.

Bancroft, born to wealth and fame, was more self-indulgent. He was ambitious, perhaps especially so since his father had failed to meet the Bancroft family's high standards. But he was also intellectually lazy. His laziness was not that of one who allows hobbies or busywork to crowd out scholarship. Bancroft dedicated long hours to his discipline; few read as widely or wrote more. But he did not dig very deeply into his subject. While Bancroft possessed the critical intelligence of his grandfather, he lacked the industry and tenacity that had led George Bancroft to mine the unpublished manuscripts of two continents while writing his history. He preferred the light exercise of raising questions to the heavier labor of answering them; the excitement of exposing shortcomings in "mathematical chemistry" to the drudgery of mastering its tools.

Why did the phase rule hold strong appeal for Bancroft? Because it was independent of hypotheses, perhaps; because it was general and of potential practical value, yes; but most especially because its study and use did not demand that Bancroft stretch himself beyond the limits within which, by education and disposition, he felt comfortable. It fit easily into that constellation of habits, values, and attitudes that Bancroft had acquired as a young man. Research involving the phase rule did not demand mathematical sophistication; nor, at the turn of the century, did it require great experimental exactitude. The problems of heterogeneous equilibria had been so little studied that a chemist with some cunning and a grasp of a few basic principles and techniques could quickly move to a position of prominence in the field.

If, however, these were the only merits of the phase rule, it is unlikely that Bancroft would have found or kept any disciples or that the phase rule would have found a secure, albeit secondary, place in physical chemistry. It offered more than an empty promise of a royal road to knowledge; it was of value in research. Despite his sometimes wobbly judgment and predilection for exaggeration, Bancroft was canny enough to realize this. He not only pontificated about the phase rule; he also sought to use it as a research tool.

Here its power is best displayed in the construction of equilibrium or phase diagrams of specific systems, diagrams that display so much information about the solubilities, freezing points, melting points, vapor pressures, and other characteristics that one commentator has called the phase rule "the basis of one of the finest 'filing systems' ever invented."[65] With an accurate phase diagram in hand, a skillful student could survey in a single glance the complex permutations that substances undergo in response to changes in physical conditions. And by studying such diagrams it quickly becomes evident that a mere handful of forms suffice to describe a multitude of seemingly diverse systems; Roozeboom, for example, showed that it was possible to depict the temperature-composition relations of most condensed systems of two components using just five basic types of diagrams.[66]

During the early twentieth century the use of phase diagrams became common in a number of fields. For the student of metallurgy, petrology, or ceramics, the phase diagram was a boon, since it constituted a handy reference encyclopedia that could be consulted quickly to obtain information about the nature of a given alloy or silicate system. For the metallurgist or petrologist engaged in research, the phase diagram, when available, proved capable of yielding important clues to the constitution, properties, and thermal history of materials. And for chemical engineers, such diagrams came to serve as useful guides in the design of procedures for distilling, purifying, or processing materials as diverse as petroleum and iron.[67]

Bancroft and his students devoted considerable effort to the construction of these diagrams, especially during the years from the turn of the century to World War I. By its nature, such research resists easy description. Unlike Noyes, who used many techniques to seek an answer to a single question of great theoretical importance, Bancroft and his collaborators applied a single technique to the analysis of a multitude of diverse issues, most of which had limited theoretical significance. To describe this work in detail would be tedious, but to neglect it entirely would be to ignore the core of Bancroft's science. By passing over many of the applications, however, and focusing on the technique itself, as exemplified in one or two pieces of research, it is possible to acquire a fair idea of the nature of Bancroft's work while avoiding an uneconomical and pointless recapitulation of papers and events. It was, after all, the technique rather than the particular results that appealed to Bancroft and that he sought to teach his students.

THE USE AND ABUSE OF THE PHASE RULE

Among the many heterogeneous systems that Bancroft and his students studied during the years prior to World War I, those to which they devoted greatest attention were systems composed of solutions of metals in other metals: alloys. Few substances received so much public attention at the end of the nineteenth century, and few were so poorly understood by scientists. Developments in warfare and industrial technology had created an enormous interest in metals with improved properties: greater impact, heat, or corrosion resistance, better tensile strength or ductility. The occasional discovery of a new alloy that better withstood the shells of dreadnoughts or the heat generated by a cutting tool stimulated much the same enthusiasm that the development of new plastics elicited in the 1950s and that recent innovations in ceramics and composite materials excite today. Yet the working of metals and the making of alloys was still much more an art than a science. Chemists, who once had been intimately involved with the refining and working of metals, had come to consider such work outside their domain.[68] Alloys generally seemed to be mixtures rather than true compounds; moreover, they were not soluble in ordinary solvents, and their working more often involved mechanical and physical operations than chemical action. The indifference of chemists was matched by their brethren in physics. The upshot was that the study of these materials was often left to the practical sort: engineers who, for the most part, had little formal schooling in science.[69]

In the years just before the turn of the century, however, a new generation of researchers began to study metals. The diffusion of knowledge of Gibbs's work on heterogeneous equilibria and advances in the techniques of producing and measuring high temperatures afforded workers new theoretical and experimental tools for analyzing alloy systems.[70] At the same time, the growth of physical chemistry as a discipline helped create both journals that would publish theoretically sophisticated work on alloys and institutes in which such research could find support. In France, Le Châtelier, whose interest in metallurgy began in the 1880s, focused on the chemical constitution and physical properties of iron and steel. In Holland, Roozeboom later used Le Châtelier's data to construct the first phase diagram of systems composed of iron and carbon (steel). In Britain, C. T. Heycock and F. H. Neville investigated brass, bronze, and other alloys. In Russia, N. K. Kurnakov took up the study of phase relations in alloys in 1903, the same year that Gustav Tammann, Ostwald's successor at Dorpat and later Nernst's successor at Göttingen, commenced a systematic study of the alloys of the common metals.[71]

Bancroft, therefore, was one among a number of talented scientists who perceived opportunities in the study of metals around the turn of the century. Just as the electrolytic dissociation theory and law of mass action had put the procedures of analytical chemistry on a more rational foundation, so Bancroft, and others who shared his interest in alloys, hoped to see the phase rule and phase diagram bring reason and order to the practices of metalworkers.

Bancroft's work on alloys commenced in 1902, when he and a recent graduate of Cornell, E. S. Shepherd, undertook the study of brass and bronze; later, Bancroft and his collaborators studied other alloys of aluminum and tin, aluminum and copper, and steel.[72] Like Noyes, Bancroft obtained support from the Carnegie Institution of Washington for his research. The support given Bancroft was not as generous as that given Noyes, nor was it as enduring; Bancroft never developed the close relationship with the administrators of the Carnegie Institution that Noyes enjoyed. Nevertheless, the five to fifteen hundred dollars Bancroft received each year between 1902 and 1911 assisted him in purchasing equipment and securing assistance. And, like Noyes, Bancroft would also dip into his own pocket to subsidize his research; prior to the Depression, he always turned his salary back to Cornell and sometimes paid assistants with personal funds.[73]

Bancroft's ultimate aim in these investigations was to determine how such mechanical and physical properties as hardness, ductility, malleability, tensile strength, conductivity, and corrosion resistance varied with the composition and thermal history of alloys.[74] Metalworkers had long known that both the composition and structure of alloys influenced their properties. Varying the proportions of copper and zinc in brass, for example, has dramatic effects on the hardness and ductility of the product. But changes in the crystalline structure of alloys caused by different heat treatments could also have significant effects, although the effects might differ from alloy to alloy. So, for example, heating bronze to a red heat and quenching it in water softens and strengthens that alloy, but the same treatment does not strengthen brass. Brass is rendered more ductile by a regimen of annealing, heating, and gradual cooling, but bronze's ductility is diminished by the same process. Traditional methods of chemical analysis could yield data about the ultimate composition of alloys, and microscopic scrutiny of their etched surfaces could reveal much about crystalline structure. But neither of these techniques could provide much insight into the conditions under which a given alloy assumed a given structure.

A phase diagram, however, could often provide such insight. By interpreting the freezing-point curve on the diagram with the aid of the phase rule, conclusions could be drawn regarding the composition and crystalline form of the alloy at elevated temperatures. The real task was to perform and interpret the many experiments necessary to reveal the structure of a phase diagram.

Figure 4.1, taken from a paper by Bancroft's assistant, E. S. Shepherd, shows the first step in this process, namely, the construction of a diagram depicting how the freezing point of an alloy varies with composition and temperature.[75] The alloys whose equilibrium relations are depicted in this diagram are mixtures of copper and zinc: the brasses. The line beginning at A and ending at F represents the liquidus, that series of temperatures and compositions at which crystals begin to separate from molten alloy as temperature is slowly lowered. The lines bB, nN, cC, dD, and eE denote temporary halts in the rate of cooling that occur at constant temperature across a range of compositions. In the case of line bB, for example, brasses containing between 62 percent and 72 percent copper by weight were found to exhibit a temporary pause in their rates of cooling just under 900 degrees. The isothermal lines intersect the liquidus at points B, N, C, D, and E, thus dividing it into six branches.

The significance of these experimental findings and their graphical representation becomes clearer through the application of the phase rule. Each of the isothermals, or horizontal lines representing arrests in rate of cooling, suggests the presence of systems in a nonvariant state. Since brass is a two-component system, nonvariance would require, according to the phase rule, the simultaneous presence of four phases at equilibrium. One of these phases must be the vapor (not represented in Shepherd's diagram), a second must be the homogeneous liquid melt, and the remaining two must represent distinct sol-

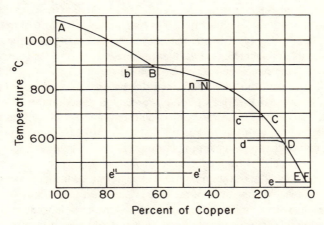

FIGURE 4.1. The freezing-point curve of mixtures of copper and zinc, redrawn from E. S. Shepherd, "The Constitution of Copper-Zinc Alloys," *Journal of Physical Chemistry* 8 (1904): 422.

ids. The six branches into which the liquidus curve is divided may then be taken to represent the conditions of univariant equilibrium existing between each of six solid phases and the liquid and vapor phases.

The phase rule, in brief, imposed constraints on the interpretation of experimental data, such as those drawn from studies of cooling rates. On the basis of the phase rule it may be deduced that there are only a very limited number of ways in which a liquid phase formed by two solid components can solidify into one or more solid phases, and that, likewise, there are a limited number of ways in which a transformation in or between these solid phases may take place. "The relation of the phase rule to practical metallography," wrote Cecil Desch, one of the leading British authorities on metals,

> is something like that of formal logic to scientific discovery. No new scientific fact is ever likely to be discovered by the aid of formal logic, but it is sometimes useful to state a process of scientific reasoning in syllogistic form as a means of detecting fallacies. In a similar way, fallacies in the expression of equilibria in complex systems may be detected by applying the strict logic of the phase rule.[76]

A system containing two components, such as alloys of copper and zinc, cannot exist in more than four phases simultaneously. If experimental data indicate the presence of five or more phases at any one moment, then the data are incorrect or have not been interpreted properly. If the data indicate the presence of an isothermal line on the phase diagram, then this must be taken as strong evidence of the existence of a nonvariant system at that temperature and marks the site as one worthy of more intensive investigation through use of analytical methods.

Figure 4.2, also taken from Shepherd's paper, depicts the complete set of liquid and solid phase relationships for brass alloys.[77] The form of the liquidus curve and the positions of isothermal lines remain unchanged, although some points have been relabeled: line nN has become c_1C, line cC has become d_2D, etc. Greek letters have been added to represent each of the six solid modifications or phases of the alloy. A dashed line has also been included to depict the solidus curve, that series of compositions and temperatures at which the transition from liquid to solid state becomes complete with falling temperature.

The forms of the lines separating regions on this diagram, it should be emphasized, were derived from experimental observations. The freezing-point or liquidus curve and the isothermal lines were determined through measurement of cooling rates. The positions of the isothermal lines, in turn, provided a clue to the regions where, at lower temperatures, boundaries between solid phases of the alloy might be found. The precise location of these boundaries, such as the line separating alpha brass from a solid solution of alpha and beta brass, could usually be determined only through the microscopic analysis of samples quenched at various temperatures. So, for example, a sample of brass of

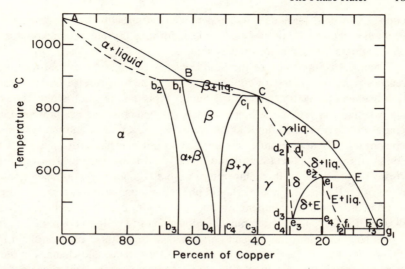

FIGURE 4.2. A phase diagram of brass, redrawn from E. S. Shepherd, "The Constitution of Copper-Zinc Alloys," *Journal of Physical Chemistry 8* (1904): 423.

known composition might be raised to a certain temperature, held at that temperature for a time sufficient for equilibrium to be established, and then suddenly cooled. Since the velocity of the process by which the internal crystalline structure of one type of alloy is transformed into another is typically very slow, quenching will usually fix the structure normal to the elevated temperature. Microscopic analysis could then be used to determine which types of crystals were present. By repeating this procedure at a number of temperatures and on samples of various compositions, the rough forms of lines such as b_2b_3 and b_1b_4 could be plotted.

Once constructed, a diagram such as that reproduced in Figure 4.2 could be used to follow the equilibrium relations of any brass alloy through changing temperature conditions. Thus an alloy containing approximately 68 percent copper by weight exists above the liquidus in two phases, liquid and vapor, and hence is bivariant. Both temperature and composition may vary independently without any alteration in the number of phases. When the temperature falls to the liquidus, however, crystals begin to form. With the appearance of the new solid phase, the system becomes univariant. Temperature may continue to decrease, but changes in temperature must be accompanied by changes in the composition. The first crystals to emerge from the melt are of the alpha variety and are rich in copper; their composition corresponds to that of a point on the diagram at the intersection of the dashed solidus curve and a horizontal line (not shown) drawn from the saturation point of a 68-percent-copper alloy on the liquidus. As temperature continues to fall, the composition of the liquid and solid phases change; later-forming crystals become poorer in copper as the mother liquor itself is depleted.

When the composition of the liquid reaches point B and that of the solid reaches point b_2, the rate of cooling slows because heat is released by a reaction between alpha crystals and the liquid. From this reaction a new type of crystal emerges, beta brass. Since crystals of alpha brass are also present, four phases coexist (vapor, liquid, alpha brass, and beta brass), and the system is nonvariant. The temperature remains fixed until one phase disappears. The compositions of the beta and liquid phases also remain fixed. Thus all beta crystals that appear at this temperature have a composition corresponding to point b_2 on Figure 4.2; and the liquid phase has a composition corresponding to point B. When the temperature falls below 900 degrees, the liquid phase disappears. The solid, which must be 68 percent (by weight) copper, is a heterogeneous mixture of alpha and beta crystals. Since three phases (again including the vapor) are present, the system is univariant, and changes in temperature entail changes in composition. As temperature continues to fall, changes in the composition of alpha crystals may be followed along line b_2b_3 and changes in beta crystals along b_1b_4. If cooling is sufficiently slow to permit equilibrium to continually reestablish itself, then the 68-percent-copper alloy will cross line b_2b_3 near a temperature of 800 degrees, where beta crystals in the mixture will tend to rearrange themselves into the alpha form.

With a phase diagram in hand, Bancroft and his students undertook to correlate such physical properties of alloys as their ductility, tensile strength, and corrosion resistance with their structure and thermal history. But like other workers in Europe, Bancroft and his colleagues soon discovered that generalizations were hard to come by. A phase diagram could alert users to expect changes in crystalline form at certain compositions and temperatures, and crystalline form had a great influence on physical properties. But the structure of crystals was not the only determinant of physical properties; such other factors as the size of crystals and their distribution in the solid also played roles, and the phase diagram was of little use in forecasting these. More important, the correlation of physical characteristics with certain structures and compositions was not the same as developing a theory that explained such findings. Here Bancroft and his associates made little progress.[78]

Bancroft hoped that systematic study of alloys and the construction of a library of phase diagrams would revolutionize the working of metals. "It might be thought," wrote Shepherd, "that technical men would gladly support investigations of this nature, so directly are the results applicable to technical problems." But, as Shepherd acknowledged, "such is not the case." At least it was not very often the case prior to World War I. Aside from giving Bancroft free samples of aluminum and steel, industrial firms did nothing to aid Bancroft. Nor did they rush to hire his students; few whose careers are readily traceable entered the metal industry prior to World War I.[79]

The conservatism of an industry that has often been slow to change its ways may have been partly responsible. Firms like U.S. Steel faced little competition and had ample opportunity to reduce costs and increase profits by using larger and better-integrated units, by improving plant design and the flow of materials, and by substituting machinery for labor. As one observer noted in 1907, "what Pittsburgh excels in is not discoveries but the scale on which the discoveries of others are applied." Not until the 1920s did American firms engaged in the production of metals by traditional means begin systematically to invest significant sums in research and development.[80]

But, as Shepherd's words suggest, it was not so much the industrialists as their technical experts who were skeptical of Bancroft's work: the metallurgists, works managers, and chemists responsible for analysis and quality control. Very little is known of these technicians. While many no doubt had formal training in the physical sciences, few had Ph.D.s. Even Bancroft's qualitative physical chemistry must have seemed esoteric to many of these men. But it would be a simplification to attribute their skepticism entirely to ignorance. At the turn of the century even academic metallurgists, those who did possess strong credentials in science and who taught at universities and technical institutes, were divided in their opinions of the phase rule. While some shared Bancroft's enthusiasm for it, others were indifferent. Rudolph Ruer, for example, one of Germany's leading experts on metals, made extensive use of equilibrium diagrams in his influential textbook of metallography, but gave little attention to the phase rule. The equilibrium diagram, he stressed,

> originates in the collective arrangement of evidence furnished by the individual experiments . . . I have accordingly made no extended use of the phase rule. When this is adopted as a basis for the discussion, much abridgment is of course possible. On the other hand, there is on the part of the beginner a certain disinclination to use the phase rule—not without good reason in my opinion, for although it does, indeed, furnish a general view of possible equilibria and a means for their classification, it is less serviceable as a key to the understanding of individual cases.[81]

Not only were Bancroft and others of his generation unable to frame a general theory linking the structure and composition of alloys with their properties, they were also unable to bridge the great gulf separating the laboratory from the industrial forge. Unlike the physical chemist, metalworkers rarely dealt with pure materials and seldom treated alloys in such a way as to ensure equilibrium at all stages of processing. Ironically, the complaint that Bancroft lodged against those who studied dilute solutions—that they were treating ideal systems of limited practical importance—could also be leveled at his own research. The study of pure materials at precisely controlled temperatures over extended periods of time bore no closer relation to industrial practice than did the study of very dilute solutions.

The work of Bancroft and other students of phase equilibria did not revolutionize industrial practice, but it was not devoid of value. Gradually during the 1910s and 1920s metallurgists and those involved in the study of similarly complex materials such as ceramics and cement came to recognize this. Phase diagrams, for example, helped explain why practices that had evolved through long experience with metals, such as quenching, had significant effects on metallic properties, even if they could not predict the precise nature of those effects. Since the spontaneous changes that metals undergo are always adjustments toward equilibrium, phase diagrams could also help to make sense of puzzling phenomena like age-hardening and tin plague. They assisted in determining whether certain alloys formed true chemical compounds or solid solutions. And, most important, they were extremely valuable pedagogical tools. By visually correlating immense amounts of data, they saved both students and teachers from drowning in seas of facts. The study of phase diagrams—of how they are constructed and read—gradually became an integral part of metallurgical education.

Nor was it only a small band of academic metallurgists who adopted the phase rule and phase diagram. After World War I, America's major producers of metals began to build research laboratories, and the phase rule became a major tool in their work. As John Johnston, the research director at U.S. Steel, explained it,

> It now looks as if no further *appreciable* lessening of the dollar cost of steel per unit of weight is possible through the use of still larger units or of further labor-sparing machines . . .
>
> It is possible, however, to lower the cost per unit of service by improving the quality of the product, that is, by ascertaining how to make a metal precisely suited to each of the manifold uses to which it is now put. . . . This demand, which in recent years has become much more insistent, for steels of higher quality yet at no higher cost, raises many new problems which have necessitated a fundamental examination to discover what actually happens in each of the long sequence of processes from ore to finished product.

The phase diagram, Johnston added, "is absolutely fundamental to a proper understanding of how to handle the particular alloy system so as to secure the best results."[82] As has often been the case, scientific research did less to revolutionize than gradually to improve and rationalize technology.

It took two decades or more for industrial metallurgists to warm to the kind of work that Bancroft did at the turn of the century. Bancroft's colleagues in physical chemistry took even longer. Indeed, few of Bancroft's colleagues ever came to view the phase rule as more than an essentially descriptive generalization, of value only in special situations. To A. A. Noyes and G. N.

Lewis, for example, a qualitative physical chemistry seemed like a contradiction in terms, and Bancroft's habit of contrasting his phase studies and their work on solutions seemed offensive. They considered the phase rule itself a trivial consequence of the thermodynamic analysis of equilibrium. Lewis gave two pages of his *Thermodynamics* to the rule, and then excused himself from treating its applications. "This simple principle," he wrote,

> has been the starting point for the development of a large field of exact but qualitative thermodynamic study, into which it will be impossible for us to enter far, since we are presenting the science of thermodynamics with the primary purpose of making it readily applicable to quantitative and numerical calculations.

Noyes, when drafting catalog descriptions of courses in theoretical chemistry at Caltech, cautioned one of his colleagues: "I would certainly not use the term Phase *Rule*, which is an absurd practise since the rule itself is really of little practical importance, and substitute Phase Equilibria."[83]

Noyes and Lewis were hardly alone in offering such criticism. Ostwald, in a review of Meyerhoffer's monograph on the phase rule, faulted his colleague for exaggerating the significance of the rule: "the rule has the drawback," Ostwald observed, "that it furnishes no direct insight into the origins of the various equilibria, but rather, so to say, ordains them from without by decree." "This rule," wrote Walther Nernst in an account of the development of physical chemistry, "is . . . rather a reliable formula than a theory proper, and that is why from many sides we are warned not to exaggerate its value." When Bancroft, in an obituary notice on Roozeboom for the *Journal of the American Chemical Society*, likened the Dutchman's achievements to those of Lavoisier and Dalton, the editor, William Albert Noyes, asked Bancroft to withdraw the sentence. The comparison, Noyes wrote, was "a little overdrawn."[84] And so Harry Clary Jones was giving voice to a common view when, in 1913, he wrote:

> The question arises, Of what use is it [the phase rule] in chemistry? There are very different opinions on this question. A few see in it the dawn of a new day, not only for chemistry, but for natural science in general. If, however, we judge of its future by its past, it is fundamentally a convenient and shorthand method of expressing facts already established by experiment.

There can be little doubt that Jones had Bancroft in mind when he concluded:

> Looking at the Phase Rule in as broad and impartial a manner as possible, I am unable to get away from the thought that by certain men it has been an overridden hobby.[85]

With the passage of time, some of these views of the phase rule came to seem as exaggerated as Bancroft's. It was true, as Ostwald and Nernst suggested, that the phase rule simply informed its users of some of the constraints operating on systems in equilibrium without saying anything of the origins of—or forces controlling—such equilibria. And, as Jones implied, the phase rule was unnecessary when a complete set of quantitative data (temperatures, pressures, compositions, and free energies) were available for a chemical system. Nevertheless, it became more and more difficult to deny that the phase rule was of value as metallurgists, ceramists, and chemical engineers increasingly began to apply it in their work. In the study of such complex systems as alloys, cements, ceramics, and petroleum oils, complete thermodynamic data were neither available nor readily ascertained, and a knowledge of constraints could be invaluable in interpreting fragmentary experimental evidence. Even physical chemists who were squarely within the Noyes-Lewis research tradition came to acknowledge this after World War I. "With the aid of the phase rule," wrote Charles A. Kraus in 1939,

> it is possible to arrive at a knowledge of the constitution of material systems and the relation of various parts of such systems to one another without destroying the systems themselves. In many respects, it is a more powerful tool than is chemical mass analysis, which is useful only if the systems in question may be resolved into their component parts without changing their nature and composition.[86]

Yet except for Bancroft and a few other enthusiasts, physical chemists rarely gave the phase rule much attention, although they were always willing to claim credit for the advances made possible by its use. It became traditional, for example, to devote a chapter in textbooks to the rule and its applications, but often a chapter toward the end of the text, where the student was informed, more or less explicitly, that the uses of the phase rule and the intricacies of phase diagrams were really best studied in courses on metallurgy or chemical engineering.

Physical chemists often sought, and deserved, credit for having established the fundamental principles of applied fields, and they boasted of the wealth of applications their knowledge possessed. But when the interesting problems became those of application, they generally preferred to turn their sights on other targets. From the inception of their discipline, physical chemists had referred to their subject as theoretical chemistry. Even if their work was done largely in the laboratory, their experiments were directed toward resolving theoretical issues. The task of showing in detail how their results could be used by industry, many felt, belonged to others.

As among physicists, so too among physical chemists there was a tendency to place applied science on a lower plane than pure science. In part this was due to age-old social distinctions between scholar and craftsman. Even though twentieth-century America was far removed from ancient Greece, the disinterested pursuit of knowledge was still considered more honorable than work related to industry and trade, at least among those whose values were shaped in the academy and university. Applied science was socially less respectable than pure science, but it also was deemed less intellectually satisfying. It was too much an end in itself and not enough a means to a larger end. The best and most absorbing research problems were generally conceived to be those that simultaneously resolved one or more of the discipline's old puzzles and opened up one or more new ones. But for physical chemists the phase rule was not problematic; its application promised the solution of many puzzles, but they were the puzzles of others. The construction of an alloy's phase diagram, for instance, might prove illuminating to metallurgists, but it promised to shed little light on the issues that absorbed most physical chemists. As useful as the phase rule might be, to scientists like Noyes, Lewis, and Nernst it seemed a dead end.

There were contradictions in this. Many physical chemists ignored or deprecated applied science and yet boasted of the contributions their discipline was making to technical fields. They worked in an academic culture that preserved, if only in attenuated form, some of the attitudes of a preindustrial society—particularly, the tendency to draw distinctions between pure and applied learning—but they also lived in a larger industrial world in which the inventor was lionized and in which boundaries between pure and applied knowledge were becoming harder and harder to recognize.

Because science is almost universally acknowledged to be one of the most powerful agents of change in modern history, scientists are often viewed as precursors of the modern order, quick to throw off the values of the past, ever forward-looking and receptive to new modes of thought. Yet a group may disturb social and cultural equilibrium without seeking to do so. As Robert Merton observed in his study of science in seventeenth-century England, purposive social action can have unanticipated consequences.[87] No less than others in a rapidly industrializing society, scientists could find themselves caught between two worlds, acting at times in accord with the values of an older culture while giving voice to those of a new.

Most physical chemists simply lived with these tensions; few were even conscious of them. They celebrated the growth of industry and industrial technology, but generally preferred to work in universities and advised their best students to follow teaching careers. They often justified their work by referring to its ultimate utility, but usually selected research problems on the basis of their value to the theoretical structure of their discipline. They sometimes performed services for industry, but did so almost furtively, as if such work were déclassé. If physical chemists themselves took little part in the tedious labor of application, if, that is, they left it to metallurgists and chemical engineers to redeem their promises, a ready-made excuse was available in the ever-increasing specialization of knowledge and function in society.

Bancroft was something of an exception. He, more than most physical chemists of his generation, tried to bring his research into line with the claim of utility. But his effort had a curious result. His research on alloys, and on phase equilibria more generally, fell between two stools. His work was too practical to interest many of his colleagues in physical chemistry and too abstract to be of immediate interest to industrial chemists and metallurgists, at least those of his generation. By the time interest in the phase rule awakened in industrial circles, Bancroft had moved on to the study of other topics.

BANCROFT REDUX

Bancroft reached the pinnacle of his career while he and his associates were exploring applications of the phase rule. In 1903, Bancroft was promoted to full professor; two years later, he was elected president of the American Electrochemical Society; and in 1910 he won the presidency of the American Chemical Society. Bancroft's good fortune may seem surprising in view of the critical reception given his ideas by fellow physical chemists. Nevertheless, it should be remembered that Bancroft's critics, eminent as they were or soon would be, were small in number. Indeed, in the first decade of the twentieth century physical chemists themselves constituted a fairly small minority in an organization like the American Chemical Society. While the best and brightest of his colleagues were dubious about Bancroft's program for physical chemistry, there were many chemists who shared Bancroft's distaste for mathematics, for recent trends in theoretical physics, and for the seemingly esoteric interests of scientists like A. A. Noyes. By championing a version of physical chemistry that was decidedly experimental rather than theoretical, and industrial rather than academic, Bancroft may have won as many friends as he lost.

Nor was Bancroft without legitimate credentials. The *Journal of Physical Chemistry*, while not outstanding in its field, did serve a useful function by publishing articles for which there was insufficient space in the *Journal of the American Chemical Society*. Although many of his colleagues had strong reservations about Bancroft's views on physical chemistry and the direction of his research, his laboratory did produce papers in abundance, and some made significant contributions, if only to topics on the border between pure and applied science.[88] And even his adversaries respected Bancroft's intelligence, or, perhaps more precisely, his cleverness. "There's a wonderful story about Bancroft that Kisty [George B. Kistiakowsky] tells," one of Bancroft's acquaintances related to an interviewer:

> He said that one time [Hugh S.] Taylor from Princeton and Bancroft were arguing the importance of mathematics for physical chemistry. You could tell who was on which side. And Kistiakowsky said, "That blank blank Bancroft—you know, he was completely wrong and he won the argument!"[89]

Perhaps Bancroft's strongest card, however, was his success as a teacher. As is often the case in science, Bancroft's reputation rested in large part upon the achievements of his pupils. A steady stream of graduate students passed through his laboratory, acquiring tools with which to study heterogeneous equilibria. Some applied their knowledge to the study of mineral systems at the Geophysical Laboratory of the Carnegie Institution of Washington, where efforts were under way to understand the formation of igneous rocks. Others went to the Bureau of Soils of the U.S. Department of Agriculture, where soil formation was studied from the standpoint of the phase rule.[90] Still others made careers for themselves in industry: Hector R. Carveth rose to a vice presidency at the Roessler and Hasslacher Chemical Company; Homer W. Gillett, after working for a succession of federal agencies and private firms, became director of the Batelle Memorial Institute; and most famous of all was Edgar E. Teeple, whose process for separating potash from other salts in the brine deposits of California's Searles Lake made him one of the most successful industrial consultants of his generation. America's expanding universities also absorbed a share of Bancroft's students; by 1925, "Banty's men," as they sometimes called themselves, were sprinkled across the colleges and universities of America.[91] The diversity of these careers testifies to America's growing capacity to absorb experts in the early twentieth century. The day when physical chemists had to justify their existence by impersonating analytical chemists, although not over, was nearing an end.

Collectively these results were impressive, sufficient to compensate for Bancroft's idiosyncrasies and to counterbalance the steady criticism directed at him from Noyes, Lewis, and others who shared what might be called a physicalist's conception of physical chemistry. Bancroft was not universally acknowledged to be a leader in his field, but he had a following and was well-known to his contemporaries.

Success within his own university and within organizations like the American Chemical Society, however, should not be mistaken for success in shaping a discipline. The phase rule did not become an organizing principle of physical chemistry; physical chemists did not become less concerned with exact measurements; the links between physical chemistry and physics became stronger, not weaker, with time. And impressive as was the record compiled by Bancroft's graduate students, it does not bear close comparison with that of Noyes's. Whereas Bancroft's former pupils staffed government bureaus, Noyes's headed them; whereas three or four of Bancroft's former students attained high office at firms like Roessler and Hasslacher, at least twice as many of Noyes's filled key positions at firms like Du Pont and General Electric. While Noyes's students went to teaching positions at schools like Illinois and California, which were developing strong commitments to research, Bancroft's went where the opportunities for research and the incentives to publish were not so great: New Hampshire, Iowa State, Indiana, North Carolina, and Rice. None of Bancroft's students found their way into the National Academy of Sciences. In short, while Noyes's laboratory was producing the elite of the next generation, Bancroft's was producing many commoners. Bancroft won notoriety, but not genuine leadership. It is time to ask why.

Many elements may enter into disciplinary leadership—charismatic authority, access to resources, institutional power, etc.—but two factors are of cardinal importance: ideas and disciples. Some scientists lead primarily by force of intellect, as expressed through personal research and publications. They sketch out the outlines of new territories and thereby propose dramatic changes in the conceptual map of science. The ideas of van't Hoff and Arrhenius, for instance, were so novel, attractive, and far-reaching that, when encountered in a sympathetic setting, they compelled attention and respect. Others, such as Ostwald and Noyes, leave a mark by building a research school: by picking and choosing artfully among the ideas and techniques of their contemporaries, by developing research programs that offer opportunities for fruitful work, and by recruiting students with the intelligence and energy to exploit those opportunities. They succeed by exploring new territories in detail, thereby improving and sometimes revising the map of science, and by making new knowledge accessible and relevant to broad audiences.

It is tempting to label these modes of leadership with adjectives like intellectual and social, but this would be deceptive. Van't Hoff and Arrhenius had students, and Ostwald and Noyes had ideas. Nernst and G. N. Lewis were very nearly ambidextrous in the sense that they demonstrated both the ability to make fundamental contributions to the theory and techniques of their discipline and to build powerful research schools. Rigid distinctions between thinkers and entrepreneurs fail; social skills and intellectual powers are present, even if in variable proportions, in all of these scientists. Nevertheless, leaders of academic disciplines typically possess one or both of these attributes in unusual degree.

Bancroft did not. Although a clever man, Bancroft's quickness was not matched by deep insight or profound originality. He admired those, like van't Hoff, who led by dint of their brilliance, but did not aspire to such a role. He did, however, seek to create a research school and, more than that, to expand and revise the agenda of the discipline in which he was trained. But here too he fell short, at least by comparison with his contemporary, Noyes.

His failure is, in some respects, puzzling. A visitor to Cornell and MIT in 1900 would have found in Bancroft and Noyes two bright and ambitious young scientists, with similar backgrounds and prospects. Noyes showed somewhat greater attention to detail and power of concentration, but Bancroft had greater skill with a pen and was the more forceful of the two, the more likely to inspire personal loyalty. Both had an ample stock of ideas and plans and the personal resources necessary to act on at least some of them. Each had embarked on promising research and had begun to draw others into their enterprises. Their situations were roughly comparable, although Cornell probably had the advantage over MIT as a site for a research school. It was larger and wealthier, it possessed a graduate school, and its administration had a stronger commitment to research, especially research in the basic sciences, than did the administration at MIT. It was, in other words, by no means clear that Noyes would come to play a central role in the development of his discipline and that Bancroft would become a secondary figure.

Yet that is what happened. This outcome is due to differences in the way Bancroft and Noyes used their personal and professional resources. Both sought leadership in their discipline, but they approached their goal with different strategies, both intellectual and institutional. Consider first their intellectual strategies. By focusing on the properties of strong electrolytes in aqueous solutions, Noyes built squarely on the foundations laid by the Ionists. His research program, at least at its inception, grew directly out of the issues confronted by Ostwald, Arrhenius, van't Hoff, and Nernst. His methods too were those of his European teachers: thermodynamic reasoning linked with exact experimental studies. And just as his teachers had, at various points in their work, looked to physics for new techniques and ideas, so too did Noyes. He preserved their orientation, even though it eventually led him to reject some of what they had taught.

Bancroft, by contrast, embarked on a research program that was only tangentially related to the research of the founders and leaders of his discipline. While the Ionists were willing to acknowledge that the behavior of complex heterogeneous systems fell within the province of physical chemistry, they did not recognize the study of such systems as germane to the fundamental problems of their discipline. Like the physicists whom they admired, the Ionists sought to master the simple before studying the complex; they were intent on understanding as fully as possible the behavior of dilute solutions of salts in water before undertaking the study of solutions of metals in other metals, of silicate minerals in other minerals, or of concentrated solutions of salts in water. Whatever the importance of the latter to metallurgists, petrologists, and industrial chemists, it seemed premature to expect that their study would yield much return to physical chemistry. To be sure, a principle like the phase rule might be of service in the treatment of complex heterogeneous systems, especially when exact thermodynamic data were unavailable, but it was theoretically rather uninteresting. The phase rule was the basis of an efficient filing system and a logic of analysis, but it did not appear to promise new laws or new insights into the nature of matter and chemical change. Its value lay almost entirely in its application.

By choosing to make the applications of the phase rule central to his research, by opting more generally to develop a qualitative, industrial, and decidedly chemical version of physical chemistry, Bancroft cut himself off from the mainstream of research in his field. On a rhetorical level, Bancroft differed little from Ostwald and other physical chemists who spoke freely of the utility of their science and its potential for unifying chemistry by giving that science a set of general principles that might underpin its various specialties. Echoes of Ostwald's *allgemeinen Chemie* could be heard in Bancroft's message. Nevertheless, Ostwald's journal and his research program at Leipzig focused not so much on reconstructing chemistry as on creating a new physical chemistry. His research school evolved around the problem of applying physical methods and principles to the study of aqueous solutions. Its research was tied most directly to the work of van't Hoff and Arrhenius that had made solutions accessible to quantitative investigation. For this reason, many chemists could simply refer to members of Ostwald's school as "Ionists." Ostwald may have talked about creating an *allgemeine Chemie*, but it was a *physikalische Chemie* that in fact resulted from his work.

Most of Ostwald's students took enthusiastically to this work on the border between chemistry and physics, but Bancroft could not. The obstacles were diverse. His grandfather had stressed the importance of applying one's knowledge, and Bancroft found it difficult to see the utility in lavishing attention on "slightly polluted water."[92] His teacher at Harvard, Josiah Parsons Cooke, had impressed upon Bancroft the importance of order and a sound and comprehensive classificatory scheme for good pedagogy, and yet the open-ended research of his contemporaries and the disturbing discrepancies between the ideal and real behavior of solutions seemed to complicate rather than simplify science.[93] Bancroft himself had a distaste for mathematics and mathematical physics that his education in the United States did nothing to erase; he was ill-prepared to master the thermodynamics of the previous generation, never mind the molecular-kinetic science of his own.

His solution, as we have seen, was to forge a qualitative version of physical chemistry that was easily taught, that might find immediate application in industry, and that was largely independent of modern physics. Congenial as this vision was to Bancroft, however, it stood in conflict with the aims and values of most of his fellow physical chemists. Bancroft had imposed upon himself a challenge that Noyes did not face, that of persuading other physical chemists that his research was centrally important to their discipline.

The ways in which Noyes and Bancroft mobilized their resources were no less different than their intellectual strategies. When Noyes launched his research laboratory, he did not wait for students to present themselves; he searched for the promising young chemists both in America and in Europe and offered them attractive pay and working conditions. He made room both for those who already possessed doctorates and research experience, such as Lewis and Bray, and for those who had yet to earn a graduate degree, such as Kraus and Tolman. Noyes ran his laboratory not as a degree-granting program but as a research institute. And this research institute lay at the center of Noyes's life and thought: it absorbed his money, time, and creative energies.

Bancroft's attitude toward his laboratory seems, by comparison, almost casual. His letters and papers contain few references to his graduate students and their laboratory work. He made no special efforts to recruit exceptional students, nor did he seek to create niches for postdoctoral fellows at Cornell. Chemists in Bancroft's program were, with few exceptions, there for degrees; they were part of the chemistry department, not members of a quasi-independent and elite corps. Some had special gifts for research, but many did not. In short, Bancroft's materials, his disciples, were not as fine as Noyes's. Perhaps it could have been otherwise. Both Bancroft's resources and willingness to devote them to his discipline were as great as Noyes's. At the turn of the century it had been within his means to construct a research laboratory similar to Noyes's. But Bancroft chose to use his resources differently. He invested not in men and equipment but in paper and ink—his *Journal of Physical Chemistry*. Year in and year out, that journal absorbed money and, perhaps as important, Bancroft's time and energies.

The return on investment was meager, far less than Bancroft had imagined when he founded it in 1896. It did provide Bancroft with an outlet for his own writings and the papers of his students. Gradually there also developed around the *Journal* a loose network of friends and former students willing to contribute papers and, occasionally, money: Louis Kahlenberg and his students, who became notorious in the first decades of the twentieth century for their outspoken attacks on the electrolytic dissociation theory; J. M. Bell and Hector Carveth, who shared Bancroft's passion for developing applications of the phase rule; Eugene C. Bingham of Lafayette, T. B. Robertson of California, and J. W. McBain of Stanford, who shared Bancroft's later interest in colloid chemistry; and others, such as Herman Schlundt of Missouri, S. W. Young of Stanford, and Frank K. Cameron of the U.S. Bureau of Soils, who for one reason or another felt that their work was unappreciated elsewhere. By 1918 Bancroft's journal enjoyed a paid circulation of approximately five hundred.[94]

Nevertheless, at no point during Bancroft's tenure as editor was the *Journal* self-supporting, and at no point did he enjoy the luxury of an ample backlog of papers. Just as Bancroft had to pay deficits from his own pocket, so too he found it necessary to write articles and reviews simply to flesh out issues that otherwise would have been embarrassingly thin. Over ten percent (83/762) of the articles published in the *Journal* during its first twenty years were written by Bancroft himself. Inevitably, the quality was uneven. As time passed, he increasingly resorted to publishing articles that were patchworks of long quotations on various topics that caught his interest. Although he occasionally made original observations in these peculiar review articles, most were little more than space-fillers.

The *Journal of Physical Chemistry* made Bancroft a visible figure; but not all readers liked what they saw. Absent from the subscription list of the *Journal* were the names of some of America's prominent physical chemists: G. N. Lewis, Arthur B. Lamb, W. C. Bray, and William D. Harkins, among others. Arthur A. Noyes did have a personal subscription but apparently believed

Bancroft's journal might be dangerous to immature minds; after moving to Pasadena, he ordered it banned from Caltech's library.[95]

Despite the burdens, Bancroft persevered as editor. Like an investor who cannot cut his losses, Bancroft was forever seeing glimmers of hope. The chemists who used the *Journal* as their principal outlet for papers encouraged him in his optimism. The only other American journal giving physical chemistry significant space was the *Journal of the American Chemical Society*. This, many of Bancroft's supporters believed, was firmly under the control of "the Tech crowd."[96]

On several occasions, William Albert Noyes, the general editor of the *Journal of the American Chemical Society* between 1902 and 1917, sought to negotiate a merger between his journal and Bancroft's. Noyes sought this combination as a part of his wider effort to unify American chemical publications under the auspices of the American Chemical Society, an effort motivated in part by his belief that particularist sentiments within new specialties posed a threat to the integrity of the Society. But Bancroft repeatedly snubbed these advances as his allies spoke darkly of the dangers of a "chemical publishing trust" and warned him that "with the passing of the J.P.C. something will be lost. Some of us will have trouble saying what we want to say and in our own way."[97] In particular, they feared that A. A. Noyes, G. N. Lewis, and their associates, who were already largely responsible for refereeing publications in physical chemistry for the *Journal of the American Chemical Society*, would control the product of a merger. As one of Bancroft's partisans wrote:

> Lewis informed me that the Journal [of the American Chemical Society] not only could not take all of the papers now offered but could not even handle all of the physical chemistry, so that a judicious selection of the "best" articles is now necessary, and this selection, as a matter of course, must be done by A. A. Noyes and himself.[98]

These fears of a publishing "trust" smack of a paranoia that is not uncommon among those who are marginally competent in their callings; some of Bancroft's supporters surely would have had difficulty meeting stricter standards of concision, accuracy, and relevance. But competence was not the only issue. By challenging the research priorities of the Noyes-Lewis school, Bancroft declared himself exempt from many of the criteria that Noyes and Lewis used in judging quality. Where they valued quantitative precision, Bancroft admired the simple qualitative experiment; "accuracy to another decimal place," he wrote, "is not yet the real goal of chemistry."[99] Where Noyes and Lewis believed that contemporary physical research would enrich their science, Bancroft lacked the mathematical skill and inclination to pursue such studies, believing them, in any case, too esoteric to benefit chemists. Whereas Noyes and Lewis came to believe that an understanding of atomic and molecular structure would be an invaluable aid in the study of processes of chemical change, Bancroft remained skeptical. When, in 1928, one of his colleagues asserted that

> without a knowledge of the present-day theories in regard to the nature and structure of the atom, it is impossible for the student to comprehend such topics as the cause of ionization, the nature of amphoteric electrolytes . . . , hydrogen ion concentration, Donnan's membrane equilibria theory,

and a variety of other topics, Bancroft responded:

> Hydrogen ion concentrations and amphoteric electrolytes were studied before anything was known about the composition of the atom. We can only marvel at the wonderful ingenuity of Donnan in deducing a relation which, by definition, he could not comprehend.[100]

It was evident to most physical chemists that A. A. Noyes or G. N. Lewis would have an editorial perspective very different from that of Bancroft. Neither shared Bancroft's aim of developing the "qualitative" side of physical chemistry; neither believed the phase rule to be of broad significance or of general interest to chemists. These differences were already evident in the contrast between the coverage given physical chemistry in the two journals. Between 1896 and 1901, for instance, nearly a third of the papers in Bancroft's journal dealt with applications of the phase rule; during the same period, the *Journal of the American Chemical Society* published not a single article on the topic.

Bancroft's decision to invest his money and time in a journal rather than a research laboratory was, therefore, not simply an arbitrary choice. His intellectual strategy dictated this allocation of resources. By articulating and promoting an agenda for physical chemistry that was at odds with the dominant tradition in the discipline, Bancroft imposed upon himself the burden of maintaining the *Journal of Physical Chemistry*. It was the one forum in which he, and other physical chemists who dissented from orthodoxy, could freely argue their case.

Bancroft never lost his interest in the phase rule, but in the years around World War I the focus of his research shifted. He was discouraged perhaps by his colleagues' indifference to his research and attracted by the prospect of beginning afresh in a new field of work. Bancroft's drift away from the study of heterogeneous equilibria, however, was hastened by events beyond his control. In 1912 the Carnegie Institution of Washington, as part of a general effort to reduce its support of individual scientists, suspended its support for Bancroft's work on alloys. Four years later a fire destroyed Bancroft's laboratory and equipment. And in 1917 Bancroft, like Noyes, Lewis, and many other scientists, took leave from his teaching and research to participate in the war to end all wars.

But while Bancroft's interests changed, his approach to science did not. As we shall see in a future chapter, he would remain a dissenter until the end.

CHAPTER 5

Physical Chemistry in the "New World of Science"

THE PROSPECTS facing physical chemists at the beginning of the twentieth century were exhilarating. The conceptual breakthroughs made by van't Hoff and Arrhenius in the 1880s had been the prelude to a wave of new ideas and discoveries in the region between chemistry and physics. Some, such as Nernst's theory of electromotive force or his later heat theorem, were made by workers within the research tradition of the Ionists. Others, such as the discovery of the electron, radioactivity, the radioactive-decay series, and the diffraction of X-rays by crystals, were made by scientists who, like J. J. Thomson, Henri Becquerel, Ernest Rutherford, Frederick Soddy, and Max von Laue, emerged from different intellectual traditions. But whatever their backgrounds, the paths of these scientists seemed to converge in the broad region between chemistry and physics. Soddy was known as a physical chemist, as was Rutherford during the early portion of his career. Thomson's works were read as avidly by physical chemists like Noyes and Lewis as by physicists. And X-ray diffraction, although discovered in a physics institute, was applied by both physical chemists and physicists. As is often the case when knowledge expands rapidly, boundaries were mobile, but physical chemistry seemed to have as strong a claim to the strategic crossroads of modern science as any other discipline.

The freedom to traverse the borderlands between chemistry and physics, the excitement that came with being present at the beginnings of a new discipline, the possibility of making fundamental discoveries—a possibility made all the more real by the recent history of such breakthroughs—all this contributed to making physical chemistry a deeply satisfying discipline for its devotees. Satisfying too, especially in a country so sparsely dotted with facilities for scientific research as America, was the construction of institutions where none had existed before. The process typically was slow; Ostwald's American pupils had to serve the needs of their departments and universities before the needs of their discipline and this usually meant heavy teaching responsibilities in subjects like analytical chemistry. But as they gained seniority, and as the universities in which they worked grew in size and wealth, physical chemists often could shed duties extraneous to their interests and concentrate on their own subject.

By 1905 or 1910 the results were becoming apparent. American laboratories were beginning to produce important papers, most of those papers were being published in American journals, and the great majority of young physical chemists were taking their graduate degrees at home rather than abroad. Foreign students were beginning to study physical chemistry at American universities, and even the founders and leaders of the discipline, van't Hoff, Ostwald, Arrhenius, and Nernst, were finding it worthwhile to cross the ocean and lecture in America. At the International Congress of Arts and Sciences, a scholarly extravaganza held in conjunction with the St. Louis World's Fair of 1904, physical chemists, both American and European, held an important place on the program.[1]

Despite the satisfactions of a career in physical chemistry and the growing strength and visibility of the subject in academic circles, there was nevertheless a great disparity between the discipline's promise and its achievement at the beginning of the twentieth century. Ostwald and his disciples had stressed the broad utility of their new science; exploration of the borderland between chemistry and physics, they had asserted, would benefit other branches of chemistry, other sciences, and industry. Yet, except for their well-publicized contributions to the rationalization of analytical chemistry, physical chemists had delivered little on these promises. Their work as yet had had little impact on workers in other fields. Physical chemists played an insignificant role in industry. Their one or two outposts in the scientific bureaus of the federal government were exposed and vulnerable. And even in universities their place was far from secure: many American colleges and universities had yet to add a physical chemist to their chemistry departments, and where the subject was taught it often drew few students. For physical chemists and a few scientific sympathizers, physical chemistry occupied a central position in the conceptual map of science. But for many other scientists, for university presidents, and for those businessmen who knew something of science, it was a small and obscure province populated by a strange breed who held unaccountably strong attachments to curiosities like the phase rule. At the turn of the century, physical chemistry was a discipline that was satisfying many of the intellectual needs of its practitioners, but few of the needs of society.

By 1920, much had changed. Specialists in fields like petrology, biochemistry, chemical engineering, and pharmacology were drawing heavily on physical chemistry for ideas and techniques. Physical chemists were much sought after by business firms and were in command of many of America's premier laboratories for industrial research. They had been entrusted by the government with responsibility for some of the country's most critical projects during the first world war. And in chemistry departments of universities large and small, physical chemists were represented, often in positions of leadership. Although it is impossible to quantify such judgments, it is probably fair to say that no specialists in the early 1920s enjoyed opportunities more diverse and numerous than physical chemists. During the preceding two decades, their science had been woven into the fabric of American society. This chapter is about some of the many and various ways in which this occurred.

THE INTEGRATION OF SCIENCE AND THE INDUSTRIAL ORDER

In following the prewar work of Noyes and Bancroft, we saw how two physical chemists made places for themselves and their science in America and how, through both their work in building institutions and in developing their science, they helped create opportunities for others. But the success of physical chemistry in the first decades of the twentieth century cannot be explained by reference to one or two individuals, no matter how able they may have been as scientists, teachers, and publicists. Nor can it be explained solely on the basis of the efforts and achievements of that entire generation of physical chemists. Their talent and initiative were essential ingredients, but little could have been achieved in the absence of opportunities to exercise their skills.

As we have seen, by 1900 opportunities had appeared in America's universities, where physical chemists proved able to adapt themselves to the special needs of those institutions, especially their need for teachers of elementary chemistry. This, however, was not a prescription for indefinite growth. Had the teaching of elementary subjects remained the only significant role for physical chemists, Ostwald's students could easily have sunk into an obscure mediocrity, their enthusiasm for research choked by the daily routine of undergraduate instruction. And their discipline could easily have remained a satellite orbiting about a European center, dependent on Europe's laboratories for both graduate training and intellectual stimulation. But other opportunities did appear. Their emergence was due less to developments internal to the discipline of physical chemistry than to changes in American society and its institutions. Casual readers of American history will be familiar with the main elements of this story: a transformation of American business that entailed changes in the volume and speed of production, in patterns of trade, and in the size and complexity of firms; changes in the rhythms of daily life associated with urbanization, mass production, and technological change; the emergence of new professional organizations whose experts not only serviced the new industrial economy but also sought to alleviate its social dislocations; and the expansion of higher education, fueled both by new wealth and by new opportunities for those with education.[2]

A few statistics illustrate the magnitude of these changes. During the lifetimes of Arthur A. Noyes and Wilder D. Bancroft the percentage of the American population living in cities more than doubled, the number of professional and technical workers increased by a factor of ten, and the number of workers who described themselves as chemists grew by a factor of fifty-eight. During their teaching careers, the nation's manufacturing output quadrupled and the number of university degrees conferred per million of school-age population quintupled. And between 1898 and 1918, while Noyes and Bancroft were building their research schools, production of petroleum increased by more than a factor of six; steel, by nearly a factor of five; and output of technologically sophisticated products like electrical equipment and fine chemicals grew at a far faster pace.[3]

Although many of these trends and developments had their beginnings in the mid-nineteenth century, it was not until the twentieth century that they began to impinge in direct and impressive ways upon the academic culture of which physical chemists were a part. That it took so long is indicative of the distance that separated the realms of science and technology in nineteenth-century America. Cheap raw materials, plentiful infusions of foreign capital, and the skill and resourcefulness of mechanics, inventors, and entrepreneurs had been far more important to the expansion of American industry and commerce in the nineteenth century than elaborately educated engineers and scientists. Through most of that century, higher education and the institutions that provided it were the concerns of members of the learned professions and those with old wealth, many of them the scions of the eighteenth-century merchant aristocracy. These custodians of culture viewed education more as a source of personal satisfaction, polish, and virtue than as a font of utilitarian skills or pecuniary profit. Those who were narrowly focused on the pursuit of wealth were apt to consider such education as irrelevant or even inimical to their goals. Where the academic and industrial cultures met, as in engineering schools, powerful conflicts could and often did develop between educators and businessmen, teachers and students, and among teachers themselves over where to draw the line between theoretical and practical learning.[4]

Yet the very existence of engineering schools and their rapid proliferation at the end of the nineteenth century testify to a developing consciousness among both educators and businessmen that cooperation was in their mutual interest. Educators recognized potential patrons in industry's barons and potential students in the households of the growing middle class of managers, bookkeepers, and clerks; they feared that their colleges might become entirely obsolete if they ignored the needs and desires of a rapidly changing society. They also believed in the benefits of a college education, even if it was not the classical education they had enjoyed. The college and university could help civilize this new America if it could train those who would guide its continued economic expansion.[5]

Businessmen, for their part, faced the task of producing increasingly complex products by more and more complicated processes in ever-larger organizations. In such technology-intensive industries as the manufacture of electrical equipment, even an inventor and entrepreneur as quick and ingenious as Edison found himself dependent on experts with formal scientific training.[6]

But growth of business interest in scientific education was not simply a technology-driven necessity. Andrew Carnegie and John D. Rockefeller did very well for themselves with only sporadic help from academically certified experts in iron and steel or petroleum chemistry. Yet both ultimately dedicated significant shares of their riches to the support of science. Their interest in science, and that of their contemporaries in American business, was the product of motives as complex as the businesses they built: a quasi-religious belief in progress, a concern with public relations, a desire to immortalize their names, and, perhaps most universal, an ambition to acquire the social legitimacy associated with patronizing high culture.[7] New elites almost always absorb some of the values of those whom they displace or join, and the newly rich entrepreneurs of the late nineteenth century were hardly exceptions. Whether or not an investment in academic science paid or promised to pay a monetary return, it surely would confer approbation and honor.

And so, gradually, the ties between America's business and academic elites, which had been very loose in the 1870s and 1880s, tightened. Entrepreneurs' occasional gifts to colleges and universities became larger and more frequent, consulting arrangements between business firms and academic scientists became more common, engineering schools proliferated, and new applied sciences like electrical engineering and, a little later, chemical engineering made their debuts. It would be an exaggeration to suggest that the union of these two cultures became complete, or to say that one party became a handmaiden to the other; their distinct histories had endowed capitalists and educators with different values and tastes, and neither party could entirely forget that their interests, while overlapping, were not identical. Nevertheless, the gulf that separated the businessman from the academic had begun to narrow by the end of the nineteenth century, and it would narrow yet further in the early twentieth century with the creation of philanthropic foundations like the Carnegie Institution of Washington and Rockefeller's General Education Board and with the establishment of the first American laboratories for industrial research.

The impact of these new institutions on physical chemistry was not immediate. The General Education Board, for instance, made enormous grants to universities, but generally to assist medical colleges institute reforms or to help universities increase endowment. It was, of course, expected that some benefits from these infusions of capital would trickle down to individual disciplines, as indeed happened, but the principal aim was to strengthen schools and not disciplines—to improve American science and medicine by making institutions less dependent on tuition income.

At the Carnegie Institution of Washington the focus was on improving disciplines rather than on aiding universities, and physical chemistry was among the disciplines to benefit. As we have seen, the bulk of the Carnegie Institution's appropriations for chemical research went to physical chemists, and Carnegie support proved critical to the work of both Arthur A. Noyes and Wilder D. Bancroft. Other beneficiaries included T. W. Richards and Harry

Clary Jones. Yet the few thousand dollars that the Carnegie Institution of Washington gave physical chemists each year was dwarfed by the hundreds of thousands it spent on such ventures as the Laboratory for Experimental Evolution at Cold Spring Harbor and the Mount Wilson Observatory above Pasadena. The administrators of the Carnegie Institution sought to provide for projects that universities could not afford to undertake or to support work in important fields that were ill-supported by existing institutions. Chemical research did not require monumental equipment, nor did the chemists who advised the trustees of the Carnegie Institution make a strong case for their science when funds were being allocated. And so chemists received minuscule sums by comparison with the purveyors of extravagant projects like the astrophysicist George Ellery Hale or the geneticist Charles Benedict Davenport.[8]

Chemists, of course, did not labor under such handicaps when it came to industrial research. Businessmen had long found uses for chemists, although more often in the management of plants and the analysis of materials than in any conscious effort to generate novelty. But when industrial firms began to build their own in-house research facilities around the turn of the century, chemists quite naturally assumed a major role. The statistics are impressive. By one estimate, the cumulative number of industrial research laboratories founded in America grew from four to over one thousand during the years between 1890 and 1930 (Table 5.1). The number of chemists employed in research in these institutions totalled more than 3,800 by 1921, approximately four times the number of chemists teaching in American colleges and universities.[9] Practically unknown in America in 1890, the research laboratory became *de rigueur* for progressive American firms in the span of a generation.

The motives behind this rush to research were as complex as those that prompted barons of industry to favor pure science with millions. From the

TABLE 5.1
The Formation of Industrial Research Laboratories, 1890–1930

Year	Cumulative Number Founded to Date
1890	4
1895	22
1900	49
1905	100
1910	180
1915	310
1920	519
1925	728
1930	1030

Source: Arnold Thackray, Jeffrey L. Sturchio, P. Thomas Carroll, and Robert Bud, *Chemistry in America, 1876–1976: Historical Indicators* (Dordrecht, 1985), pp. 345–346.

broadest perspective, research laboratories seem like logical and inevitable appurtenances of the modern corporation. Through consolidation and growth, American business firms were becoming sufficiently large to afford the luxury of permanent research facilities. The professional managers who were coming to preside over these companies were more concerned with rational planning and orderly growth than the entrepreneurs they replaced and were far more likely to see systematic research as a means toward such ends. Not only could research laboratories occasionally create the products and processes by which new markets might be created and conquered, they also promised the patents and incremental improvements by which markets new and old could be defended. The logic seems compelling, and over the long run it was. Few laboratories, once established, were later written off as failures.[10]

Seen up close, however, all manner of accidents could play roles in the decision to create particular institutions—the whims of influential individuals, the pressure to solve especially critical problems in operating systems or to eliminate bottlenecks in production, fear of foreign competition, the urgent need to find substitutes for raw materials or imported products cut off during World War I.

Nor can the broad movement toward industrial research be understood entirely in terms of economic imperatives. The rapid material progress of America, and of western societies more generally, had a dark side that thoughtful observers found troubling. Even before World War I, there were prognosticators who feared that industrial nations were entering an age of ever more intense competition for resources and markets. Englishmen counted the years until Britain's coal reserves would be exhausted. Americans, within and without the incipient conservation movement, warned their countrymen that America's bountiful resources were not infinite. And citizens of all industrial nations were concerned with trade: preserving those markets they had and winning new ones for their finished goods and products. Competition for resources and markets would intensify in the twentieth century; upon this all agreed. And the nation whose businessmen, scientists, and engineers could wring the greatest possible benefit from resources—who could refine and improve old productive processes and develop new ones—this nation would prosper, at the expense of others. These forebodings, distressingly consonant with readings given the second law of thermodynamics and the theory of natural selection by the popular press, lent urgency to the message of those who, like the Boston consulting chemist Arthur D. Little, preached the gospel of industrial research.[11]

There was a kind of sympathetic reverberation between the gospels of industrial research and of physical chemistry. Both taught that the future would belong to the efficient, that the study and improvement of the processes by which matter is transformed would reap rewards. It is surprising, therefore, that fewer than one in ten productive physical chemists worked in industry during the five years prior to World War I, nearly half of them at General Electric, and that physical chemists like Wilder D. Bancroft, who were favorably disposed toward cooperation with industry, found such a feeble response.[12]

It is difficult to know how to interpret this disparity between the anticipated and the actual. Part may be due to the use of publication as a criterion for defining the population of physical chemists. Physical chemists in industrial employ were less visible than their academic colleagues since they had fewer incentives to publish and, in some cases, may have been discouraged from publishing. Nevertheless, using a less stringent criterion for identifying a sample of physical chemists does little to enlarge the pool of industrial workers; only a handful of the students who earned Ph.D.s under Bancroft and Noyes prior to World War I took full-time positions in industry. The more likely explanation is a reticence on the part of Ph.D.-holders to enter low-status industrial jobs and a lingering suspicion of highly specialized scientists among industrial employers. An example of the former may be found in John Johnston, a physical chemist fresh from postdoctoral research at Noyes's laboratory in Boston and a stint at the Geophysical Laboratory of the Carnegie Institution of Washington. "I take no job unless they are prepared to treat me decently in the matter of conditions of work: for instance, I am for no 7 a.m. to 6 p.m. business, nor a job with Thanksgiving and the 4th of July as my annual vacation," Johnston wrote in 1911.

> I hardly know whether I prefer University work to industrial work: it would depend nearly altogether on the particular conditions of the job. I am inclined to think that I favor the former though, because one meets more congenial people, is more independent in the matter of research work and other things, and is on the whole better treated.[13]

Businessmen could hardly have found attitudes such as Johnston's congenial, and examples of resistance to the hiring of "highfalutin" doctors are legion, especially in older industries like the production of glass and metals. In 1913, for example, when some of the directors of the Owens Bottle Company suggested that the firm hire an expert chemist, M. J. Owens, the president and founder of the company insisted on hiring an experienced batch mixer. For Owens and many of his generation there was no substitute for practical experience. But even in industries in which innovations in products and processes were of critical importance, there was a strong inclination to vest authority over research in the hands of those who had proved their mettle in the plant rather than in the university. When, for example, Standard Oil of New Jersey sought to initiate systematic research into methods of cracking petroleum, it turned first to Edgar M. Clark, a refinery manager for Indiana Standard who, although lacking formal schooling past grammar school, had invented an extremely profitable continuous cracking process.[14]

There are, of course, examples of industrial research laboratories that were organized by, and put under the direction of, academic scientists. The first and most famous of these was General Electric's. Impressed and intimidated by advances in the technology of lighting, the directors of General Electric felt compelled to build up their firm's own research capacity to compete. Since some of the most promising of the innovations of the 1890s had been the work of Europeans with training in physical chemistry, the firm focused on acquiring expertise in the new discipline. By promising high pay, opportunities for rapid advancement, and exceptionally liberal privileges, GE was able to build a stable of outstanding physical chemists: Willis R. Whitney, William D. Coolidge, Irving Langmuir, Colin G. Fink, and Saul Dushman.

Yet even in this Harvard of industrial research, academic and industrial values clashed. As the historian George Wise has skillfully shown, it took years before the research laboratory was able to escape the pressure to show immediate returns on investment; it was only by undertaking certain engineering and production problems that it survived long enough to achieve a secure place in the corporate scheme. And if the company's directors were uncertain about hiring scientists from the ivory tower, scientists were no less uncertain about enlisting in the corporation. Whitney, Coolidge, and Langmuir all had deeply ambivalent feelings about industrial work when first invited to Schenectady; had their academic prospects been brighter, it is unlikely that any of them would have accepted GE's offers.[15]

Once at GE, these scientists discovered that the corporate world was not as bad as they had imagined and in fact offered many advantages over academic life; all developed fierce loyalties to their company. But they also discovered that harnessing their learning to the needs of industry was by no means simple. The ionic theory of electrical conduction that had been so fruitful in the study of solutions proved of little value in treating the conduction of electricity in gases and solids—the phases of matter that were most important in light-bulb manufacture. Like their academic colleagues at MIT, the scientists at GE gradually realized the need for a fuller understanding of the structure and motions of atoms and molecules. A few, Langmuir chief among them, were able to prosecute research that both enlarged scientists' understanding of these topics and use that new understanding to improve technology. But the ability to develop ideas that were important both by disciplinary and corporate criteria was rare.

Empirical work—the trial-and-error testing of ideas and materials—played the dominant role in most of the innovations that came from the GE laboratory during its first two decades. And for the most part, the real value of an academic training often turned out to be the knowledge of techniques and the habits of work that it imparted: the ability to locate information quickly in the scientific literature, to design and use equipment, to plan, execute, and interpret experiments, and to measure systematically and precisely. Formal training in a laboratory science reinforced and enlarged the skills that an earlier generation of inventors had acquired by tinkering with machines and chemicals in their attics and basements. As Coolidge acknowledged when describing his discovery of a commercially practical technique for making tungsten ductile, "we were guided, in the main, by the experiment itself, rather than by metallurgical knowledge."[16] Theory could sometimes be a tool, as Irving Langmuir demonstrated, but his integration of theoretical and practical knowledge, of disciplinary and mission research, was the exception rather than the rule in the early years of industrial research.

The hiring policies of industrial laboratories acknowledged this reality. There was little reason to hire Ph.D.-holders when the skills sought were possessed by students with less training. Where pioneering research on new products and processes was involved, industrial research laboratories often engaged one or two senior scientists with Ph.D.s in the basic sciences and then hired a bevy of junior workers who might possess a baccalaureate or perhaps a year or two of graduate training to assist and serve as extra hands. Where the critical issues involved questions of plant design or scaling-up, firms increasingly turned to workers who possessed training in the basics of both chemistry and mechanical engineering and who specialized in the problem of moving from production by the test tube to production by the tank car, i.e., chemical engineers.

The exponential growth of corporate laboratories suggests that industrial research was a sudden and revolutionary development in American business, and the rhetoric of both scientists and the proponents of industrial research make it seem as if physical chemists would be logical beneficiaries of this movement. In fact, there were many kinds of industrial research, ranging from glorified quality control to Langmuir's Nobel-prize-winning work in surface chemistry, and there were many kinds of industrial researchers. In early-twentieth-century America, relatively few scientists in corporate employ held Ph.D.s, in physical chemistry or in other sciences. Their number was growing, but not nearly as rapidly as the proliferation of industrial research laboratories or panegyrics of industrial research would suggest.

The economic and social dislocations caused by World War I helped to alter this picture by catalyzing changes in both the structure of American industry and in the values of businessmen and scientists. On the eve of the war, the United States possessed an enormously powerful, but unevenly developed, industrial sector. American firms manufactured glass for countless bottles, windows, and inexpensive lenses, but little high-resistance glassware and high-quality optical glass. They produced huge volumes of bulk chemicals like sulfuric acid, alkalies, and chlorine, but were dependent on Europe for most fine chemicals: synthetic dyes, aromatics, and organic reagents. In general, American industry excelled in the mass production of relatively simple commodities, but showed serious weaknesses in lines of business in which products were comparatively complex, sold in small volumes, or made by technologically sophisticated processes.

The Allied blockade of German shipping isolated America from its most important sources of these high-technology goods. At the same time, the war brought a flood of orders to American firms, from both domestic and foreign customers, for products made in small quantities, if at all. Shortages and fears of shortages of essential materials like nitrates, industrial alcohols, and gasoline exacerbated the sense of crisis and stimulated crash programs to develop new sources and to maximize production from old. The war precipitated not only a vast expansion in productive capacity but also furnished America's chemical and chemical-process industries with powerful incentives to develop along new lines.[17]

To catch up in fields in which Germany had traditionally held hegemony demanded the deployment of scientific and engineering talent on a scale unprecedented in American history. Chemical engineering, a profession that was practically unknown in 1890, was among the chief beneficiaries. Possessing some knowledge of both physical chemistry and mechanical engineering, and taking as their focus the distillation columns, filtration beds, and other man-made apparatus of chemical industry, members of this new breed of engineers described themselves as experts in the scaling-up of processes from the laboratory to the plant. Between 1910 and 1920 enrollments in chemical-engineering programs grew by a factor of six, as many American businessmen realized that these were exactly the skills they needed to enlarge production.[18]

But American business needed more than ingenuity in design and in solving problems of scale. On the eve of World War I, chemists and engineers in several countries had begun to develop a cluster of exciting new technologies entailing the use of extreme temperatures and pressures: the production of synthetic oil by the hydrogenation of coal or heavy petroleum hydrocarbons, the synthesis of industrial alcohols from carbon monoxide and hydrogen, the cracking of natural petroleums to enhance the yield of gasoline, the production of industrial gases via the liquefaction of air, and the synthesis of ammonia from its elements. With the exception of petroleum cracking, which was pioneered in the United States, German firms had far more experience with these technologies than their American counterparts. It was through these technologies that Germany obtained the synthetic oil, rubber, and the nitrates that were essential to its war effort. With control of sea lanes, the Allied powers did not have such an urgent reason to find substitutes for traditional sources of rubber, oil, and nitrates. But the U-boat made such control tenuous in 1916 and 1917. Moreover, as supplies of raw materials dwindled and costs rose, there was ample reason to use these new technologies to supplement production by more traditional processes, even in an ostensibly neutral country like the United States.[19]

Reconstruction of the techniques of German firms from such fragmentary sources as patent applications and published descriptions and the adaptation of those processes to American needs and conditions called for a broad spectrum of skills. Chemical engineers were best equipped to handle many of these problems: the development of efficient techniques of heat transfer and the design of high-pressure reaction vessels, for example. Others, however, such as the evaluation of equilibrium constants, were best treated by physical chemists. Not only did they have knowledge of the theoretical principles that governed chemical reactions under extreme temperatures and pressures, they also had laboratory experience with such conditions and with the catalysts that were usually necessary to enhance yields. Such experience was in short supply.

And so, as American industry cranked up production, and as the American government drifted from neutrality toward belligerence, scores of new opportunities arose for physical chemists both in industry and in the government agencies that worked with industry on these technical matters: the Bureau of Mines, the Chemical Warfare Service, and the National Bureau of Standards. And no longer was it merely the castaways from academe who considered such positions. The preparedness campaign made such work seem like a patriotic duty; even physical chemists as firmly committed to academic ideals as A. A. Noyes put aside their chalk to enlist in the war effort.

These scientists-turned-modern-day-armorers did not find the transition easy. They had to discover for themselves the lesson that Willis R. Whitney and William D. Coolidge had earlier learned at General Electric: science is not easily translated into technology. They were unaccustomed to spending large sums of money, to working under deadlines, to thinking in terms of tons rather than grams, and to dealing with others who wanted to know how rather than why. They were also handicapped by some of the very shortages that they sought to remedy: in fine chemicals, glassware, and precision-made industrial products. The first Haber process plant in the United States, a twenty-million-dollar facility in Sheffield, Alabama, proved far more valuable as a source of experience than as a source of fixed nitrogen—not so much because American chemists could not master the principles of the process, but rather because the plant was plagued by leaky gaskets, unreliable compressors, and impurities in materials. It operated sporadically for only a few months in 1918 before being shut down.[20]

Nor did the nation always make the best use of its chemical talent. The seasoned Charles A. Kraus and the young Farrington Daniels spent much of the war struggling to find a solution to the problem of the fogging of the eyeglasses of gas masks—an important problem for any wearer of such apparatus, but one that was more appropriate for engineers than for two outstanding physical chemists. G. N. Lewis was even less fortunate. Invited to join the Chemical Warfare Service with the rank of major, Lewis was promptly shipped to

France, where he eventually became responsible for training gas officers for the field. "The war," he told a friend, was a great disillusionment.

> I had always supposed that G.H.Q. was the place where the commanding General sat with the more intelligent and more experienced officers to plan the major strategy of the war. I thought they would be studying in all detail the ways in which warfare was being modified, and could be modified, through the agency of such new instruments as aviation, tanks, and lethal gases. . . . I thought at least there would be something like one of our research seminars where the war would be discussed in all its phases. I found in the whole Army nothing approaching this ideal. G.H.Q. was filled with young officers recently out of West Point who were considered not quite good enough to command troops, all enmeshed in a great entanglement of red tape from which they emerged occasionally to send me repeated telegrams asking why the divisions were not being supplied with gas masks for pigeons. Since we never had had any gas masks for pigeons, although the item appeared in one of the early lists of war materials, and since, moreover, there were no pigeons, this was a very hard question to answer, especially as my pride as an author required me to give a new answer each time.[21]

Illusions were punctured when the war threw scientists into close contact with military officers and government officials. Scientists were appalled by what they took to be inefficiencies in the army and in the federal bureaucracy more generally. Waste was inevitable given the size of the government before the war and the exigencies of rapid mobilization, but the identification of government with incompetence deflected more than a few political progressives toward the right—a development that helps account for scientists' enduring skepticism about the value of federal support in the postwar era.

The war may have precipitated some doubts about the competence of the federal government, but it also helped dissolve many of the suspicions that had for years divided scientists and businessmen. Mobilization led to the creation of scores of meeting grounds where businessmen, engineers, and scientists met, mixed, and worked together on urgent tasks—in committee rooms in Washington, in the production plants where new processes were installed, and on the building sites of such projects as the government's massive plants for the production of nitrates. Scientists and businessmen discovered that they shared common backgrounds and professional outlooks and a common desire to see more efficient and rational procedures introduced into American government and industry. They also perceived common adversaries: politicians and old-fashioned military men and bureaucrats.

Friendships formed during this experience would have consequences after the war. The 1920s were marked by an easy cooperation between scientists like Robert A. Millikan and George Ellery Hale and "enlightened" men of affairs like Herbert Hoover, Owen D. Young, George Rosenwald, and Gano Dunn.[22] Consequential too would be the scientists' new experience of power and affluence. The war gave many scientists their first taste of work that was deemed of demonstrable and immediate importance to the general community and their first experience in spending money freely. A. A. Noyes, who before the war had managed a laboratory whose budget amounted to a few thousand dollars a year, found himself suddenly advising the War Department on the expenditure of tens of millions of dollars for nitrate plants. Arthur B. Lamb, who had been a professor at Harvard, was made director of defensive chemical research of the Chemical Warfare Service. Richard Chace Tolman, who had moved from MIT to California and then to a professorship at Illinois, was tapped as associate director and then director of the Fixed Nitrogen Research Laboratory—a government facility created just after the war to perfect techniques for nitrogen fixation and to develop new uses for nitrates. Its budget—$300,000 a year—was far more than that of any academic chemistry department.

Like the proverbial doughboys who found it hard to go back to the farm, so some scientists found it hard to return to the routine of classes and the narrow horizons of small-scale academic research after the excitement of wartime Washington. Their understanding of the scientists' role in American society had changed, as had the scope of their ambitions. For some the war stimulated dreams of new laboratories and research budgets unprecedented by prewar standards, for others it awoke hopes of fruitful research in wealthy and enlightened corporations. Most took away an enhanced appreciation for the dignity and difficulty of applied research: the problems of industrial production were often neither as boring nor as simple as they had once imagined.

The war and immediate postwar years affected the attitudes of businessmen toward scientists no less than it affected the attitudes of scientists toward businessmen. For many firms, the decade 1914–1924 had brought a succession of crises: first a desperate scramble to meet demands for new products and to boost production of old during a period of ever higher prices and recurring shortages, then a conversion from wartime to peacetime production, and finally an ongoing struggle to hold on to markets that had fallen to them by default during the war. Sophisticated processes that had been practically unknown in American industry prior to the war had quickly become commonplace. Businessmen, struggling to incorporate exotic new technologies into their production processes, found in academe a little-tapped pool of talent. Scientists might speak an unfamiliar language and have some annoying habits, but they had experience with the behavior of matter under extremes of pressure and temperature and with the substances and machines of the laboratory. Moreover, as businessmen acquired more experience with science and scientists, they discovered that a knowledge of theory was not necessarily antithetic to inventive skill. Of course, the problems of building a cracking still or hydrogenation chamber could not be solved by a touch of the magic wand of thermodynamics, specialists in reaction kinetics could not accurately forecast the rate and yield of reactions under industrial conditions, and students of catalysis were woefully equipped to predict the effectiveness and durability of the substances used to promote chemical reactions. Nevertheless, a sound understanding of physico-chemical theory could truncate the process of trial and error experimentation. As businessmen worked with scientists to achieve common goals their attitudes softened; a grudging respect and sometimes even an enthusiasm for the scientist and his learning displaced contempt.

The outcome is evident in the statistics. The number and size of industrial research laboratories expanded during and after the war, and the role of Ph.D.s in industrial research grew dramatically. Whereas industry employed fewer than one in ten productive physical chemists on the eve of World War I, during the postwar decade industrial employment grew to close to one in five. Many of those who entered industrial employment were giving up promising careers in universities to do so (see Table 5.2). Not only were more good chemists enlisting in business, but also more of their academic colleagues were doing consulting work than ever before.

During times of peak business demand for physical chemists, those who remained wed to purely academic ways voiced fears about the future. Shocked to learn that his friend John Johnston, the same chemist who had sniffed warily at industrial work before the war, had left a professorship at Yale to become research director at U.S. Steel, G. N. Lewis warned that American industry was consuming its own seed grain. "The industries are gradually draining the

TABLE 5.2

Movement of Productive Physical Chemists from Academic to Industrial Employment, 1915–1925

Chemist	Academic Position	Industrial Position	Year
C. W. Bennett	asst. prof., Cornell	Douglaston Labs	1915
Francis Frary	asst. prof., Minnesota	Oldbury Electrochemical/Alcoa	1915
W. W. Holland	assoc. prof., Johns Hopkins	Standard Oil/Indiana	1916
W. A. Taylor	instr., Wisconsin	du Pont	1916
E. P. Wightman	prof., Richmond	Parke Davis/Kodak	1917
H. N. McCoy	prof., Chicago	Lindsay Light Co.	1917
N. E. Loomis	asst. prof., Purdue	Standard Oil Indiana/Standard Oil N.J.	1918
Marks Neidle	asst. prof., Pittsburgh	Sterling Varnish Co.	1918
F. S. Mortimer	assoc. prof., Iowa	National Analine & Dye	1920
Harry A. Curtis	prof., Northwestern	International Coal	1920
E. B. Spear	assoc. prof., MIT	Goodyear	1920
Graham Edgar	prof., Virginia	Ethyl Gasoline Corp.	1924
A.F.O. Germann	asst. prof., Stanford	Lab Products Co.	1925
C. O. Henke	ast. prof., Indiana	Newport Co.	1925

Source: Survey of authors of five or more articles on physico-chemical topics; *American Men of Science*.

universities of many of their best scientists,'' wrote Lewis to the Dean of Yale's Sheffield Scientific School in 1927;

> They not only offer better salaries, but in many cases they even promise a greater freedom to the individual. Universities still think that they are above the universal law of supply and demand, and instead of meeting the industries in battle they retreat timidly. . . . every few months I am horrified by the calamity that is overtaking the scientific departments of American universities which will ultimately affect most deleteriously the industries themselves. . . .[23]

While Lewis worried about the long-term fate of American universities and industries, his fellow physical chemist at MIT, F. G. Keyes, worried about the consequences of this boom for his department and the young chemists in it. Unbridled consulting, he told the president of MIT,

> is bound to become more and more pernicious in its effects on the quality of the Institute staff, particularly those who should devote extra time to scientific research. . . . At the present time, one can easily observe that most of the younger members of the Institute staff are dissipating the most formative portion of their lives in working out in their spare time various small applied science problems for pecuniary reward, instead of devoting themselves to the pursuits of pure science. . . .[24]

Fears that a strong industrial demand would spell disaster for science in America were surely exaggerated. The very best industrial laboratories were capable of supporting important work on fundamental topics, and an Irving Langmuir was probably as productive at General Electric as he would have been at a good university. In the far more numerous laboratories where such opportunities were absent, high salaries proved a weak incentive for imaginative chemists. After two years at a print works in Cranston, Rhode Island, D. McIntosh, a Canadian physical chemist, developed such an abhorrence for cotton cloth that he was ready "to go to one of the South Sea islands, where the natives do not even wear a rag around their loins."[25] Most disgruntled industrial scientists settled for a job in a government agency or university.

More important, the presence of scientists like Lewis, Keyes, and A. A. Noyes in influential positions meant that the ideals of disinterested research were hardly without advocates, even in the business culture of the 1920s. They and their colleagues in other fields helped see to it that the needs of academic science received attention from university presidents, philanthropists, and foundation officials.

And they did receive attention. American colleges and universities benefited from the prosperity of the 1920s no less than did business firms, and as their general incomes grew so too did funds for salaries and research. Despite serious erosion during the inflation following World War I, academic salaries and benefits rebounded quickly and later managed to stay within hailing distance of those in industry.[26] Fellowship programs, such as those established by the Rockefeller Foundation in 1919 and the Guggenheim Foundation in 1925, also helped keep the best minds in academe. Consider the case of Worth Rodebush, a student of G. N. Lewis at Berkeley who had drifted from a wartime position with the Bureau of Mines into a job with the U.S. Industrial Alcohol Company—a major producer of alcohol and such chemicals as acetone and acetic acid. It was late in 1918, the universities were in turmoil, the purchasing power of an instructor's salary was steadily dwindling, and it was still both patriotic and manly to put one's brains and, if need be, shoulders into the task of production. Rodebush was attracted by the high salary, although he found it "worth every dollar of it to live in Baltimore," where he was stationed. But, as he told Lewis, "Now that I am in the game . . . I must say that I find research in large scale production very interesting. I want to stay with it until I learn as much applied chemistry as possible and apply some myself if I get the opportunity."[27] In Lewis's eyes, however, industry was no place for an intelligent and creative physical chemist:

> I have been a little worried ever since you entered industrial work at the idea that you might continue indefinitely in that kind of thing, although I feel sure that you will ultimately be better satisfied with academic life and pure research, providing that conditions in the universities are made at least tolerable.[28]

To make those conditions more tolerable for Rodebush, Lewis helped him to win one of the first Rockefeller-funded fellowships awarded by the National Research Council. Rodebush found the liberal stipend and the opportunity to pursue research unfettered by other responsibilities irresistible, and he promptly resigned his position in Baltimore. Few physical chemists would turn down such a prestigious and exciting opportunity for a job in industry, no matter the difference in income. And most who accepted such national fellowships remained in academic life. Rodebush used the opportunity to complete research that would earn him a full professorship at Illinois in 1924.

Despite the growth of industrial positions for physical chemists during and after World War I, and the increasing number of physical chemists who availed themselves of those opportunities, physical chemistry remained an essentially academic field. The bulk of the papers published each year came from university laboratories; textbooks cast hardly a glance at industrial applications; and the very best young physical chemists continued to show a preference for the professorial life. Even in a society distorted by war, wholesale inversions of habits and values are exceptional. But amidst the continuities, there was change. The war did accelerate technological and attitudinal changes under way since the late nineteenth century. New production technologies demanded highly specialized expertise; greater contact between businessmen and scientists, while not obliterating all distinctions between their values, did erase some of their worst prejudices about one another. Industry began vigorous efforts to recruit young doctors of science like Rodebush, and when scientists like Rodebush chose between academe and industry it was not so much from ignorance and suspicion as from a rational weighing of the assets and liabilities of both kinds of careers. While not losing all traces of independence, America's academic and industrial cultures had moved far toward integration.

For physical chemists this integration brought the direct benefit of a new range of jobs in industrial research. But it also brought many indirect benefits. For every Ph.D.-holder in physical chemistry employed by industry, several chemists or chemical engineers were hired who, while lacking graduate training in physical chemistry, had taken one, two, or more courses in the subject. These anonymous scientists and engineers were the rank and file of industrial laboratories. In their earlier incarnations as students, they constituted an important reason for the improvement of academic conditions for physical chemists during the early twentieth century. Universities and engineering schools, hard put to meet the demand for chemical training, devoted large new resources to their departments of chemistry and chemical engineering. And some of these resources found their way to physical chemists, both because they had made themselves so valuable in the teaching of elementary courses and because they possessed knowledge and skills for which others were discovering new and practical uses.

Chemical engineers were especially important here. Physical chemistry found a place in the curriculum in chemical engineering shortly after chemical engineering began to emerge as a distinct field of study. By 1902, on the urging of Arthur A. Noyes, a course in theoretical chemistry was made a requirement for students of chemical engineering at MIT. Other schools gradually followed suit. By 1920, a semester of physical chemistry was nearly a universal requirement for students of chemical engineering, and textbooks, such as Walker, Lewis, and McAdams's influential *Principles of Chemical Engineering* (1923), spoke as freely about equilibrium constants as they did about distillation columns and evaporators. Come the 1930s, chemical engineering bore much the same relation to physical chemistry that electrical engineering bore to the physics of electricity and magnetism: the field was suffused with the language and concepts of the basic science. Their service role in the education of chemical engineers gave physical chemists a captive audience and a claim on resources wherever chemical engineering was a popular course of study.[29]

Shortly after the end of World War I, a number of America's prominent scientists, A. A. Noyes among them, collaborated to produce a book entitled *The New World of Science*. Part history and part propaganda, the authors celebrated, and often embroidered upon, their contributions to the Allied victory. Yet despite the self-serving exaggerations, these scientists perceived an essential historical reality of the war: it had greatly multiplied and strengthened the filaments connecting American science to American society. Esoteric expertise, such as that possessed by physical chemists, had become relevant, not simply to the fellow cognoscenti, but to businessmen and engineers in a wide variety of industries. Industrial work, once the last resort of impecunious scientists, had become a respectable career, even for those who could entertain other options. And whether or not scientists accepted industrial employment, they lived and worked under conditions that the broad new industrial demand for their services had helped create. World War I had accelerated the processes leading to a convergence of America's academic and industrial cultures; it had, as the title of the scientists' book suggested, ushered in a new world of science.[30]

THE INTEGRATION OF PHYSICAL CHEMISTRY AND OTHER SCIENCES

The evolution of new ties with industry was one avenue for the growth of physical chemistry in the early twentieth century. But it was not the only direction in which expansion occurred. For while physical chemistry was evolving, so too were other sciences. As physical chemists in America were securing niches for themselves in universities and business firms and building laboratories and journals, other specialists, no less committed to their fields, were doing much the same. Growth in close proximity often led to a convergence of concerns and techniques. In many cases, practitioners of other specialties discovered in physical chemistry tools that were critically important to their own work. In some cases, entire fields became so suffused with physical chemistry that disciplinary boundaries became difficult to discern.

These ties between physical chemistry and other sciences, their origins and elaboration, are well worth study. The long-term prospects of a discipline are dependent not only on the skill with which its members till their own field, but also on the contributions that they are able to make to those who cultivate others.[31] This is how constituencies are built at colleges and universities and how intellectual capital is amassed. With disciplines, as with nations, commerce is essential to prosperity.

The early twentieth century presents several examples of this interpenetration of physical chemistry and other disciplines. We have already noted the growing role of physical chemistry in metallurgy and chemical engineering, and in a future chapter we shall discuss physico-chemical approaches to biology. The size, origins, and character of these fields, however, make the task of reconstructing their commerce with physical chemistry exceedingly difficult. Biochemistry, linked as it was to medicine, underwent an expansion that was probably as rapid as that of physical chemistry itself during this period. It was, as the historian Robert Kohler has skillfully shown, internally fractured into diverse and competing schools of thought. While some had strong affinities for physical chemistry, others were more closely related to organic chemistry or physiology. Chemical engineering likewise was a discipline that sprouted quickly from many roots: physical chemistry, but also industrial chemistry, electrochemistry, mining and metallurgy, mechanical engineering, and even electrical engineering. To disinter one of those roots entails the excavation of many others. Metallurgy was smaller and less promiscuous in its relations with other sciences than biochemistry or chemical engineering; with a few notable exceptions, however, Americans did not much cultivate metallurgical theory until the 1920s. The strongly practical orientation of this discipline in the United States meant that the earliest and most consequential interactions with physical chemistry generally developed in Europe.[32]

Nevertheless, it is important to examine in some detail the ways and means by which fruitful exchange developed between physical chemistry and its neighbors, and there is one science whose history reveals a relatively clear and clean example of this process: petrology. Unlike metallurgy, petrology was a field in which a significant number of Americans attained prominence in the early twentieth century. The volume of petrological work done in America was sufficient for trends to become evident, and those trends were similar to, roughly contemporaneous with, and yet not entirely derivative of those in Europe. Hence, by focusing on the evolution of relations between petrology and physical chemistry in the United States we are not examining developments that were atypical or notably backward. And yet, because petrology was a smaller discipline than biochemistry or chemical engineering, it is not forbiddingly difficult either to trace some of the paths by which ideas and techniques moved from one discipline to the other or to assess their impact.

With the benefit of hindsight, it seems obvious that physical chemistry would find fruitful uses in the study of rocks. As early as 1851, Robert Bunsen had observed that magmas were solutions and might be treated by the same principles that governed the behavior of salts in water. Minerals, for example, did not simply crystallize out of magmas in the reverse order of their fusibilities, but seemed to show complex solubility patterns resembling those of aqueous solutions. In 1857 the French petrographer, J. Durocher, drew another analogy when he noted similarities between magmas and alloys. If one could understand the principles governing the solidification of metallic melts, which could yield alloys with different physical properties depending on the rate and conditions of cooling, might not that understanding be extended to rock melts? Both Bunsen and Durocher were seeking to comprehend the inaccessible and complex in terms of the familiar and simpler. Their suggestions, however, bore no direct fruit, for although aqueous solutions and alloys might be simpler than magmas, they were little understood by chemists.[33]

The remarkable series of conceptual developments of the 1880s associated with the genesis of physical chemistry seemed to afford the understanding that Bunsen and Durocher lacked. Insofar as physical chemists had revolutionized the chemists' comprehension of solutions and heterogeneous equilibria, they had also transformed the basis of the geologists' understanding of the chemistry of rock formation and alteration. The Ionists themselves were among the first to appreciate the geological significance of their ideas. Both van't Hoff and Arrhenius, for instance, turned toward geological issues in the 1890s. Such applications were not only intellectually challenging, they also helped legitimize the Ionists' ideas and strengthen the still fragile institutional supports of their discipline.[34]

But a transmission of ideas depended as much on the receptivity of geologists as on the assertiveness of physical chemists, and geologists generally were ill-prepared to follow—never mind make use of—recent developments in physical chemistry. And despite the advances of physical chemistry, there was still a great disparity between the intellectual resources of physical chemists and the needs of geologists. Magmas were more complex than dilute aqueous solutions or alloys and far less accessible to study. Whereas the chemist or metallurgist could check his conclusions against the behavior of solutions or alloys in the laboratory, the geologist did not possess facilities to duplicate the natural conditions of rock formation.

The transmission of ideas and methods from physical chemistry to geology was, therefore, a slow and halting process, and the path and pace of the transmission varied a great deal from country to country. The geologists of France and England were rather slow to adopt a physico-chemical approach. But geologists in Germany and in countries on the orb of German scholarship were far more receptive. In Germany, geologists collaborated with van't Hoff in his studies of the Stassfurt salt beds; in Norway, J.H.L. Vogt undertook experiments on the thermal behavior and composition of slags; in Russia, both A. Lagorio and F. Y. Loewinson-Lessing drew on physical chemistry to account for the processes of differentiation of igneous rocks; in Austria, C.A.S. Doelter established a laboratory for the experimental study of the behavior of silicate minerals at high temperatures.[35]

Three factors appear to have been crucial in leading to this pattern: the presence of a vigorous group of petrographers interested in the problems of the classification and genesis of rocks; the existence of a community of chemists acquainted with recent developments in solution theory and chemical thermodynamics; and the existence or creation of institutions within which geologists and chemists might mix and collaborate. Where these conditions were present, research on the borders between geology and physical chemistry flourished; where one or another of these conditions was absent, such work was slow to develop.

The importance of these conditions is well illustrated by the American case. By the mid-1890s, a large and growing cadre of petrographers was active in America. Some had learned their science at home, but at the core of this group were scientists who had acquired in Germany knowledge of the structure, texture, and composition of rocks. Beginning in the 1870s, a stream of American students had gone to Heidelberg, Leipzig, and other German universities to study under Ferdinand Zirkel, Harry Rosenbusch, and other German masters of petrographic techniques. Although most went in order to learn the intricacies of rock classification and microscopic analysis, many took advantage of their stay in Germany to attend lectures on chemistry as well.[36]

After returning to the United States, many joined the U.S. Geological Survey, where their primary duty was to identify and map districts with economically interesting minerals. But as long as the Survey was young, growing, and well financed, they also had freedom to study broader questions. The eruptive districts of the American west, for instance, were excellent sites for study of the origins and differentiation of igneous rocks, and the ancient rocks of the Lake Superior region offered tempting materials for work on processes of metamorphism and ore formation. By the early 1890s several American petrographers had begun to make good use of these opportunities.

These young petrographers had much in common with the even younger physical chemists who were beginning to trickle back from Germany in the 1890s. They had studied in the same institutions, they had both faced the challenge of adapting themselves to German scientific culture, and now both were confronted with the task of readapting to American institutions. Both petrographers and physical chemists saw themselves as practitioners of new methods in an old science; both were engaged in the dual tasks of reforming old institutions and creating new. But petrographers and physical chemists were united by more than simply these similarities of experience and situation. They also gradually discovered that they shared intellectual concerns, that many of the problems of petrogenesis and metamorphism were amenable to physico-chemical treatment.

Petrographers, for example, were concerned to understand the mechanisms by which a single body of cooling magma, such as might be found in a laccolith or dike, could give rise to several types of rocks or mineral assemblages. Geologists more generally were concerned to understand the relationship between magmas. It was clear that the chemical composition of magma varied greatly from one region to another and even within regions; the nature of the magma produced by a single volcano, for example, could change drastically over time. Did these differences denote the existence of many chemically distinct reservoirs of magma below the surface? Or, was it possible that all magmas were derived from a single homogeneous substance? If so, exactly how did that unity give rise to diversity? Was it the result of a differentiation within the liquid itself, or was it the result of the absorption of country rock into the magma, or was it a by-product of the cooling process—a fractional crystallization?[37]

Questions such as these were just being formulated in the 1880s and 1890s; traditional field evidence could offer little guidance to their solution. Nor could physical chemists instantly produce answers to such questions. It took no less effort and ingenuity to make physical chemistry relevant to petrological problems than to industrial. Nevertheless, those acquainted with both the geological issues and recent developments in physical chemistry were at least in a position to recognize some of the complexities of these petrological problems. And they had access to analogies that might guide their research. So, for example, just as the solubility of a salt in water is affected by the presence of other salts, so the fusibility of a mineral might be affected by the presence of other minerals. Just as pressure will alter the solubility of salts in water, so too it might alter the solubility of minerals in magmas. And just as slow cooling can result in supersaturated aqueous solutions, so too it might result in magmas supersaturated with one or more components. Hence, the order in which minerals crystallized out of a magma could be affected by pressure, by the composition of the magma, and by the rate of cooling. Physical chemists were developing a sophisticated understanding of how these variables affected equilibrium. Even though their knowledge was based principally on the study of aqueous solutions in laboratory vessels, it took only some imagination to see that the principles that governed aqueous solutions in laboratory vessels might also extend to inaccessible magmas in the earth.

Shortly after 1890, liaisons began to be established between these two specialties. The first American contacts were not made in emulation of developments abroad; rather they were independent of, and roughly simultaneous with, similar contacts then being established in Norway, Austria, and Russia.

Two petrographers, Joseph Paxson Iddings and Charles R. Van Hise, were the most influential of those who established these contacts in America. They were not extremely original scientists, nor were they at ease in the chemical laboratory. Rather they were synthesizers who cast their nets wide. Initially, they were interested more in broad and qualitative theory than exact measurement, but gradually both came to believe that cooperation between physical chemists and geologists would yield answers to the perplexing problems of rock formation and metamorphism.

Iddings appears to have been the first American geologist to draw attention to the geological implications of the Ionists' work. He was one of the many American petrographers who had studied under Rosenbusch and who had then joined the young U.S. Geological Survey. During the 1880s Iddings spent his summers in the field, amidst the volcanic rocks of Yellowstone Park, and his winters in Washington, where he struck up a friendship with Frank W. Clarke, the chief chemist of the Survey and a man with wide knowledge of the new physical chemistry. Both of these experiences proved important for Iddings.[38]

His field work gave Iddings experience with eruptive rocks that exhibited a fairly smooth and continuous gradation of mineral and chemical composition. Mineralogical and chemical analyses of samples from these beds, together with reports of similar deposits elsewhere, led Iddings to suggest that the igneous rocks of any region are so intimately related that they must have originated in a common source—some single magma that could give rise to various kinds of igneous rocks through a process of differentiation. Several European petrographers had already noted textural and structural evidence for the existence of discrete petrographic provinces. Iddings's contribution was to stress the chemical relatedness of rocks in these provinces, their chemical consanguinity.[39]

Iddings's work led directly to two questions: In what form did siliceous substances exist in molten magmas? And by what process could a wide variety of distinct but chemically related rocks be derived from a single, homogeneous source? It was here that physical chemistry became important. In his first paper on the crystallization of igneous rocks, written in 1889, Iddings had been unable to go beyond his predecessors in treating these issues. He considered magmas to be saturated solutions of silicate molecules from which mineral species crystallized as temperature and pressure conditions changed. As to the exact nature of the compounds in solution and the details of differentiation, he confessed ignorance.[40]

In his next paper on the subject, written in 1892, Iddings adopted a far more assertive tone. Whereas he had earlier been without suggestions as to how to proceed, he now advanced a new hypothesis regarding the condition of substances in magmas and posited a mechanism for differentiation. In large part this new confidence derived from his discovery of the papers of van't Hoff and Arrhenius. His attention was drawn to Arrhenius's work, he tells his reader, through the suggestion of his colleague at the Survey, Clarke.[41]

Iddings had already concluded that chemically homogeneous magma could produce very different minerals depending upon the conditions of cooling. Analyses of rocks drawn from two locations in the Yellowstone region had shown that although these samples were composed of the same elements in the same proportions they differed markedly in mineral composition. Iddings had even tentatively suggested that minerals might not retain their molecular integrity in the molten state, but might decompose into simpler units which could shift about independently of one another and enter into several associations, depending upon the physical conditions prevailing during crystallization.[42] Arrhenius's theory of electrolytic dissociation provided Iddings with a powerful support for this idea by suggesting to him that magmas might belong to that large class of solutions in which molecules of the solute dissociate into ions.[43]

Arrhenius's work also suggested to Iddings a method of testing this hypothesis. If the analogy between magmas and aqueous salt solutions was valid, then magmas should conduct an electrical current. In experiments on the electrical resistance of fused rock magmas, conducted with Carl Barus, a Survey physicist who had studied under Friedrich Kohlrausch, Iddings found persuasive evidence for his conclusion. Not only did molten rocks conduct a current, but the changes in conductivity seemed to vary with temperature and ionic concentration in ways markedly similar to aqueous solutions. "Looking at our results as a whole," the authors concluded,

we find them trenching in a novel way on the solution theories of Arrhenius, Ostwald, and van't Hoff. It is difficult to withhold one's assent from the proposition, that the ions of a molten magma are largely present in the dissociated state. . . . To the extent of our enquiry, the behavior of molten rock magmas is in its nature quite identical with that of any aqueous or other solution, the difference being one of solvent.[44]

Although Iddings's belief in the exactness of the analogy between aqueous solutions and magmas was based on only a few laboratory experiments, his work was an important advance both because it put new content in Bunsen's old suggestion that magmas were solutions and because it went some way toward explaining how different mineral assemblages could emerge from one magma. A magma was not an association of definite mineral species; instead of being composed of molecules, a magma was composed largely of ions, and these could combine in many different ways depending upon the physical conditions under which cooling proceeded.

As to the mechanism of differentiation, Iddings once again drew on the work of a physical chemist—in this case, van't Hoff. Van't Hoff's theory of osmotic pressure provided a theoretical basis for an empirical principle discovered in the early 1880s by the physicist C. Soret: in a solution in which there is a temperature gradient, molecules of the solute will tend to concentrate in the cooler portion. In a cooling body of magma contained within an irregularly shaped cavity within the earth's crust, significant variations of temperature are to be expected. Such variations, such as between the walls of a dike and its center, Iddings suggested, would suffice to cause a diffusion and differential sorting of constituents.[45]

Iddings could point to no laboratory evidence to support this contention, and later workers would find many flaws in his ideas. In particular, the time necessary for the Soret effect to cause a significant diffusion of materials in a large body of viscous magma was found to be enormous, making it unlikely that it plays as important a role in differentiation as Iddings believed.[46] But he was conscious of both the tentativeness of his reasoning and the need for experimental studies of petrogenesis. He closed his paper of 1892 with an admonishment:

The complexities of a compound solution that exists only at extremely high temperatures and experiences the pressures to which rock magmas have undoubtedly been subjected may long remain beyond the reach of direct investigation. Still the steady advance of experimental physics offers great possibilities in this direction. Until the establishment of definite knowledge concerning the nature of molten magmas we must proceed along the lines of analogy by applying to them such laws as may be found applicable to solutions that exist at lower temperatures and pressures.[47]

One clear implication of Iddings's papers was that geologists should pay heed to developments in the borderland between chemistry and physics. They should make use of the analogies physical chemistry had to offer and seek to foster research that eventually might obviate the need to rely on analogy—that is, research on molten silicates themselves. Not long after these papers were written, Iddings's colleague, Charles R. Van Hise, came to the same conclusions by a somewhat different route.

Although Van Hise was an exact contemporary of Iddings, his education was, in most respects, inferior to that of his colleague. Van Hise did not study in Germany; his formal training was as a mechanical engineer at the University of Wisconsin. He learned his geology in the field, first as an assistant in the state survey of Wisconsin, where he became acquainted with the use of the microscope, and later as a geologist and division chief in the U.S. Geological Survey. Although Van Hise taught at his alma mater throughout the 1880s and 1890s, and later became its president, his professional life was oriented toward the Survey, where his primary responsibility was to study the iron and copper districts of the Lake Superior region.[48] A lesser intellect might have been satisfied to prepare useful but essentially descriptive reports on the topography and minerals of the area, but Van Hise had unusual discipline and tenacity. His descriptive work led him, during the late 1890s, to study the causes of ore deposition and the larger subject of metamorphism. Frustrated by his inadequate preparation in the physical sciences, Van Hise embarked on an ambitious campaign to educate himself in the principles of physics and chemistry. "During the past five years," he wrote in 1902,

> in order to handle the problems of geology before me, I have spent more time in trying to remedy my defective knowledge of physics and chemistry and in comprehending advances in these sciences . . . than I have spent upon current papers in geology; and with, I believe, much more profit to my work.[49]

During the course of this ambitious program of intellectual reconstruction, Van Hise discovered the writings of the Ionists and was so taken by them that he hired a young physical chemist, Azariah T. Lincoln, to tutor him in the new science. The fruits of his labor appeared in a series of papers on ore deposition and metamorphic processes and in his massive *Treatise on Metamorphism*, an encyclopedic work that some of his colleagues took to be the last word on the subject.[50]

Van Hise's treatment of metamorphism was based on the field observation that young rocks are often marked by numerous fissures, faults, and joints, whereas older rocks, uncovered by surface erosion, show many signs of folding and flexure but few of fracture. He ascribed the difference to the physical conditions under which deformation occurred. Young rocks were deformed near the surface in what Van Hise called a zone of fracture. Here stress might result in rupture, but incumbent pressure was insufficient to cause fissures, once opened, to be closed. Ancient rocks were deformed beneath this zone, in a second region that Van Hise called the zone of flow. Here pressure was greater than the strength of any rock, and hence fissures were impossible. Deformation in this lower region resulted from the plastic flow of rocks.

Although Van Hise attributed the gross deformation of rocks to physical causes, he believed that their alteration was a chemical or physico-chemical problem. Thus he maintained that a different set of chemical reactions characterized the alteration of rocks in each zone. In the zone of fracture, reactions typically occurred with the expansion of volume and the liberation of heat: oxidation, carbonation, and hydration. In the lower zone, these reactions were reversed, as pressure rather than temperature became the factor controlling chemical change. In both zones, Van Hise insisted, the alteration of rocks took place chiefly through the agency of water and the mineralizers that water carried in solution. In the zone of fracture, meteoric waters circulated through fissures, dissolving and depositing metals and other materials as a result of variations in solubility arising from changes in temperature and pressure. In the zone of rock flow, minute quantities of trapped water acted to maintain rocks in a plastic state through the continuous solution and deposition of rock material.[51]

The names of van't Hoff, Arrhenius, Ostwald, and Nernst dot the pages of Van Hise's work. His understanding of the effects of temperature and pressure on chemical reactions and of the roles of water and ionic equilibria in metamorphic processes was derived largely from his reading of the works of these physical chemists. "The working out of the principles of metamorphism," he later wrote,

> was a physico-chemical problem. The handling of the problem of rock alteration with fairly satisfactory results was possible because of the rise of physical chemistry. Had this science not developed within the past score of years, it would not have been possible to have gone far upon the problem. . . .[52]

Van Hise believed that he had sketched out the major features of a general theory of metamorphism, but he recognized that many of the details of rock alteration were as yet uncertain and recognized the existence of a vast field for research in the study of the formation of igneous rocks. Greater clarity and understanding, he thought, would require experimental studies by scientists with a surer knowledge of physical chemistry than he had been able to acquire.[53]

Despite differences in education and experience, Van Hise and Iddings both came to the view that a closer collaboration between physical chemists and geologists would be necessary, and both perceived advantages in conducting such collaborative work in the laboratory. But formidable obstacles confronted such an undertaking. Facilities did not exist for approximating and controlling the conditions under which rocks were formed.[54] Advances in the technology of producing and measuring high temperatures and pressures suggested that such conditions might be realized, but a laboratory equipped to study molten silicate solutions would be an expensive affair, requiring air and water compressors, high-pressure bombs, gas and electrical furnaces, and prodigious quantities of platinum for reaction vessels. A fairly large staff with a variety of special skills would be needed to design and supervise delicate experiments that might run for hours, days, or even weeks. And the entire enterprise would be directed toward assembling data with no demonstrable economic benefit. Despite retrenchments in the 1890s, the Geological Survey was the one plausible site for such a project, and it was through the Survey that Van Hise and Iddings chose to work.

The opportunity came in 1900, when the director of the Survey, Charles D. Walcott, appointed Van Hise chairman of a committee to study relations between the geological and chemical divisions of the agency.[55] The Survey had sponsored a limited program of geophysical investigations during the 1880s and early 1890s, of which the work of Barus and Iddings was representative. During a budget crisis in 1892, however, the laboratory in which this work had been done was abolished and chemical work restricted to routine analyses.[56] By the turn of the century, chemists in the Survey, together with laboratory-minded geologists like Van Hise, had grown restive with this role. The report of the Van Hise committee, not surprisingly, recommended that chemical research in the Survey be greatly expanded and that time and equipment be made available for research on problems such as had caught the attention of Van Hise and Iddings.[57]

Later that year, a laboratory of chemistry and physics was organized within the Survey under the direction of George F. Becker, a scientist with strong credentials in chemistry, physics, and geology. It became the kernel from which a geochemical tradition grew in America. Becker drew together an impressive staff whose members had backgrounds in several disciplines: the analytical chemists F. W. Clarke and W. F. Hillebrand; Arthur L. Day, a Yale-trained physicist who had worked on high-temperature thermometry at the Physikalisch Technische Reichsanstalt; and E. T. Allen, a young chemist from Johns Hopkins who had a strong interest in the new physical chemistry. They would later be joined by F. E. Wright, a petrographer trained at Rosenbusch's institute; J. K. Clement, a physical chemist who had worked under Nernst at Göttingen; and E. S. Shepherd, Bancroft's former student and research assistant.[58]

The first assignment of Day, who had responsibility for organizing physical research within Becker's laboratory, was to determine the melting points of rock-forming minerals. His progress, however, was slow, since he had little equipment and a small budget. Long delays in obtaining apparatus left Day with much free time, and he used it to read extensively.[59] Not surprisingly, given that he was surrounded by chemists and petrographers, Day turned toward the literature on petrology and physical chemistry. This reading, which included van't Hoff's work on solid solutions, J.H.L. Vogt's work on the crystallization of igneous rocks, and H.W.B. Roozeboom's papers on the application of the phase rule to steel and alloys, left Day convinced that the future in petrology would belong to physical chemists. His enthusiasm was that of a convert: "an extraordinarily effective weapon has come into the service of petrology," he later wrote,

> the full power of which cannot at once be understood or appreciated. We refer to the methods and established generalizations of physical chemistry. The older science of chemistry has made such strides under these new theories of solutions that we really have little more to do than to apply them ready-made to our own problems, like a smooth and powerful machine tool of guaranteed effectiveness.[60]

Under Day's leadership, the multidisciplinary staff developed an ambitious interdisciplinary program for the experimental study of rock formation. "Our plan," Day and Allen wrote,

> was to study the thermal behavior of some of the simple rock-making minerals by a trust-worthy method, then the conditions of equilibrium for simple combinations of these, and thus to reach a sound basis for the study of rock formation or differentiation from magma. Eventually, when we are able to vary the pressure with the temperature over considerable ranges, our knowledge of the rock-forming minerals should become sufficient to enable us to classify many of the earth-making processes in their proper place with the quantitative physico-chemical reactions of the laboratory.[61]

Part of this program was realized at the Survey. Between 1901 and 1907 the chemist Allen, the physicist Day, and the petrographer Iddings completed a rigorous study of the thermal properties of the plagioclase feldspars. These feldspars, composed in varying proportions of albite ($NaSi_3O_8$) and anorthite ($CaAl_2Si_2O_8$), were the most abundant of the rock-forming minerals; but their nature was a matter of controversy. Some believed that albite and anorthite were capable of forming a series of true chemical compounds; others believed that they formed eutectic mixtures, that is, that each substance lowered the melting point of the other until a minimum temperature was reached at which both substances crystallized in constant proportions. Despite marked differences between the chemical composition of albite and anorthite, others thought them capable of forming an isomorphous mixture in which the physical properties changed continuously with composition.[62] Debate over these issues, however, took place in the absence of reliable knowledge of the physical properties of the plagioclases. Day and Allen sought to remedy this deficiency.

Using artificial minerals to guarantee purity and thermal methods of measuring melting points to avoid the ambiguities inherent in optical determinations, Day and Allen determined the melting points of a series of plagioclases of differing composition.[63] They then interpreted their data using the resources of physical chemistry. Their diagram relating the compositions and melting points of plagioclases, reproduced below (Figure 5.1), revealed no evidence of the formation of new components (compounds of definite proportions between albite and anorthite). Nor did it show evidence of a eutectic point, a mixture with a melting point lower than that of either of the pure components. Rather, the melting point of albite was steadily raised, and that of anorthite was steadily lowered, by the addition of the other substance, although the melting points did not fall on a straight line. After trying, unsuccessfully, to interpret these findings in terms of van't Hoff's laws of perfect solutions, Day and Allen turned to Roozeboom's work on heterogeneous equilibria. Here they discovered that the Dutchman had already described a general category of systems into which the plagioclases fit: that in which the components were miscible in all proportions and in which the melting points of mixtures lie on a continuous curve joining the melting points of the pure components without maximum or minimum. Called Type I systems by Roozeboom, such isomorphous mixtures, he had hypothesized, would have an equilibrium diagram of the general form described in Figure 5.2. The important point to note in this diagram is the gap between the liquidus and solidus curves, which denotes the range of temperatures across which solid and liquid phases can remain in equi-

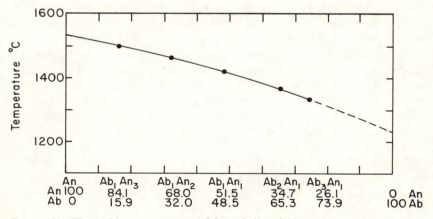

FIGURE 5.1. The melting temperatures of the soda lime feldspars according to Day and Allen. Figure redrawn from Arthur L. Day and E. T. Allen, *Isomorphism and the Thermal Properties of the Feldspars* (Washington, D.C., 1905), p. 60.

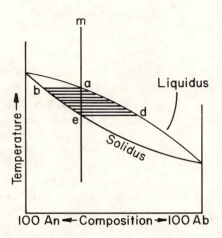

FIGURE 5.2. Equilibrium diagram of soda lime feldspars. Redrawn from Arthur L. Day and E. T. Allen, *Isomorphism and the Thermal Properties of the Feldspars* (Washington, D.C., 1905), p. 68.

librium. There is, in other words, no single melting point for such mixtures: a cooling melt will begin to solidify at one temperature and become entirely solid at another. Day and Allen, who had already noted the extreme difficulty involved in specifying from thermal data a single melting point for their mixtures, were satisfied that their convex melting-point curve represented the upper or liquidus line in an equilibrium diagram of the form in Figure 5.2.[64]

This conclusion had both petrological and petrographical implications. It threw light on both the process by which differentiation could occur in large masses of molten plagioclase and the process by which crystals of plagioclase could develop the zonal patterns that were commonly visible under a microscope. Starting, for example, with a mass of plagioclase feldspar of composition *m*, crystals will begin to form at temperature *a*. The first-forming crystals will have composition *b*, that is, they will be rich in anorthite, and the liquid will have composition *a*. As the temperature falls, the composition of both the crystals and liquid will change, following curves *be* and *ad* respectively, as both grow richer in albite. When the temperature reaches *e* the solidification will be complete.

Day and Allen realized that this process could yield a solid either of uniform or variable composition, depending upon the rate of cooling and other physical considerations. For example, Day and Allen suggested that if the early-forming crystals remained stable, the final solid would contain crystals showing a range of compositions; by contrast, if the liquid of composition *a* undercooled to temperature *e* before crystallization commenced, then all crystals would have composition *e*.[65] Likewise, differences in the rate of cooling could explain the puzzling zonal structure found in many thin sections of the feldspars, where the outer zones were almost always more sodic than those they enclosed. Once again, Day and Allen hypothesized that if the process of crystallization proceeded without undercooling, the composition would change continuously and the centers of crystallization would be expected to be richer in calcium and the periphery richer in sodium. If there were undercooling, the crystals would lack a zonal structure.[66]

Charles D. Walcott, the director of the Geological Survey, called Day and Allen's study "one of the most important contributions to geologic physics ever printed," and not without reason.[67] Their study was important in several respects. It was the first fully quantitative investigation of the cooling of silicate melts. Its authors introduced many techniques that later became standard features in such research, for instance, the use of thermal rather than optical methods to determine melting points and the use of artificial minerals to guarantee purity. And the authors managed to resolve several important issues in petrology and petrography: not only had their results ruled out the formation of compounds and eutectics among the plagioclases, but they also threw light both on the process by which differentiation occurred among plagioclase feldspars and on the process by which feldspar crystals assumed a zonal structure.

More important than the details of their results, however, was the path they had followed. Day and Allen had examined their findings in light of recent developments in physical chemistry and, within physical chemistry, they had found Roozeboom's work on heterogeneous equilibria to be of special value. By studying silicate systems under controlled laboratory conditions and analyzing their results with the aid of the phase rule and phase diagrams, Day and Allen had taken a crucial step in the direction suggested by Iddings, away from crude analogies and toward a quantitative and theoretically satisfying understanding of petrogenesis. Their work proved an exceedingly influential precedent.

The research program begun by Day and Allen at the U.S. Geological Survey expanded and prospered at a new institution, the Geophysical Laboratory of the Carnegie Institution of Washington. Authorized in 1905 and occupied in 1907, the Geophysical Laboratory quickly attained international leadership in the laboratory study of minerals and rocks.[68] During its planning, Van Hise, Iddings, and other American petrographers played a crucial role in guiding the trustees of the Carnegie Institution toward the support of the physical chemistry of petrogenesis. These petrographers saw physical chemistry as a source of ideas and techniques, but also as a model of the kind of borderland science that they themselves sought to create. The success of physical chemistry seemed to validate their own ambitions. "Until recently," Van Hise wrote,

the natural sciences and physical sciences have been handled as if almost independent of each other. The ground between has been largely neglected. The occupancy

of this ground is certain to lead to important results. The order of results to be expected is illustrated by the great advances which have recently come from occupying
the middle ground between astronomy and physics, and between physics and chemistry. For a long time astronomy and physics were pursued as independent sciences.
The great discoveries of astro-physics have shown the advantages of their combination. Chemistry and physics for a long time were pursued as independent sciences.
The rapid rise of physical chemistry has shown how wonderfully fruitful is the
ground between the two.

Exploration of the "great, almost untouched territory" between chemistry,
physics, and geology, Van Hise suggested, was bound to yield discoveries as
important and unanticipated as those made by astrophysicists and physical
chemists.[69]

Van Hise, Iddings, and members of their generation were midwives at the
birth of the Geophysical Laboratory, but the institution belonged to a younger
generation, one with a much firmer command of both laboratory procedures
and physical chemistry. Arthur L. Day was appointed its first director; E. T.
Allen became his chief chemist. They brought with them to the new laboratory
others with whom they had worked at the U.S.G.S.: the petrographer Wright
and the physical chemists Clement and Shepherd. Using this group as a nucleus, Day expanded the staff by hiring other petrographers and physical
chemists, among the latter, students of both Wilder D. Bancroft and A. A.
Noyes.[70]

Following the procedures established at the U.S.G.S., these scientists went
about the task of methodically collecting the data necessary to construct phase
diagrams of mineral systems of two and three components. The results were
of cardinal importance to geochemistry and petrology. Their data enlarged the
chemist's knowledge of silicates and, more generally, of the behavior of substances under high temperatures and pressures; their equilibrium diagrams
served as the basis for N. L. Bowen's theory of the evolution of igneous rocks,
which, though not unchallenged, has remained a cornerstone of teaching and
research in petrology and geochemistry since it was advanced in 1922.[71]

Wilder Bancroft believed that the successes of the Geophysical Laboratory
vindicated his conception of physical chemistry: "the Geophysical Laboratory
at Washington," he wrote, "might well be classified as a Phase Rule research
laboratory"; it is, he wrote elsewhere, "a wonderful example of what can be
done with the phase rule as an instrument of research."[72] In a sense he was
correct; the phase rule played an important role in the analysis of the behavior
of silicate systems, and phase diagrams were the immediate goals of much of
the Geophysical Laboratory's research. But, as Bancroft was wont to do, he
both simplified and exaggerated. While the phase rule and the visual depiction
of equilibrium relations were important, they were not the only tools used at
the Geophysical Laboratory. And the Laboratory's goal was not simply to gain

a qualitative comprehension of equilibrium relations among rock-forming minerals, but, insofar as was possible, to win a precise quantitative understanding. Toward that end, the laboratory's staff soon discovered that it was important to command not only the uses of the phase rule, but also the thermodynamics upon which the phase rule was based.

Although two of Bancroft's former research assistants were included in the original eight-man staff of the Geophysical Laboratory, the physical chemists who were hired subsequently were taken from programs in which a greater emphasis was placed on the mathematical and quantitative side of physical chemistry. Prominent among these were students of A. A. Noyes: John Johnston, R. B. Sosman, Eugene Posnjak, and F. H. Smyth, all of whom worked in Noyes's research laboratory in the years before World War I. Indeed, the most illustrious member of the staff of the Geophysical Laboratory, Norman Levi Bowen, had studied physical chemistry under Noyes, Lewis, and Bray before going to Washington, where he learned to construct a phase diagram from Bancroft's former student, Shepherd.[73]

Like his teachers at MIT, Bowen did not rest satisfied until he had gained thorough mastery of his subject. A brief example here may suffice. In one of his earliest papers, Bowen returned to the system studied by Day and Allen, albite-anorthite. Although his predecessors had established that albite and anorthite formed a continuous series of solid solutions, they had been able neither to specify the exact form of the liquidus and solidus curves nor treat these solutions by means of the laws of solutions. Bowen solved the first problem by using a technique that Shepherd and Rankin had imported from metallurgy into experimental petrology: that of quenching samples to fix their compositions and structures. He solved the second by comparing data thereby attained with the predictions of an equation that the Dutch thermodynamicist van Laar had derived from the solution laws. The fit was nearly perfect; both liquid and solid solutions of albite and anorthite obeyed the laws of perfect solutions.[74] It became possible, therefore, to predict the compositions of solid and liquid phases in equilibrium at a given temperature between the liquidus and solidus curves either by inspection of a phase diagram constructed directly from experimental data or by solving a general equation based on a quantitative thermodynamic argument.[75]

Bowen, in other words, creatively used tools he had acquired during his studies with Noyes and Lewis at MIT and during his work with Bancroft's former pupils at the Geophysical Laboratory. In his later work, Bowen relied on simple graphical methods whenever possible, but when circumstances called for the use of analytical techniques, he was prepared to use them.[76] While he used the phase rule as skillfully as any scientist, he did not rest content with it, and his enormous impact on petrology was due to his thorough grasp of thermodynamics and his commitment to quantitative reasoning. Not all of his co-workers at the Geophysical Laboratory shared Bowen's skills, but his work, generally recognized as the most important of his era in igneous petrology, inevitably served as a model toward which others aspired. The Geophysical Laboratory, therefore, was something more than Bancroft described—not simply a Phase Rule research laboratory, but a laboratory of applied physical chemistry. Its success was not so much a vindication of Bancroft's program for physical chemistry as it was a sign of the capacity of the discipline as a whole to follow through on its early promise.

By 1920, a review of elementary physical chemistry had become a common feature in the textbooks of geochemists and petrologists, and the phase diagram had become as indispensable to practitioners of those sciences as maps were to garden-variety geologists. Even though field-workers sometimes complained about the necessity of adding another arrow to their quiver, or grumbled about the arrogance of young Turks who would make geology into a laboratory science, the vocabulary and tools of physical chemistry permeated petrology rapidly. When the world's leading authorities on the problems of petrogenesis convened in London in 1924, it was the physical chemistry of rock formation that dominated their discussion, and it was the Geophysical Laboratory that was acknowledged to be the pacesetter in such work.[77]

If physical chemistry left an imprint on petrology, the adoption of physicochemical tools by petrologists had a reciprocal influence on physical chemistry. When leading geologists referred to petrology as "the natural history branch of physical chemistry," or described physical chemistry as lying "at the very centre of modern petrological research," they repaid physical chemists with the strongest of currencies: their esteem.[78]

But physical chemists gained more than honor from their commerce with geologists. The Geophysical Laboratory created positions for seven or eight physical chemists and perhaps an equal number of petrologists who, like Bowen, had done coursework in physical chemistry while completing degrees in their own specialty. These are small numbers, but through the influence of the work of Day, Bowen, and their colleagues, opportunities were created elsewhere as well. Geologists interested in petrology and petrography soon discovered that a course or two in physical chemistry could be essential to their professional success, thus creating a modest, but nonetheless important, constituency for teachers of physical chemistry in colleges and universities.

More interesting, because it illustrates how roundabout the path from academe to industry could be, was the influence of the Geophysical Laboratory in stimulating an industrial demand for physical chemists. None of the founders of the Laboratory had envisioned its work as having immediate economic importance. But soon after it opened, members of its staff began to make contributions of industrial significance—spin-offs from their basic research into the properties and behavior of silicates. Day and Shepherd developed a practical method for the production of quartz glass; others made discoveries of potential importance to the cement and ceramics industries. Their work was little noticed by businessmen until World War I. Then, under the pressure of shortages of materials, especially optical glass, first Bausch and Lomb and then both the Spencer Lens Company and Pittsburgh Plate Glass opened their workshops to scientists from the Geophysical Laboratory. A phase diagram was worked out for the system composed of silica, lead oxide, and alkali oxide, the principal components of most high-quality German glasses; systematic research was begun into the relationship between the composition and properties of glasses; studies were made of the annealing process, which drastically reduced the time necessary for cooling; improvements were introduced into the design of pot furnaces, the control of temperatures, the stirring of molten glass, and the inspection of products. American firms, which were entirely dependent on imported optical glasses in 1914, produced 675,000 pounds of such glasses during the war, 95% of it in plants under the direction of scientists from the Geophysical Laboratory. By the close of the war, America was exporting optical glass to Europe.[79]

After the armistice, several members of the staff of the Geophysical Laboratory helped firms organize ongoing research laboratories to consolidate these gains and to pursue leads that had been uncovered during the war. It was during this period, too, that makers of ceramics and Portland cement began to turn to physical chemistry. With phase diagrams depicting the equilibrium relations of the two or three earthy oxides that, in most instances, constitute ceramic materials, it became possible to predict the temperatures at which melting began and ended as well as the amount of the body that could be melted at any given temperature. Similar advantages accrued to those who possessed equilibrium diagrams of Portland cement, a three-component system consisting of oxides of aluminum, silicon, and calcium. The upshot was that a variety of new positions opened up in universities, government agencies, and business firms to relate the basic knowledge of silicate chemistry being developed at the Geophysical Laboratory to the practical concerns of industry.[80]

And so, by processes as unplanned as those of international trade, intricate relations between physical chemistry and other sciences were elaborated in the early twentieth century. In the case of commerce between physical chemistry and petrology, the Geophysical Laboratory was an especially important site. Not only was it a place where physical chemists and earth scientists met, exchanged ideas, and collaborated in pushing the experimental study of petrogenesis toward its limits, it also, quite serendipitously, became a conduit through which physical chemistry entered the glass, ceramics, and cement industries.

But petrology was only one among several sciences in which physical chemistry came to play an important role in this period, and the Geophysical Laboratory was but one among many institutions that served physical chemists and their trade partners as intellectual bourses. Physical chemists cooperated with chemical engineers in industrial firms and in the experiment stations of the U.S. Bureau of Mines. They collaborated with biochemists at institutions like the Rockefeller Institute for Medical Research and the Department of Physical Chemistry of the Harvard Medical School. They worked with pharmacologists and bacteriologists in the newly organized research laboratories of the ethical drug firms and even, occasionally, in hospitals and research clinics. In short, in the 1920s physical chemistry became a coordinate science in ways that Ostwald, Bancroft, and their contemporaries had only begun to imagine at the turn of the century.[81]

The Integration of Physical Chemistry into the University: Chemistry at California

Universities did not assume an especially prominent role in the elaboration of relations between physical chemists and other scientists during this era. This seems strange. The word "cooperation" fell as easily from the lips of academic scientists as from those in government or industry. Many academics were drawn to study topics that fell between departments. And despite the disruptions caused by the war and postwar inflation, America's universities were sufficiently prosperous to afford experimentation. Yet, as often as not, universities were followers and not leaders in bridging disciplinary boundaries; their administrators greeted new fields with much the same enthusiasm that western powers showed the Bolsheviks.

Their caution is understandable, and not simply because of the need to conserve resources. In institutions devoted to the achievement of a specific goal, be it the understanding of petrogenesis or the production of glass, disciplinary boundaries often receive little respect. Workers at the Geophysical Laboratory cared little about each other's pedigrees; once the Laboratory's mission had been defined, the skills of particular specialists became so many tools to be judged by their performance. Universities, by contrast, are organized around disciplines; boundaries are constantly at issue whether it be in the allocation of positions, facilities, and money, or in the design of the curriculum. No matter how committed individuals might be to the ideal of the seamlessness of knowledge, the structure of the institution could frustrate their ambitions.

This conservatism was nothing new. Universities had not hastened to create a place for physical chemistry when it was a fledgling. They took physical chemists in because extra hands were needed to teach swollen enrollments in introductory chemistry. Likewise, in the 1910s and 1920s, curricular and research initiatives in topics on the borderland between physical chemistry and other sciences often came only in response to pressure, from industry, foundations, or students. Where such pressures were strong, as in the case of chemical engineering, new academic structures appeared expeditiously; where the pressures were weak, the response could be very slow; not until the 1930s, for instance, did an American university build facilities for research in experimental petrology.[82]

Conservatism, of course, has its virtues. Even though American universities greeted initiatives in interdisciplinary and multidisciplinary research with caution, they offered conditions that were unexcelled for work within established disciplines, and by 1920 physical chemistry had become such a discipline. Between a quarter and a third of all graduate students in chemistry during the 1920s did their research in physical chemistry, and roughly the same proportion of all papers published in American chemical journals fell into that category (see Table 5.3 and Table 2.2). An integral part of introductory courses in chemistry, required in many places of majors in chemistry, metallurgy, and chemical engineering, recommended by some medical schools, and increasingly important to majors in geology and biology—physical chemistry had many constituents on university campuses.[83] It had become integrated into the structures and activities of American universities no less than into American industries.

Nowhere was this more striking than at the state universities of the midwest and west. These schools, some of which had been smaller than the large urban high schools of the east in 1890, and little more sophisticated, grew strong enough to compete with the strongest private universities by the 1920s. The universities of California, Illinois, Minnesota, Michigan, and Wisconsin each enrolled more full-time students than Harvard or Columbia by 1924; and by 1932, according to one survey, more eminent scientists taught at California than at Yale, and as many taught at Illinois as at Johns Hopkins.[84]

TABLE 5.3
Graduate Research Students in Chemistry, 1924–1932

	Year				
Research Area	1924	1926	1928	1930	1932
General and Physical	240	343	406	520	595
Colloid	69	58	86	78	81
Catalysis	51	31	27	44	52
Subatomic and Radiochemistry	20	21	18	39	35
Electrochemistry	56	45	38	68	70
Photochemistry	24	25	22	44	57
Total Physical Chemistry and Allied Fields	460	523	597	793	890
Total All Fields	1,700	1,882	2,071	2,795	3,348
Physical Chemistry as % of All Fields	27	28	29	28	27

Source: Clarence J. West and Callie Hull, Journal of Chemical Education 5 (1928): 882–884; 7 (1930): 1674–1675; and 10 (1933): 499–503.

Chemistry departments, with their multiple service roles, enjoyed special prestige and attention within these universities; and physical chemistry, with its polyvalent ties to other specialties, did especially well. In Madison, the specialty found a special friend and patron in Charles R. Van Hise, who presided over the University of Wisconsin from 1903 until his death in 1918. Enamored of physical chemistry during his earlier studies of metamorphism, Van Hise appointed Louis Kahlenberg head of the chemistry department in 1908. The department would remain in the hands of physical chemists for the next fifty years. At Minnesota, S. C. Lind, who had studied under A. A. Noyes at MIT and then taken a Ph.D. at Leipzig, was Director of the School of Chemistry from 1926 to 1935. And both Michigan and Illinois sought to hire physical chemists as department heads during this period, although without the success of their sister institutions.[85]

It was California, however, that was preeminent among these institutions in physical chemistry. There, under the direction of G. N. Lewis, a department was built that would produce 290 Ph.D.s, five Nobel laureates, as many National Research Fellows as Harvard, and professors for dozens of American colleges and universities.[86] The history of that department reveals, far better than statistics alone, the ways and means by which physical chemistry moved from the periphery to the center of academic chemistry in the early twentieth century.

At the turn of the century, California, like many state universities, had emerged from its beginnings as an all-purpose preparatory academy and college, but had not yet become a center of learning and research of the first rank. With a full-time enrollment of nearly two thousand students in 1899, its student body was already approaching the size of Cornell's, but only a handful of its students were seeking advanced degrees; in 1899 it awarded only two Ph.D.s. Benjamin Ide Wheeler, the vigorous philologist who was named president of the university in that year, arrived from Cornell with the decidedly ambitious dream of elevating standards while accommodating ever-larger numbers of students—and to accomplish these tasks in a community that was thousands of miles from the centers of intellectual life and in a region that would soon be rebuilding from one of the most devastating earthquakes in history.[87]

Berkeley's natural charms and the generous support of the local business community, especially Phoebe Apperson Hearst, enabled Wheeler to work this near-miracle. Ludwig Boltzmann, the distinguished Austrian physicist who was brought to Berkeley in the summer of 1904 as part of the effort to put Berkeley on the intellectual map, was struck by the beauty of the campus and the energy of its builders. "The university . . . ," he wrote, "is the most beautiful imaginable. It consists of a kilometer-square park with trees which must have seen centuries—or is it millennia—who can say. In it beautiful buildings with modern conveniences are already proving grossly inadequate. But new structures are going up—there is both money and space."

Boltzmann warned his readers not to drink the local water, and bemoaned the absence of alcoholic alternatives (among Wheeler's favorite themes was the importance of clean living), but on the whole he was sanguine about Berkeley's prospects. "Yes," he concluded,

> America will accomplish great things. I believe in these people, even though I observed them in an endeavor at which they do poorly, namely when dealing with integral and differential calculus in a seminar in theoretical physics. They did roughly as well at that as I in all the jumping over ditches and running up and down hillocks, which one can't avoid on the Berkeley campus.[88]

Physical chemistry was not high among the priorities at California. During the 1890s, when Cornell, MIT, Harvard, and Johns Hopkins were appointing their first physical chemists, the Berkeley chemistry department was struggling to organize elementary instruction in inorganic and analytical chemistry—the courses that hundreds of would-be engineers and medical doctors took. Come the new century, however, California, like Chicago, Illinois, and many other American universities, made room in its chemistry department for a practitioner of the new art. Perhaps it was the unsolicited advice that Wilder Bancroft sent Wheeler: "Some day it will come over you with a rush that no university is complete without a professor of physical chemistry"; perhaps it was the publicity that Jacques Loeb had recently reaped for physical chemistry through his discovery of artificial parthenogenesis, or the appearance of courses in physical chemistry in Stanford's catalog. Or perhaps it was simply that Berkeley's College of Chemistry, with 171 majors, seemed ready for expansion and diversification.[89] In any case, in 1901, Wheeler launched a search for new instructors, and on the shopping list was a physical chemist.

The emissary Wheeler sent east described the sort of chemists Berkeley sought: "We want young men who have taken their Doctor's degree and have already shown themselves capable of carrying on original investigation. Above all things, we want gentlemen. Young men with whom it would be pleasant to be intimately associated. To men of the proper qualifications and the proper spirit, I think the place will be very attractive. We shall try to give them an opportunity to continue their research work . . ." This message, especially the last phrase, betrayed an attitude that must have seemed ominous to the young chemist at Harvard who was ranked first among the finalists for the position, G. N. Lewis. Nor did the other physical chemists on the shortlist, two of whom were working with Bancroft at Cornell, accept a position. Instead, Wheeler and his chemists settled on Frederick G. Cottrell, a young man who had the dual advantage of foreign study and local connections. Born in Oakland, Cottrell had been known around Berkeley's chemical laboratory since he had entered the University of California as a freshman in 1893. "Tall, well-built, dark hair, straight piercing brown eyes, a friendly smile, a winning manner, one never forgot him,'' was one teacher's description. After graduating, Cottrell had taught in the Oakland high school for three years and then gone to Germany, where he studied under both van't Hoff and Ostwald.[90]

Cottrell pleased his former teachers, now colleagues. He did not attract many students—indeed, his enrollments rarely broke into double digits—but his course offerings were numerous and more sophisticated than those which his older associates felt comfortable teaching. And he was always able to lend a hand with the large introductory classes, which drew upwards of three hundred students.[91] Cottrell, legendary for his energy, also made time for occasional pieces of research. None of this work made much impact on his discipline, but some of it proved profitable, especially an electrostatic "scrubber" he invented for reducing particulate emissions from smokestacks. Such moonlighting, one suspects, enhanced Cottrell's standing among his colleagues, some of whom had commercial ambitions of their own and all of whom viewed chemistry as a decidedly practical science.[92]

Like another physical chemist on the other side of the continent, A. A. Noyes, the young Cottrell harbored dreams of making scientific research pay for itself. Unlike Noyes, however, Cottrell had an enduring interest in industrial technology and, it would seem, a restlessness of mind and spirit that made sustained basic research difficult. Rather than using the income from his patents to build an institute in which he himself might work, Cottrell turned U.S. rights to his invention over to an institution of his own devising, the Research Corporation. A peculiar hybrid between a philanthropic foundation and a bank, the Research Corporation would furnish scientists with seed money for research and in return receive profits from patent rights if, in fact, the research resulted in patentable inventions. Convinced that basic research often does lead directly to invention, Cottrell believed that the Corporation's stable of profitable patents was bound to grow over time, and hence too the resources available for reinvestment in basic research. Cottrell himself, however, neither drew on his bank nor took an active role in managing it. He put the Research Corporation in the hands of a group of eastern scientists, engineers, and industrialists and, at about the same time, left the University of California to accept a job as chief physical chemist with the U.S. Bureau of Mines.[93]

Cottrell's departure in 1911 gave Benjamin Ide Wheeler opportunity to reorganize the College of Chemistry—something that had been on his agenda for some time. California's chemists had ingratiated themselves with local businessmen by performing analyses and other chores, but the staff published little and attracted few graduate students. And even among undergraduates the College had a low standing. While enrollments in elementary courses continued their inexorable growth, fewer students majored in chemistry in 1908–1909 than a decade earlier. No other college in the university had suffered such a decline.[94] Chemistry, it seems, was becoming a subject that students took only because other departments required it of them.

Wheeler resolved to use the occasion of Cottrell's departure to appoint a chemist at the senior level, someone who might infuse the College of Chemistry with vigor and lend it prestige. Since Cottrell's departure stripped Berkeley of its only physical chemist, it was natural to seek a replacement from that same specialty. Wheeler's first choice was George Augustus Hulett, who had recently won promotion to full professor at Princeton. Like Cottrell, Hulett had done his graduate work at Leipzig. An able if not brilliant man, Hulett was respected both for his teaching and for the precision of his experimental technique. At Princeton, he was making a start at building a graduate program, but earlier he had taught at the University of Michigan, so the special demands of teaching at a large and still growing state institution were not entirely foreign to him.[95]

Wheeler pursued his quarry relentlessly for six months, promising him a new laboratory, ample support for his research, and freedom from the obligation to teach underclassmen. "You may be sure," Wheeler wrote, "that we shall be glad to develop the department of chemistry as fast as its teachers make it incumbent upon us. It has been heretofore an excellent department as a teaching department for undergraduates. We know it has not been developed in the higher ranges. We want it done."[96]

Hulett, for his part, was intrigued by the opportunity. He was interested in the intersection of chemistry and industrial technology, especially the mining and metals industries, and was eager to try to mobilize industrial support to build up a graduate program. Earlier in 1911, he had gone so far as to suggest that Arthur D. Little, the famous consulting chemist, be made head of Princeton's department of chemistry in order to attract industrial attention. These ambitions, however, excited little enthusiasm among administrators at Princeton. At Berkeley, the prospects for such an experiment looked much brighter.[97] Negotiations broke down, however, in September 1911, when Wheeler, on the advice of the Dean of the College of Chemistry, Edmund O'Neill, refused to grant Hulett authority to hire one or two young physical chemists. "Physical chemistry," Hulett replied,

> is the scientific and fundamental side of the science and it is naturally the leader in advanced work. In the last two decades it has been the stimulus to most other lines of chemistry but there must be sources of energy as well as stimuli to get results . . . I had . . . asked how a program of adding one or two good men in a year or so would be viewed. I take it from Professor O'Neill's letter that the staff does not deem this necessary.

California's program in chemistry, Hulett suggested, could not meet Wheeler's expectations if Cottrell were simply to be replaced without further changes.[98]

Although Wheeler had not bagged his game, he did learn a great deal from the chase, both about the expectations of established chemists from the east and the inadequacies of his own chemists. Within two months of receiving Hulett's final answer, Wheeler visited Boston to speak with another candidate—G. N. Lewis. The meeting went well, for after it Wheeler spared no effort to secure Lewis for California. Lewis had just been promoted to the rank of full professor at MIT, but was nonetheless ready to make a move. He had gained administrative experience, confidence in his executive abilities, and a taste for management by running the Research Laboratory of Physical Chemistry while Noyes was Acting President of MIT between 1907 and 1909. Now Noyes was back in charge of the Laboratory, and Lewis was again his loyal, but somewhat junior, associate. There was little hope of winning his own institute at MIT, since the school was embarking on the expensive task of building a new campus. Nor were relations within the chemistry department all they might have been. Colleagues were jealous of the independence, free time, and research facilities available to Noyes and his crew and wondered why an institute of technology needed so many theoretical chemists. And some were cultivating relations with industry in ways that alarmed Lewis. Unlike Hulett, Lewis believed that industrial money would be poisonous to the best traditions of academic research. In short, the future of the Research Laboratory of Physical Chemistry did not look as bright as its past. So, when Wheeler invited Lewis to follow up their conversation with a visit to Berkeley, Lewis accepted, if a bit cautiously:

> I have had the opportunity since your visit to Boston of talking with several men who have known the University of California and its chemical department. They all seem to agree that this is at present one of the weakest of your departments, and that it will need considerable reorganization before a satisfactory graduate department can be developed.[99]

After visiting Berkeley, Lewis reiterated his conviction that major reforms would be necessary. Lewis's list of demands was lengthy: a new laboratory, funds for new equipment, a five-thousand-dollar salary, a mechanic, glassblower, bookkeeper, and administrative assistant, and a fifty-percent increase in budget to cover the cost of new instructors. But Wheeler and the regents of the University agreed to all of them without cavil. Rather than giving the senior faculty in chemistry veto power over the conditions of the new appointment, as he had done during the negotiations with Hulett, Wheeler ceded complete control of the College of Chemistry to Lewis.[100]

Some observers believed Lewis was making a serious error. His relations with the chemists already at California were bound to be rocky, and Wheeler had been known to promise more than he could deliver. Moreover, Berkeley's enormous distance from the centers of scientific activity made it seem, to some, like a form of exile little less severe than Lewis had earlier imposed on himself by going to Manila. Even long-time Californians like Edward S. Holden, the director of the Lick Observatory, could find the isolation less than splendid. "Heavens—what fun it must be to live where you can talk to someone," he wrote to a friend in Washington after touring the east in 1896. Despite Wheeler's best efforts to stock his university with talented and interesting scholars, the situation had improved little in the subsequent fifteen years. Even Wheeler's prize catch, Jacques Loeb, lured to California by a research professorship in 1902, had grown disillusioned with the dream of building a "Woods Hole of the West." In 1910 Loeb had resigned his position to accept a post at the Rockefeller Institute. Not surprisingly, Loeb and his one-time co-workers at Berkeley, W.J.V. Osterhout and T. Brailsford Robertson, were among those who questioned Lewis's judgment. Robertson was still on the faculty at Berkeley in 1911, but Osterhout, like Loeb, had recently left for a position on the east coast. "I had a long letter from Robertson," Loeb informed Osterhout early in 1912:

> He told me incidentally that everybody in Berkeley was sorry for poor O'Neill, while Robertson bestowed his sympathies on Lewis. I will send you the announcement of Lewis's coming in the San Francisco Chronicle. It frankly states that he is coming purely in order to get five thousand dollars salary and deplores the fact that a man like O'Neill, who is considered one of the most prominent chemists in the country, should be dethroned by an outsider. They also mention that O'Neill is the author of a number of valuable books.
>
> It did me good to read these statements because they alleviate my homesickness for Californian climate and sea-urchins a little. [101]

Lewis quickly proved the skeptics wrong. During the year or two after arriving at Berkeley, he had engaged a bevy of new instructors and assistants, many of them physical chemists he had known at MIT: William C. Bray, Richard Chace Tolman, Joel Hildebrand, Harry Wheeler Morse, G. E. Gibson, Merle Randall, and Paul V. Faragher, among others. A temporary laboratory, already under construction when Lewis arrived, was finished late in 1912, in time to greet the new arrivals from the east. Three years later a state bond issue financed the construction of a permanent chemical laboratory, a large building with superb facilities for physical chemistry. "I shall be disappointed," Lewis told Wheeler, "if ours doesn't become the foremost chemical laboratory in the country." [102]

The relationship between Wheeler and Lewis was not without tension. Wheeler conceived of chemistry as an essentially utilitarian science that would contribute mightily to California's economic development, and he cast Lewis in the unlikely role of alchemist: the one who would, like a scientific Sutter, discover new gold in California's hills. "California," Wheeler wrote, "has everything imaginable in the way of raw materials, organic and inorganic, common and rare, heavy and precious, and it only awaits some one to work up this raw material. The other essentials of chemical manufacturing, viz., power and fuel, are here in enormous quantities; . . . It needs some one to point out the possibilities and to guide the experiments, . . . The appointment of Professor Lewis is the first step in this direction.''

Lewis, for his part, was committed to disciplinary—not industrial or political—goals. ''Notwithstanding its heavy responsibility in undergraduate instruction,'' Lewis wrote, ''the Department regards the training of men in research as the most important of its functions.''[103]

By research, Lewis meant basic research. He did not tolerate the conduct of industrially sponsored investigations in his College and his attitude toward technical chemistry and chemical engineering was frosty. During the years of American neutrality in World War I, Lewis recognized that an enormous new industrial demand was being created for chemists' services, and he acknowledged his university's obligation to meet that demand. But, like Noyes, Lewis believed that an education in pure chemistry was superior to one in technical chemistry or chemical engineering, even for those whose destination was an industrial rather than academic career. When in 1916 the College of Chemistry inaugurated a new curriculum in chemical engineering, Lewis described its graduates as ''well qualified to act as privates in the chemical industries.'' But ''the officers,'' he suggested,

> will, as a rule, be chosen from those who have received the Doctor's degree. Of the nine men who have received the Ph.D. degree in chemistry during the past two years, the majority have entered upon academic work, but several have accepted important positions in industrial enterprises. It is hoped that this desirable balance between the pure and applied chemistry may be maintained.[104]

For Lewis, a desirable balance meant that in his College graduate instruction would have stronger claims on resources than undergraduate and that instruction at all levels would be directed toward conveying an understanding of the fundamentals of chemistry. It also meant that the majority of Ph.D.s, especially the brightest of them, would be expected to enter teaching rather than industrial careers, which is exactly what happened.[105]

This was not precisely what Wheeler had envisioned when he hired Lewis, but it was hard to argue with Lewis's success. Undergraduates responded well to a streamlined curriculum that eliminated much descriptive material and stressed chemical principles. Between 1912–1913, the year Lewis arrived, and 1918–1919, the percentage of undergraduates majoring in chemistry doubled. More important, the number of graduate students in the College of Chemistry grew by leaps and bounds. By 1915 Lewis could boast that ''over 150 men from other institutions have this year made places in our laboratory.'' Some were seeking to satisfy a new state requirement that high-school instructors take a year or more of graduate work in a university belonging to the Association of American Universities; Stanford and Berkeley were the only institutions west of Minnesota that met this qualification. Others were no doubt seeking to acquire skills that might be marketed to industries that were beginning to prosper off of Europe's misfortune. A few may have been preparing themselves for entry into the nation's increasingly selective medical colleges. But some were doctoral candidates who in earlier years would have matriculated at Johns Hopkins, Harvard, or a European university.[106]

These changes—in staff, facilities, graduate and undergraduate enrollments—together had an effect that was synergistic. The growth in undergraduate enrollments created a demand for more teachers, while the presence of large numbers of talented graduate students made it possible to expand the instructing corps rapidly and inexpensively by adding graduate teaching assistants to the payroll. The availability of such teaching assistantships in turn made graduate study at California more attractive to those who had to support themselves. And the burgeoning graduate program proved an effective remedy to the loneliness that had driven Loeb from Berkeley a few years earlier. With evening colloquia, advanced seminars, visiting lecturers, and—as one chemist put it—"frequent arguments with other men in the laboratory," there was little opportunity to feel isolated. Lewis's lieutenants, who had initially been drawn to Berkeley by the promise of a new laboratory and the opportunity to work with Lewis, soon found themselves in one of the largest and strongest chemistry departments in the nation.[107]

The euphoria that often accompanies unexpectedly rapid success pervaded the laboratory. Even the undergraduates who filled the lecture halls and laboratories struck newcomers as belonging to a special breed. "The most satisfactory feature of our work here is the attitude of the students toward it," wrote Bray the year after he arrived. "Although the great majority of them are not going to specialize in chemistry, the attitude of indifference which is met with so often in the East is conspicuous by its absence."[108]

Impressed by the progress made at Berkeley and eager to emulate it, the University of Illinois in 1916 offered full professorships and hefty raises to three of Lewis's junior associates: Bray, Hildebrand, and Tolman. Lewis was unable to counter Illinois's offer with more than assurances of eventual promotion. Nevertheless, only Tolman agreed to move; Bray and Hildebrand were willing to accept lower rank and salaries to stay at California. "All of us here are engaged in conducting a large-scale experiment in methods of teaching chemistry, and I personally am very much interested in watching the working out of this experiment. The work of organizing our courses," Bray added, "—the hard work—is practically over and we are now free to devote a large portion of our time to research."[109]

Given the constitution of the faculty, it is hardly surprising that physical chemistry was at the heart of Berkeley's new program in chemistry. Undergraduates met it repeatedly during their studies. Even in the first semester of the introductory course, students were given "thorough training in the ideas of concentration and equilibrium, neutralization, hydrolysis, solubility, etc., paying attention chiefly in the laboratory to the phenomena of aqueous solutions . . .'' And it was upon this foundation that all subsequent work was based. Research too was strongly focused around problems that were central to physical chemistry: the determination of the free energies of formation of common substances, the behavior of solutions, the structure of molecules, and the nature of chemical bonds—in short, the topics upon which Lewis and his associates had worked at MIT. When he arrived at Berkeley, Lewis had made a point of insisting that all professors in the College of Chemistry carry titles of ''Professor of Chemistry,'' rather than ''Professor of Physical Chemistry'' or ''Professor of Organic Chemistry.'' This, Lewis explained to Wheeler, would preserve ''all possible mobility in the department.'' What it really did was allow Lewis the freedom to ignore other branches of chemistry as he appointed one physical chemist after another to his staff. To Lewis, the science of chemistry was largely synonymous with physical chemistry.[110]

When Lewis accepted a position at the University of California in 1911, there was only one active physical chemist in the College of Chemistry: Cottrell. By 1919, Berkeley was home to ten scientists who had published five or more articles in the field—twice as many physical chemists as were active at any other American university. Although precise statistics are not available, it is likely that no other institution in the world was producing as many Ph.D.s in physical chemistry during the late teens and twenties as Berkeley; few rivaled the quality of its graduates. Nor was Lewis's impact confined only to chemistry. As the historian Robert Seidel has observed, Lewis was largely responsible for instigating Berkeley's rise to prominence as a center for study and research in physics. Like Noyes, Lewis was convinced that a program in physical chemistry could not long prosper without frequent commerce with physics, and he missed no opportunity to secure for California talent in that subject.[111]

Still young enough to be exuberantly optimistic about the future, yet experienced enough to avoid most of the errors of enthusiasts, Lewis had displayed a consummate mastery of the art of building and managing a research school. He had exciting ideas and an ample fund of projects to enroll the hands of students. He knew his field so intimately that he could quickly identify scientists whose skills would complement both his own and those of each other, and he was so widely respected that he had little trouble recruiting those he had selected. His reputation, large and still growing, gave him the leverage to extract all he wanted from a president and regents who were desperate to see not only a chemistry department reformed, but an entire university vivified. In rebuilding California's College of Chemistry, Lewis no doubt benefited from his apprenticeship under Noyes, but in many ways the student had eclipsed the teacher: the scale of Lewis's success at Berkeley was simply much grander.[112]

Much of the credit for this achievement must be given Lewis, but there was no single key to the growth of California's College of Chemistry. Whereas Noyes had built his research laboratory in the face of indifference from colleagues and administrators, Lewis worked under generally favorable circumstances. Accomplishments in research mattered far more to Wheeler at California in 1911 than to Pritchett at MIT a decade earlier. In part this reflected the steady increase in value that university presidents were attaching to the scholarly attainments of their faculties. In part it was due to differences between the two institutions. MIT at the beginning of the twentieth century was small, poor, and facing uncertain prospects. Its trustees and faculty were divided over the direction in which it should move: whether to stress basic science or applied, whether to stay in Boston or to build a new campus. Some even wondered if it had a future. MIT did not command the resources to compete with both engineering schools and research universities, to pursue both the goals of utility and pure research.[113]

The University of California, by contrast, was large and growing; its future looked as bright as that of the state it served. It could afford simultaneously to build schools of agriculture and engineering and centers of research in the basic sciences. Its critical need was not so much money or buildings as it was respectability. "It is not the most pleasant thing after one has been boasting about our great university out there, to be asked who we have on the faculty," wrote one alumnus who had gone to Johns Hopkins. "It is always safe to mention LeConte, but having done this it is always best to change the subject and talk about the weather."[114] Such a gambit might work for a student; it could not satisfy a Benjamin Ide Wheeler.

Respectability could be won only through scholarship. And scholarship, Wheeler gradually learned, came dear. Repeated failures to attract and retain scientists of national reputation taught him that stars required supporting casts, that exceptional talent demanded exceptional freedom. And so while Pritchett and his successors at MIT viewed Noyes's laboratory as an ornament, Wheeler viewed Lewis's as a cornerstone.

Lewis's skill and local circumstances both contributed to making Berkeley a center for physical chemistry. But they were not the only ingredients. Physical chemists at other universities were prospering as well, albeit with somewhat less spectacular results. At Wisconsin, undergraduate enrollments in physical chemistry tracked those in metallurgy, chemical engineering, and industrial chemistry. When the wartime boom stimulated a rapid expansion of those programs, enrollments in physical chemistry doubled. In response to this demand, as well as to the ever-increasing need for teachers of introductory chemistry, permanent new appointments in physical chemistry were made; by 1925, half a dozen physical chemists taught in Wisconsin's chemistry department.[115] At Minnesota, it was medical students who supplied a major stimulus. Rather than incorporating some elementary instruction in thermodynamics and the solution theory into its course in biochemistry, the Medical College of the University of Minnesota required all first-year students to take an undergraduate course in physical chemistry. This requirement, instituted during World War I, helped enrollments increase several-fold in the span of three years. The results were predictable: new appointments to the faculty, additional laboratory space, and growing influence in administration.[116]

Few schools commanded resources as rich as California's, and none had a talent greater than Lewis's, but a rising tide raises all ships. Physical chemistry had become integral to so many fields that it could prosper amid adverse conditions. As more and more graduate schools began to look for it on the transcripts of applicants, even small colleges and normal schools added physical chemistry to their course offerings.[117] By the end of the 1920s, the possession of a physical chemist had become the mark of an up-to-date chemistry department.

For physical chemists, service had brought recognition and even a measure of power. What had been an esoteric specialty practiced by a handful of young and very junior academics had burgeoned into a universally recognized branch of chemistry whose leaders chaired departments at America's most distinguished universities and directed research laboratories in industry and government. Those just beginning their careers often could select, rather than merely settle for, their first job. Those who had patiently worked their way up the academic ladder, chemists like Noyes, Lewis, Cottrell, and Hulett, commanded a choice of attractive positions. The days of twenty-five-hour teaching schedules, of grueling work in analytical laboratories, and of buying apparatus with personal funds were fading memories for older members of the profession and utterly foreign to bright young students. The era in which physical chemists had to be jacks of all trades had given way to an era in which they could pursue their interests with remarkable independence. The integration of physical chemistry into American institutions—in the largest sense a by-product of the integration of science into an interdependent, industrial culture—had not fettered these scientists, but had freed them.

CHAPTER 6

From Physical Chemistry to Chemical Physics

INTELLECTUAL COMMERCE spelled prosperity for physical chemistry: new jobs, influence in universities and other institutions, funds for equipment, and new students. Prosperity in turn brought freedom—freedom to choose among employers and freedom from severe financial constraints on research. And it also brought new knowledge. The 1920s saw conceptual changes in the science as momentous as those of the 1880s. Physical chemistry in the postwar era was, in every sense of the word, richer than it had ever been.

But wealth too has its problems, different though they be from those of poverty. Commerce with other disciplines brought fruitful exchange but also conflict. When geologists, chemical engineers, metallurgists, and biochemists delved into physical chemistry, they brought with them their own questions and concerns. Geologists and metallurgists were apt to be far more interested in the phase rule and its applications than were most physical chemists. Biochemists were far more interested in the large and complex materials of living cells than inorganic ions. Tensions were inevitable, especially when physical chemists and their counterparts held not only different interests but different values.

Nor did conflict arise only on the interfaces between physical chemistry and other disciplines. Pressure toward specialization and opportunities for fragmentation were as much concomitants of expansion as new jobs and laboratories. When the borderland between chemistry and physics was sparsely settled, physical chemists had been expected to be jacks of all trades. Few of Ostwald's students restricted themselves to one research topic. After leaving Leipzig, Noyes wrote papers on reaction kinetics and solution theory, colloids and catalysis, not to mention his forays into industrial chemistry and extensive work in analytical chemistry; Lewis roamed across the entire landscape of chemical thermodynamics, but also found time for papers on relativity theory and the structure of molecules; Bancroft worked on electrochemistry and photochemistry, phase equilibria and surface chemistry. Each developed special expertise in one or two research areas but published on other topics when targets of opportunity presented themselves. Like their teachers, Ostwald, Arrhenius, van't Hoff, and Nernst, these scientists had some of the fearlessness and breadth of learning that characterized natural philosophers of an earlier age. This versatility proved an important advantage, and was sometimes a necessity, in the undifferentiated job market of the 1890s.

The 1920s produced a few physical chemists with similar wide-ranging interests: Harold C. Urey, whose enthusiasms included the chemistry of isotopes and the chemistry of the moon, and Linus Pauling, who developed an easy familiarity with all manner of chemical bonds and most of the techniques for studying them. But the great majority of their colleagues specialized, as did most graduate programs in physical chemistry. Cornell and Minnesota became known for the study of colloids, and Wisconsin for its work on reaction kinetics; California and MIT became centers for work in chemical thermodynamics; Caltech the school to study physical methods of structural analysis; and Princeton a center for the study of catalysis and photochemistry. By the mid-1920s, students of X-ray analysis, chemical thermodynamics, and colloid chemistry might all identify themselves as physical chemists but have little else in common. Physical chemistry was beginning to resemble the Holy Roman Empire—a constellation of duchies and principalities which, while continuing to acknowledge some historical relationship, were no longer parts of a functional whole.

In truth, physical chemistry had never been a tidy entity. Its roots were too diverse, its ambitions too grand. The excitement caused by the work of van't Hoff and Arrhenius on solutions, the dominance of Wilhelm Ostwald's research school at Leipzig, and the necessity of presenting a common front to skeptics had endowed the discipline with a certain unity in the 1890s. But it was a unity that time did not respect. Come the 1920s, the simple ideas of van't Hoff and Arrhenius had become far more complex, Ostwald had long since retired, and skepticism had yielded to esteem. When opportunities for physical chemists began to diversify, centrifugal forces that had long been developing threatened to fracture the discipline into a dozen new specialties.

To say that specialization brought a crisis of identity to all physical chemists would be an exaggeration. Some accepted the process with equanimity. To be sure, it made teaching, textbook writing, and communications more difficult, but such seemed the price of progress. But many physical chemists, even some in the vanguard of the new specialties, found the process alarming. Change brings obsolescence, and few enjoy watching their hard-won knowledge and skills lose value. No less important were the unreasoned loyalties that institutions and hallowed traditions often inspire. As the population of physical chemists grew and as their relations with neighboring sciences became more numerous and intricate, as the weight of past learning became greater and the journals and textbooks more congested, it became harder and harder to discern just what physical chemistry was or if there was such a thing as physical chemistry at all.

Superimposed on this process of specialization were other changes, most important among them a polarization over the role of mathematics and physics in chemistry. Classical physical chemistry was born out of the effort to apply new physical concepts and tools to chemical problems. Physics, however, did not remain static in the ensuing decades, and as it changed it confronted physical chemists with an ongoing series of choices about how much and how far to stretch their own science in response. New physical concepts and techniques presented opportunities and were essential to many branches of physical chemistry in the 1920s. But they also demanded new skills of the physical chemist: acquaintance with ideas, instruments, and mathematics that had no role in the physical chemistry of their teachers. Some physical chemists greeted the new physics of the twentieth century with eagerness; out of their work there emerged in the early 1930s a new vision of the borderland between chemistry and physics that, while quite different from that of the Ionists, proved even more capacious. Other physical chemists, unable to adapt to the new ideas and methods or unwilling to alter their mental maps of science, threw up ramparts against the new physics and cultivated their science, insofar as possible, as if the nineteenth century had never ended.

Interdisciplinary competition, specialization, polarization—these are words that carry little meaning unless rooted in the specifics of historical action. In the chapters that follow we shall illustrate them through reference to the lives of Arthur A. Noyes and Wilder D. Bancroft and the work of their research schools. Not only do their careers exemplify two extreme responses to twentieth-century physics, they also illustrate, much better than abstract words or statistics, the multiple ways in which physical chemistry grew richer and more complicated in postwar America.

THE ONCE AND FUTURE KING: A. A. NOYES FROM MIT TO CIT

For Arthur Amos Noyes the new era commenced with a wrenching choice: to remain at MIT or to follow the example of his friends Hale, Lewis, and Bray and go west. The decision was probably the hardest of his life. Boston was his home. Both his parents were now dead, but he had settled into habits both comfortable and comforting. He lived in Back Bay, in a modest but respectable house on Hemenway Street together with two Irish housekeepers, Mary and Margaret, well known to several generations of his students. MIT had moved from Boston to Cambridge in 1916, and so too had Noyes's laboratory, but it was still an easy walk from his home, at least in spring and autumn. His co-workers at MIT included his childhood friend, Mulliken, school chums like Goodwin, Talbot, and Gill, and protégés like Miles Sherrill and Arthur A. Blanchard—former students, now colleagues. He owned a yacht, named *The Research*, in which he cruised with students and colleagues from Marblehead to Bar Harbor. The islands, lights, and currents of these waters were as familiar to him as the streets of Back Bay.

Had conditions permitted, Noyes might have remained contentedly constant in these habits to the end of his life. But the changes through which his country and his discipline were passing in the early twentieth century brought upheavals to MIT and to Noyes. The integration of science and industrial technology put a premium on engineers and engineering education and thrust upon MIT a leadership for which it was not fully prepared. And while the growing commerce between physical chemistry and other disciplines brought Noyes welcome new students, it also brought conflict. The year that meant returns to home and ''normalcy'' for many Americans, 1919, was the year that Noyes broke his decades-long association with MIT, left his research laboratory, and migrated to Pasadena, California, there to begin anew his effort to build a scientist's utopia.

Noyes's decision, although difficult, was timely. He had realized the potential of his situation at MIT. The Research Laboratory of Physical Chemistry would never be more than what it was during the days of Lewis, Bray, Tolman, and Kraus, and it would likely be much less. But in California there were opportunities to build on broader foundations. Noyes made good use of these opportunities, and the research school he built proved as important to the physical chemistry of the 1920s and 1930s as the Research Laboratory of Physical Chemistry had been to that of the 1900s and 1910s. The field of molecular structure, tentatively reconnoitered at MIT, was invaded in force, and the results so transformed physical chemistry that some sought to rechristen the subject ''chemical physics.''

It is important to understand how Noyes went about building his new school, and to see how that school, by using new material and conceptual resources, extended the trajectory of the work begun at MIT. But before doing so, it is worthwhile to pause and explore the circumstances that led Noyes to break the habits of a lifetime and move to California. The new era offered Noyes and other physical chemists great new advantages, but it also presented hazards. While Noyes's success in building a new research school at Caltech illustrates the opportunities, the events leading to his departure from MIT illustrate some of the dangers.

CHEMICAL ENGINEERS, PHYSICAL CHEMISTS, AND THE STRUGGLE FOR MIT

It is easy to suppose that relations between scientists and engineers became increasingly congenial as science and technology converged in the early twentieth century. That surely did happen in some places. But proximity is a condition for friction as well as friendships, for conflict as well as cooperation. Scientists and engineers were united by growing similarities in education and methods. But even as they came to share knowledge, experience, and language, they continued to pursue different goals. Whereas physical chemistry was an essentially academic science, chemical engineering was an essentially industrial art. To be sure, a growing number of physical chemists worked in industry, and chemical engineers, in growing numbers, taught in universities. Some physical chemists made outstanding industrial employees and some chemical engineers became unusually fine and disinterested scholars. Yet students did not typically enter physical chemistry to build refineries nor chemical engineering to probe the limits of chemical knowledge. Despite some convergence between America's academic and industrial cultures, physical chemists and chemical engineers tended to measure their success differently: one in publications, the other in patents. Distinctions between the scholar and craftsman, which run deep in western culture, could hardly be obliterated by a few decades of industrial experience.

Instances of both cooperation and conflict are readily visible in relations between physical chemists and chemical engineers in the era of World War I. They were drawn together by self-interest. Physical chemists saw in chemical engineers a potentially large clientele and an opportunity to influence the world of business. Chemical engineers recognized in physical chemistry tools of industrial value. "Physical chemistry," wrote one,

> is described as that branch of chemistry which has for its object the study of the laws governing chemical phenomena. When these laws and their application to a reaction or process are once understood, it is a relatively easy matter to select the most favorable working conditions in an apparatus which operates on an industrial scale.[1]

"To be able to actually apply the laws of chemistry and to predict the course of reactions from general principles already proven," wrote another, "is a tremendous economy of both time and energy."[2]

Yet, characteristically, the engineer's emphasis was on efficiency rather than understanding. "The owners of electrochemical industries," wrote Charles Burgess, a founder of Wisconsin's program in chemical engineering,

> are not advertising for men to determine the dissociation constants at extreme dilution, or to measure solution tensions and osmotic pressures under ideal conditions, or to determine the velocity of migration of ions; but, rather, men who can operate machinery, and design and work electrolytic appliances so that a kilowatt hour may be made to do the greatest service.[3]

Burgess required students to take a semester of work with Louis Kahlenberg, Wisconsin's physical chemist. But one semester was rather too little to suit Kahlenberg, who soon launched a competitive program in industrial chemistry, and rather too much for some of Burgess's students, who grumbled about Kahlenberg's antagonistic attitude toward engineers. Chemical engineers and physical chemists grew increasingly interdependent in the early twentieth century, but at Wisconsin, and at other universities, mutual interests did not always guarantee harmonious relations.[4]

At MIT, tensions between chemical engineers and physical chemists were especially pronounced, in part because MIT was an innovator in developing physical chemistry, chemical engineering, and contacts between the two disciplines. MIT was the site of both the first research laboratory of physical chemistry in America and the first program in chemical engineering. It was also among the first schools to bring a course in physical chemistry into the curriculum of chemical engineers. Being in the vanguard brought benefits to MIT, but pioneering also had penalties. The scientists responsible for these innovations had strong personalities and sharply contrasting values and goals; they made MIT a scene of conflict as well as cooperation. Confined to a single department of chemistry and chemical engineering, the proponents of basic science, led by physical chemists, and the champions of applied science, especially the chemical engineers, competed with one another for money, space, faculty positions, students, and control of the curriculum.

But more was at stake at MIT than the resources of a single department. The skirmishes between physical chemists and chemical engineers were part of a larger campaign being waged for control of MIT as a whole. Faced with the increasing sophistication of twentieth-century technology and the growing complexity of American society, both MIT's scientists and engineers recognized that the Tech would have to adapt. The question was, what should it become? Should MIT broaden its goals by becoming a science-based university with a graduate school oriented toward basic research and an undergraduate curriculum rooted in the fundamental sciences? Or should it reaffirm its heritage by focusing on the training of engineers and cultivating work in the applied sciences? Was basic science to be a means toward an end, or should it become an end itself?

A. A. Noyes and those associated with him in the Research Laboratory of Physical Chemistry lobbied vigorously for greater emphasis on the basic sciences. "The general principle which should determine the character of our four year course of study," Noyes wrote in 1908,

> is that a liberal education be provided such as will develop character, breadth of view, and high ideals of science, and that the professional education be mainly confined to a thorough training in the principles of the fundamental sciences and in scientific method, specific engineering subjects being included only so far as the remaining time permits and as the minimum requirements of professional practice demand.[5]

While recognizing that MIT should continue to prepare students for careers in industry, Noyes and his followers held that the best way to do that was by educating them in the physical sciences, wherever possible using the problem-solving method Noyes and his colleague Miles Sherrill had pioneered. The idea here was to encourage students to discover principles for themselves by slowly working through a carefully designed series of questions. Noyes acknowledged that engineers with such an education might initially be at a disadvantage in industrial work, but he believed their deeper understanding of science and problem-solving ability would quickly allow them to compensate.

More than that, such an education would equip them to handle challenges beyond the capacity of those with narrower training:

> The engineer is trained to put in application existing methods; and it seems to me that what is wanted of the factory chemist in this country is rather the power of solving new problems and making improvements in processes—a power to be acquired far more by a good chemical training, which should include a large proportion of research and other work requiring independent thinking. . . .[6]

Noyes's position was grounded in the conviction that advances in engineering were predicated upon training in basic science and experience in research.

A second faction asserted that MIT's future lay in the applied sciences. MIT, these educators asserted, was not a university and should not endeavor to compete as such. It was a school of engineering and technology that was, in their view, uniquely situated to train the builders and leaders of industry in twentieth-century America. It could seize that opportunity by developing novel and more efficient methods of training engineers. Led by William H. Walker, an energetic chemical engineer who had as much experience in business as in education, this faction preached the dignity of the engineer's calling and the necessity of specialized instruction in engineering practice and principles. Work in the basic sciences, physical chemistry included, would constitute an important part of a chemical engineer's training, but such work should be subordinate to the goal of such education: the preparation of engineers for the world of action rather than the world of scholarship. As Walker told Noyes:

> You and I have always disagreed on one fundamental proposition:—You contend that it is enough to learn a law or theory of science, and if once learned, the application of this law or theory to the solution of problems of daily practice will care for itself. I contend that when the student has learned the law or theory, that only one half has been done; that it requires a more able mind, more experienced judgment, and more work (if I may use the term) to intelligently apply, for example, Raoult's law to the diverse and complicated conditions under which the chemical engineer must work than it does to learn Raoult's law, staged as it always is in an environment where it is quantitatively valid. . . .[7]

It was only through exposure to problems drawn from industry that students could learn the uses and limitations of theory. Moreover, it was only through practice with industrial operations and techniques that students could master production by the ton rather than by the test tube, to understand the constraints imposed by material and energy costs, to be sensitive to the potential uses of by-products, and to prepare notes and reports in such a way as to be attractive to businessmen and useful to patent lawyers.[8] "Science by itself," Walker told a colleague in the geology department, produces a very badly deformed man who becomes rounded out into a useful creative being only with great difficulty and large expenditure of time. Despite Noyes and his satellites, nevertheless I still contain [*sic*] I am in a position to prove that it is a much smaller matter to both teach and learn pure science than it is to intelligently apply this science to the solution of problems as they arise in daily life. . . .[9]

Acting on these ideas, Walker, in 1905, reorganized MIT's somnolent undergraduate program in industrial chemistry. In so doing, he transformed it from a potpourri of courses in chemistry and mechanical engineering into a unified program in chemical engineering, based increasingly as time passed on the study of unit operations and the balance book aspects of industrial practice.

The idea of organizing instruction around unit operations, such as distillation, filtration, and condensation, had been simmering for some time. The pedagogical advantages were obvious. Once taught the principles of a dozen or so basic operations, a clever student could easily comprehend how they could be combined and recombined to realize many and diverse industrial aims. No longer would it be necessary to memorize the details of scores of industry-specific processes.[10] But an education in chemical engineering, even when streamlined in this way, was expensive. The machinery and materials necessary to illustrate unit operations on an industrial scale were costly and prone to rapid obsolescence. Few universities could afford such facilities.

MIT was one of the first schools to overcome this obstacle. In 1916 Walker, together with his younger colleague Warren K. Lewis, established a School of Chemical Engineering Practice. Basically a cooperative extension program, the School sent faculty members and students from MIT to select industrial plants where they had, for part of the school year, that direct contact with production Walker advocated. Equally important, the program gave MIT's chemical engineers access to the costly facilities necessary to illustrate classroom instruction in unit operations. The School of Chemical Engineering Practice was a notable and widely emulated success.[11]

Walker was not oblivious to the call of research. Far from it. He was himself a Göttingen product and an advocate of graduate education at MIT. But for Walker the research that should hold pride of place at an institute of technology was research on the applications of science. It was in this field that Walker made many of his most significant contributions.

The institutional vehicle for Walker's ideas on applied chemical research was the Research Laboratory of Applied Chemistry. This institution, organized in 1908, was a hybrid between Noyes's Research Laboratory of Physical Chemistry and the consulting firm in which Walker had once been a partner: Arthur D. Little, Inc. Like Noyes's laboratory, Walker's was set up as a semiautonomous division of the chemistry department with its own staff and budget. Like Noyes's laboratory, Walker's received a building and a small subvention from MIT and was expected to meet operating expenses by securing outside funds. But whereas Noyes looked to his own resources and those of the Carnegie Institution of Washington for support, Walker would rely on research contracts with industrial firms and trade associations. Like a consulting firm, the laboratory would make research pay for itself.[12]

Walker and his collaborators in this project believed that such a laboratory would benefit both MIT and American industry. At MIT, it might become the focus for research work in industrial chemistry and a foundation stone for a graduate program in chemical engineering. It would also bring MIT's program in chemical engineering before the eyes of industrialists around the country, with obvious rewards for both MIT and its graduates. At the same time, such a facility would give businessmen access to MIT's staff at relatively small cost. Firms that were too small to afford their own research laboratories or uncertain about the investment could rent MIT's facilities and experts without permanent or expensive commitment. The laboratory, Walker and Little hoped, would foster many new links between industry and American universities and technical schools.[13]

The aims of Noyes and Walker were anything but compatible. Nevertheless, MIT's administration resisted making a choice between the two programs. Richard C. Maclaurin, whom Noyes had helped select as his successor in the president's office, pursued a policy of smothering differences.[14] As a former physicist he felt the attraction of Noyes's ideals; as president of MIT he felt the weight of its tradition as an engineering school and the responsibility of ensuring its fiscal health. The head of the chemistry department, Henry P. Talbot, was in a similar position: that of mediating between two irreconcilables.[15]

A rapprochement never occurred. Noyes's influence at MIT gradually waned during the years prior to World War I. It deteriorated further immediately after the war. In part this was due to the popularity of Walker's program in chemical engineering. Whereas during the years 1905–1909 the majority of MIT's undergraduates in chemistry took their degrees in basic chemistry, baccalaureates in basic chemistry were outnumbered by better than two to one by baccalaureates in chemical engineering during the subsequent five-year period. The disproportion grew even greater following the war, peaking in the 1920–1924 quinquennium when 419 chemical engineers graduated but only 52 chemists (see Table 6.1). By and large, students entered MIT in order to become engineers, not research scientists. Their tolerance for subjects like theoretical chemistry was limited to those topics that seemed directly relevant to their careers. Walker acknowledged this reality; Noyes sought to change it. Like many an educational reformer, Noyes discovered that a school cannot be much more enlightened than its students.

But shifting enrollments were not entirely responsible for the erosion of Noyes's influence at MIT. Perhaps even more significant was the support

TABLE 6.1
Baccalaureates Awarded in Chemistry and Chemical Engineering at MIT by Five-year Periods, 1885–1934

Years	S.B.s in Chemistry	S.B.s in Chemical Engineering
1885–1889	38	—
1890–1894	50	31
1895–1899	98	49
1900–1904	78	51
1905–1909	82	65
1910–1914	50	132
1915–1919	63	187
1920–1924	52	419
1925–1929	81	238
1930–1934	71	240

Source: Annual reports of the registrar in the MIT *President's Report*.

Walker enjoyed among the Institute's patrons and trustees. A transformation was under way during this period in the sources of MIT's capital funds. When Noyes established the Research Laboratory of Physical Chemistry in 1903, Boston Tech depended largely on the support of men with the names Cabot, Lowell, Peabody, Endicott, and Choate. The old Boston aristocracy held control of the governing Corporation. During the years after 1910, the influence of these families waned as George Eastman, T. Coleman du Pont, and Pierre Samuel du Pont began to take an interest in the school. When MIT moved from Boston to Cambridge in 1916, its new home was built with their dollars. Between 1911 and 1921, the du Pont family gave MIT over $1.1 million; George Eastman contributed more than $10.5 million. These grants exceeded the entire value of the endowment and plant in 1910 by a factor of three.[16]

Walker commanded the confidence of these businessmen. He had once made a living as an industrial consultant, had made signal contributions to industrial technology, and during World War I had demonstrated exceptional executive skills in overseeing the construction of the Chemical Warfare Service's Edgewood Arsenal, one of the largest and best-managed construction projects of the war. Like MIT's new patrons, Walker was an organizer and achiever who harnessed knowledge to practical goals. His vision—of a technological institute cooperating with industry in the building of a more efficient industrial society—was their vision. Noyes, who sometimes struck even his fellow scientists as "a little narrow, with something of the Boston atmosphere about him," cut a less impressive figure; his plans for MIT's future were harder to grasp.[17]

The shift in policy at MIT was apparent to advocates of basic research on the faculty. Dissatisfaction with the increasingly industrial climate appears to have played a role when Lewis led a substantial fraction of the chemistry department's staff and graduate students to Berkeley in 1912. Three years later, Noyes nearly followed Lewis's precedent when Noyes's close friend George Ellery Hale extended him an invitation to join the faculty of the Throop Technical Institute of Pasadena. Hale was on the board of trustees of Throop and ambitious to make the small school into a partner worthy of the magnificent observatory he was building atop Mount Wilson. Hale promised Noyes that he would have a splendid new laboratory and a free hand in building Throop. Noyes, however, was reluctant to leave Boston and anxious about the future of his laboratory at MIT. He agreed to spend three months a year in Pasadena, thinking, apparently, that he would be able to simultaneously build a new technological institute while reforming an old one.[18]

The plan miscarried. His absences from MIT exacerbated the resentments felt toward him by some colleagues, and his efforts to steer MIT toward the basic sciences were rebuffed. In 1916 Noyes, alarmed that the Institute's "science is being more and more subordinated to engineering work," urged Maclaurin to right the balance by giving more funds to research laboratories and fellowships, linking promotion to research, and shortening teaching hours.[19]

The disruptions caused by World War I excused Maclaurin from addressing these demands. But when Noyes and Walker returned to Cambridge in 1919, conflict erupted anew. Both had held responsible executive positions during the war in which they had better learned their own worth; neither was prepared to compromise. Complaining of Noyes's obstruction and the unreasonable demands his problem method of instruction placed upon undergraduates, Walker presented Maclaurin with an ultimatum: if Noyes did not either "surrender control of the teaching of theoretical chemistry, cease to consider himself a member of the Department of Chemistry," and "confine his activities to research work and the Research Laboratory of Physical Chemistry," or "become an inherent member of the Department of Chemistry upon the same plane and basis as other professors," Walker would resign. And if he left, he told Maclaurin, many of MIT's other chemical engineers might follow. The first alternative would cost Noyes control of the undergraduate program he had done so much to construct; the second would cost him the freedom to appoint scientists of his choice to the staff of the Research Laboratory of Physical Chemistry. Neither was acceptable to Noyes, and indeed Noyes responded with his own list of demands: preferential treatment for faculty teaching the basic sciences, more money for research work, greater selectivity in undergraduate admissions, and an end to contracts with private corporations.[20]

Maclaurin was forced to make a choice. It was an excruciating decision. Noyes was MIT's most distinguished scientist. He had spent his entire career at MIT and had been Maclaurin's predecessor in the President's office. His resignation was bound to hurt the morale of faculty in the basic sciences at MIT and raise eyebrows among educators elsewhere. What Noyes was to physical chemistry, however, Walker was to chemical engineering, that is to say, he had as strong a claim to being the father of American chemical engineering as Noyes had to being the father of American physical chemistry. Like Noyes, Walker had numerous disciples and allies on the faculty. And he brought the Institute not only honor but also money: business at the Research Laboratory of Applied Chemistry was booming. Maclaurin was anxious to increase revenues and plans were already afoot to enlarge the scope of MIT's contract arrangements with industry through a new Division of Industrial Cooperation and Research. Walker was a natural choice to lead the campaign for clients.[21]

In the end, Maclaurin opted to keep Walker. A review of Noyes's work by E. B. Wilson, professor of mathematical physics and Maclaurin's advisor, may have been decisive. A graph depicting the research productivity of Noyes and his laboratory showed a marked decline dating from 1913. "There are those who say Noyes has 'shot his bolt' and will never do much more either himself or through his students," Wilson wrote.

> The chart looks it. . . . Hale, however, thinks that Noyes may and probably will accomplish a great deal if he settles permanently at Throop. I believe the chances for a pronounced and prolonged activity at Throop are less rosy than Hale imagines, but I feel sure they are far better than here as far as Noyes is concerned.[22]

Armed with this quantitative evidence, Maclaurin appears to have concluded that Noyes's loss would be easier to absorb than Walker's. The day he received Wilson's report, he asked Noyes to withdraw from an active role in the chemistry department. This was tantamount to a request for Noyes's resignation. By the end of the year, he had it in hand.[23]

MIT's treatment of a scientist of Noyes's stature was, as the historian Roger Geiger has observed, remarkable. Maclaurin himself understood this. News of Noyes's resignation was withheld until the very eve of his departure to minimize its effect on fund-raising and Noyes was asked to accept a nonresident professorship, to blunt "Harvard criticism."[24]

Noyes's departure hurt the cause of basic science at MIT. During the next decade, MIT's standing in scientific research slipped badly. The chemistry department, once among the nation's strongest, found it difficult to find and retain bright young research scientists. Significant work continued to be done on the physical chemistry of aqueous solutions by a talented group of thermodynamicists, but work along new lines, such as X-ray crystallography, expired after Noyes left. The Research Laboratory of Physical Chemistry, entrusted to the capable hands of F. G. Keyes, found it necessary to undertake more and more consulting work to survive. By 1923 it was studying the extraction of helium from air for the Bureau of Mines, the properties of steam for the American Society of Mechanical Engineers, and was even doing research on torpedoes for the Navy. The pressure to undertake research that paid for itself was felt throughout the Institute and was bitterly resented by some members of the faculty. Even teachers of applied sciences, chemical engineers among them, eventually became disenchanted with the volume and nature of the contract research undertaken for industry.[25]

Ultimately, MIT reversed its course. In 1929, alarmed by the deterioration in faculty morale and the Institute's reputation, MIT's governing Corporation gave a physicist from Princeton, Karl Compton, the presidency and a mandate to clean house. The policies Compton instituted were strikingly similar to those Noyes had advocated a decade earlier. Intent on making the basic sciences "the backbone of the Institute," Compton aggressively sought new funds for basic research, brought new leadership into the science departments, and imposed stringent controls on consulting and contract research. But for the Research Laboratory of Physical Chemistry, the reforms came too late. The leadership it had once enjoyed had been forfeited, and despite infusions of money and talent in the 1930s, it could not regain its former eminence.[26]

The postwar era did not begin auspiciously for physical chemists at MIT. Applied sciences, like chemical engineering, had evolved into powerful competitors for influence and resources. Industrial interest had threatened the independence of the Institute and the autonomy of its scientists. But these problems were not inevitable consequences of the integration of science and industrial technology. Like a youngster of humble background invited to court, MIT lacked both income and self-assurance. It desperately needed money, for technical education was capital-intensive and its new quarters in Cambridge were alarmingly expensive. Its leaders were in disarray over how best to use its resources to educate young Americans for the new industrial world. These circumstances made MIT especially susceptible to entrepreneurs like Walker, who preached a perfect harmony between industrial and academic interests, and made it especially uncomfortable for those like Noyes, who perceived the possibility of conflict as well as cooperation.

But across the continent, in Pasadena, another kind of technological institute was being built. Its architect, George Ellery Hale, combined the entrepreneurial flair of Walker and the academic values of Noyes. In this new school, physical chemistry and other basic sciences would flourish, nourished by—but not subordinate to—America's new industrial wealth.

CALTECH: FROM BASE CAMP TO TEMPLE

When A. A. Noyes arrived in Pasadena following his final break with the Massachusetts Institute of Technology, he was met by many of the same problems he had faced at MIT twenty years earlier: building a staff, attracting graduate students, and stimulating research. Even some of his friends were skeptical about the prospects. But for Hale's observatory, Southern California was about as well known for science as New England was for oranges. And

Noyes was no longer young. Now in his mid-fifties, he was worn by years of hard work, his battles at MIT, and the trauma of leaving an institution and a community in which he had spent his entire adult life. But Noyes also possessed several advantages that he had not had at the turn of the century. A magnificent new laboratory, built with a $200,000 gift arranged by Hale, was one; his reputation for being among America's most eminent chemists was another. A third was the close relations Noyes had cultivated with the philanthropic foundations of the east.[27] Finally, Noyes was confident of firm support from the trustees and administrators of the Throop Institute. Hale and James A. B. Scherer, the president of Throop, guaranteed him a free hand in molding both the chemistry program and the undergraduate curriculum; Arthur Fleming, a lumber magnate and chairman of Throop's board of trustees, arranged a $200,000 fund for chemical research.[28] Hale, in a flamboyant display of his affection, gave his friend a Cadillac. Noyes must have found the contrast between his treatment at MIT and at Throop as striking as the difference in climate, terrain, and flora. As Hale had predicted, the change of scene was invigorating. Settling into a house one block from his laboratory, Noyes threw his energies into building a school.

His policies were largely a product of his analysis of the shortcomings of the technological institute he knew best: MIT. MIT, in Noyes's eyes, had grown too large; it had sacrificed excellence for numbers. Its engineers had grown too powerful. Its administrators had compromised the independence of the institute by leasing its scientists and facilities to private firms. The school overworked and underpaid its scientists, neglected graduate education, and treated research as a personal rather than a communal concern.

In Pasadena, Noyes used his considerable influence to steer Throop past these pitfalls. Throop, Noyes believed, should remain small. Its engineers should be subordinate to its scientists, reflecting what Noyes believed was the intellectual relation between technology and science; its scientists should not double as industrial consultants. The Institute, in Noyes's view, should focus on doing a few things well, the most important being research and advanced training in the physical sciences. High salaries and generous support for research could attract a distinguished faculty; generous fellowships could attract able graduate students. As Throop's reputation grew, applications would increase and permit greater selectivity in admissions. At every level the aim would be excellence, "to train scientific leaders, rather than to afford mass education."[29] Throop would become what MIT was not: a large version of that scientific utopia of which Noyes had dreamed as a young man and had realized on a small scale at his Research Laboratory of Physical Chemistry.

Much of Noyes's energy during the remaining years of his life was dedicated to this mission. At his suggestion, the Throop Institute adopted a new name in 1920, one that would be more appropriate for its ambitions—the California Institute of Technology. A fellowship program was begun, again at his suggestion, to attract graduate students to Pasadena.[30] Together Noyes and Hale selected and recruited the new faculty that Caltech would need to succeed.

Their largest catch was Robert A. Millikan, the physicist from the University of Chicago who had gained fame for measuring the charge on the electron. In 1917, Hale extracted Millikan's promise to spend three months a year in Pasadena by using the same strategy that had won Noyes—the offer of a new laboratory and freedom to run it. In 1921, after lengthy courting, Hale and Noyes induced Millikan to accept a full-time appointment. Approximately the same age as Noyes and Hale, Millikan shared many of their attitudes and values. He was also a superb experimental physicist, a talented executive, and an able fund-raiser—in short, the perfect choice not only to oversee the development of physics in Pasadena, but also to administer the Institute. When Millikan went west it was as chairman of the eight-member Executive Council that collectively set policy and managed Caltech.[31]

This was a job that, under different circumstances, Noyes might have had. But high office had never held much appeal for him. He had already declined offers of the presidency of MIT and the Johns Hopkins University.[32] Noyes's interest lay in science and students; the public roles of a university president demanded social skills and energies that Noyes did not possess. So long as he could have influence on policy—something that Hale and Millikan gladly gave him—Noyes preferred to remain in the background. On the Executive Council, Noyes assumed special responsibility for undergraduate and graduate courses of study. Less formally, he was a key player in the making of all manner of policies—the member of the triumvirate who knew most about education and who, at least until Millikan hit full stride in the mid-1920s, best understood the inner workings of Caltech. "Millikan," one observer later remarked, "became a great public figure, who in the minds of the people of the country represented the California Institute of Technology; but Noyes was often the one who was responsible for the policies that were announced by Millikan."[33]

George Ellery Hale, the instigator of the buildup of scientific talent in Pasadena, had intended that the Throop Institute serve as a kind of scientific basecamp for his observatory atop Mount Wilson. Astrophysical research, Hale believed, was best conducted in cooperation with terrestrial work. Physicists and chemists, for example, could assist astrophysicists in such tasks as interpreting spectral data and in improving photographic dyes. When Noyes and Millikan arrived, however, the focus of scientific research in Pasadena shifted from the mountaintop to the valley below. In 1919 Hale's telescopes and Pasadena's climate were the only reasons a touring scientist might put Pasadena on his itinerary; ten years later the telescopes and climate still exerted appeal, but so too did the laboratories and scientists clustered at Caltech.

Indeed, by 1929 Caltech clearly was among the leading three or four centers of research in physics and chemistry in the United States. Its faculty included scientists of established reputation, like Millikan and Noyes, and a larger number of promising young instructors, like J. Robert Oppenheimer, Carl Anderson, and Linus Pauling. It attracted more National Research Fellows in physics than any other university; in chemistry it trailed only the far larger departments at Berkeley and Harvard.[34]

The speed with which Caltech vaulted from obscurity to leadership had few precedents in the history of education; in the United States only Johns Hopkins and Chicago had achieved comparable fame in such a short time. Unlike Johns Hopkins and Chicago, however, Caltech was not the creation of a single patron. Its success was built on dollars from a diverse assortment of donors: local real-estate men who thought that Caltech would boost property values; power-company executives who sought technical support and engineers to help meet Southern California's massive and growing electrical needs; bankers, lawyers, and newspaper tycoons who hoped that Caltech would attract industries to a region deficient in natural resources; retired millionaires from the east who wished to adorn their adopted community with institutions of high culture; local philanthropists who envied the high tone that universities lent the San Francisco region; and, perhaps most important, the officials of the great national foundations who had confidence that Hale, Millikan, and Noyes could make a desert bloom.[35]

The job of building Caltech called for salesmanship of a high order. Paeans to pure science would hold little appeal for the self-made men who held so much of the wealth of the Los Angeles region. But too much emphasis upon the immediate material benefits of research would alienate foundation officials and would-be patrons of high culture. It would also make Caltech's scientists captives of their own promises. Millikan, Hale, and Noyes were willing to devote a portion of Caltech's resources to work on such practical problems as the long-distance transmission of energy, earthquake prediction, aeronautics, and fuels. But they had no desire to set a timetable for solving such problems or to make them central to the mission of their institute.

This dilemma, however, is more apparent to us than it was to them. Millikan, Hale, and Noyes were true believers in the notion that basic science had strong and direct links with technology. They felt no compunction about telling businessmen that investment in basic science was the surest path to the solution of pressing practical problems.[36] And when an audience seemed too obtuse to understand that the shortest route between two points is not always a straight line, they could draw upon the regional pride that was as intense in Los Angeles as the sunshine. Here Millikan proved the consummate master, perhaps because his mind, despite its undeniable power, was essentially as uncomplicated as his listeners'. "Southern California," he told one millionaire, offers probably the last chance which the United States affords for laying the foundations of our American, yes, our western civilization. I mean by this that the foundations of the other important centers have already been laid. Their building period is gone while we are just entering upon ours, and those of us, whether financiers or educators, who can build ourselves into these foundations have the sort of opportunity that comes only at rare periods to any group, and exceedingly rarely to individuals.[37]

The tone was typical of the message Millikan broadcast through the business community of Southern California. The best qualities of western civilization had somehow been distilled and concentrated as settlers had moved westward. Now, in a desert on the edge of the Pacific Ocean, a new civilization was gestating. Science and science-based technology would be its midwife and wet nurse. A contribution to Caltech, therefore, would be a contribution to the community, to the nation, and even to the race.

> It is no hard luck story that I have to tell; quite the reverse. . . . Some of the things that have already happened, not that we hope will happen, have made it certain that the foundations are here just now being laid for what is destined to be a center of commanding influence with respect to the whole Pacific area, and even with respect to the future of Anglo-Saxon civilization. . . .[38]

Angelenos, already famous for their enthusiasms, proved receptive to this pastiche—evocative at once of the doctrine of manifest destiny, Spengler's metahistorical musings, Turner's frontier hypothesis, and the ruminations of scores of eugenicists, evangelists, and hucksters. Scores of the region's wealthiest families responded to the appeal.[39]

But local support was only part of what Hale, Millikan, and Noyes sought. They also looked to the philanthropic foundations for dollars. Here talk of Southern California's destiny cut little ice. The Rockefeller and Carnegie philanthropies were looking for ideas, and these too Caltech's builders had to offer. What Hale, Millikan, and Noyes proposed to do was nothing less than to mount a joint attack on the constitution of matter. "[T]he possibilities of such a joint attack have reached an extraordinarily favorable state," they told the trustees of the Carnegie Corporation of New York.

> In each of the branches of science involved, the methods and instruments of research have advanced to a very high degree of development. . . . The application of the spectroscope to astronomy, affording the means of determining the chemical composition, distances, motions, temperatures, pressures and magnetic state of the stars, has led to many advances of fundamental importance to chemistry and physics. The rise of physical chemistry, which transformed chemistry from a largely descriptive to an exact science and revealed the fundamental role played by electrically charged particles in solution, opened another new world of thought. The extraordinary discoveries and developments in physics, particularly in the fields of radioactivity, the

electrical nature of matter, X-rays and radiation, have brought to light wholly un-expected relationships between the elements which are of the greatest significance, both from the purely scientific and the practical point of view.[40]

Millikan proposed to concentrate on the structure of the atom and on processes of nuclear transformation. He hoped to use spectroscopic techniques to test Bohr's model of the atom's electronic structure and high-voltage electrical discharges to shatter its nucleus. And he proposed to study the penetrating radiation that he later christened "cosmic rays," a radiation that, he hoped, might reveal how large atoms were built up from small—a process comple-mentary to the nuclear disintegrations of radioactive decay. Noyes's labora-tory would focus upon the structure of molecules and upon the valence elec-trons that appeared to be largely responsible for their chemical properties. In particular, Noyes proposed to use X rays to study the structure of crystalline substances and to explore the effects of thermal and other forms of radiation on chemical reactions. Hale's equipment atop Mount Wilson, designed to study the "cosmic crucibles" of the "vast laboratory of nature," would give Caltech's physicists and chemists means to study matter at temperatures and pressures unattainable on earth.[41]

Impressed by the credentials of the three authors, the resources they had already mobilized, and the scope of their plans, the Carnegie Corporation of New York awarded Caltech $30,000 a year for research in chemistry and phys-ics. It was the first of many grants that Caltech would receive. The original Carnegie Corporation grant would be renewed annually through the 1920s. The Carnegie Institution of Washington continued to supply Noyes with an-nual grants for his personal research but also helped pay for the creation of a geology department at Caltech. The Guggenheim Foundation supplied the In-stitute with an aeronautical laboratory. The General Education Board gave Caltech nearly three million dollars—more than it gave any other university— to assist Caltech in building up an endowment. Capping it all off was a grant of six million dollars in 1928 from the International Education Board to fi-nance construction of a new observatory atop Mount Palomar.[42]

Caltech had succeeded where MIT had failed; its founders had found ways to secure the large sums necessary for an institute of science and technology without mortgaging their laboratories. While Caltech accepted generous gifts from businessmen, it seldom entered contracts with individual firms. Amid Millikan's extravagant prophecies about Caltech's potential contributions to the economic development of Southern California there were very few prom-ises of specific returns on investment. In short, businessmen were enlisted in their project without being given control. And so, Caltech, far better than MIT, was able to remain focused on basic research, thereby preserving and enhancing its reputation with the major philanthropic foundations.[43] The foun-dations responded with a generosity that contrasted sharply with their niggardliness toward MIT. While showering one with grants in the 1920s, the Rocke-feller and Carnegie philanthropies ignored the other.

CHEMISTRY AT CALTECH

In Pasadena, Noyes found an important role in shaping institutional priorities and policies. But Noyes's interest in the Institute as a whole did not eclipse his commitment to chemistry. At the Gates Chemical Laboratory, named after its donors, C. W. and P. G. Gates, Noyes practiced the policies he preached in the Executive Council: an emphasis on research in both undergraduate and graduate education; stress on basic science rather than applied; and the pursuit of excellence in a few areas rather than mere competence in many. Approximately half of the budget at the Gates Laboratory was earmarked for research; in 1923–1924, $25,000 in a total budget of $54,065. Salaries and promotions were determined chiefly by publishing habits, and even undergraduates were judged largely on the basis of their potential for original work. For Noyes, the power to solve problems independently was the true mark of an educated man. Work on commercial topics was undertaken only rarely, and then only if it involved scientifically important issues. And little effort was made to achieve balance among chemistry's specialties in appointments, research, and course offerings. Unlike department heads at many other schools, Noyes did not have to seek balance. He did not contend with powerful local traditions or en-trenched interest groups, because Throop, prior to his arrival, hardly had a chemistry department. Nor was there need to cater to the interests of a large and diverse student body. Hence Noyes could focus on building strength in one field at a time.[44]

When Noyes arrived in Pasadena, his principal concern was to attain excel-lence in physical chemistry. This of course was the field that Noyes under-stood best, and it was, in Noyes's view, the foundation upon which other branches of chemistry ought to be based. But it also offered the greatest op-portunities for overlap with the work that Hale directed in astrophysics and the work that Millikan proposed to undertake in physics. Traditional disciplinary distinctions were preserved in Caltech's organization; there were, for exam-ple, separate divisions, laboratories, and budgets for physics and chemistry. But Hale, Noyes, and Millikan did not wish to see administratively and edu-cationally convenient categories mold the intellectual life of the Institute. Not only had each a lifelong record of work on topics in disciplinary borderlands, but all believed that the mission research of World War I proved the efficacy of multidisciplinary collaboration. Cooperation was more than a slogan at Cal-tech; its scientific leaders understood that nature does not heed man-made cat-egories.[45]

Even more than at the Research Laboratory of Physical Chemistry, Noyes's success would depend upon his staff. The Gates Laboratory was many times the size of the Research Laboratory of Physical Chemistry, and Noyes's personal influence would inevitably be diluted in the larger institution. More important, Noyes himself was no longer at the cutting edge of research in physical chemistry. He had largely withdrawn from experimental work. And while his interest in certain theoretical issues, especially those relating to solutions, was undiminished, he was unprepared to travel very far into the regions that younger scientists were opening up in the 1920s. He did not, for example, have the mathematical tools to explore the implications of quantum theory for chemistry or to follow in much detail the involved reasoning called for in the interpretation of crystal structure. Hence, his role at the Gates Laboratory would not be that of a proprietor, who actively engaged in every aspect of his business's operations and led through personal example; rather it would be that of a chief executive officer, who took responsibility for policy and staffing, for strategy rather than tactics.

From the outset, Noyes was firm on the type of scientist he wanted in his new laboratory. In a letter to a prospective faculty member then working at the University of Chicago, Noyes emphasized that he had no interest in anyone "who is more interested in the immediate execution of his own researches than in the general development of the research laboratory." Pasadena, he continued, "is not the place for the individualist in science or research . . ."[46] Noyes found these team players through his network of former students and research associates. His first appointments, made while he was still commuting from MIT, were of current and former students. In 1916 James Ellis, who had just completed his doctorate at MIT, was assigned to assist Hale in devising more sensitive photographic plates, and Charles Lalor Burdick, who had worked both in Noyes's laboratory at MIT and in the Braggs' laboratory in England, was dispatched to Pasadena to equip the laboratory for work in X-ray crystallography. The following year, Roscoe Dickinson, a graduate student at MIT, was sent west to initiate research with the X-ray equipment Burdick had installed. After completing his degree (at Caltech) in 1920, Dickinson stayed on in Pasadena, first as a National Research Council Fellow and then as a faculty member. During the 1920s he fell heir to the role that Coolidge and Kraus had held at MIT; he was the gifted experimentalist who could make cranky apparatus work, spot the virtues and flaws in an experiment's design, and teach newcomers the fundamentals of good technique.[47]

But the new laboratory also needed a chemist with theoretical leanings, someone who could, as Lewis had done at MIT, maintain contact with the world of theoretical physics. Noyes found his man in 1921: his old friend Richard Chace Tolman. Since obtaining his Ph.D. in 1910 at MIT, Tolman had taught at California and Illinois, and had served as associate director and director of the Fixed Nitrogen Research Laboratory in Washington, D.C. Devoted to research and possessing "a very unusual combination of experimental and theoretical knowledge," Noyes considered Tolman "a great prize." "Indeed,'' wrote Noyes, "with the possible exception of G. N. Lewis of Berkeley, there is no physical chemist in the country who ranks with him."[48] What especially appealed to Noyes was Tolman's strong grounding in mathematics and physics. Not only was he a master of thermodynamics, he also had a firm grasp on statistical mechanics and gas kinetics and was strongly interested in both quantum and relativity theory. He had made, and would make, no single contribution to science comparable to Lewis's work on the determination of free energies or valence, but his breadth of learning, his productivity, and his excellence in seminars made him indispensable. During the 1920s he became the department's "principal work-horse in graduate teaching."[49]

After Tolman's appointment in 1921, Noyes ceased to look to his MIT circle for physical chemists and began to make appointments from within Caltech's own growing corps of graduate students. Among those who entered the faculty by this route were Richard M. Badger, Linus Pauling, and Don Yost, recipients of Ph.D.s in 1924, 1925, and 1926, respectively. Each had been handpicked for graduate fellowships by Noyes and all were given that broad-based education in physics, mathematics, and chemistry which Noyes deemed essential for physical chemists. And their training did not end with the Ph.D. As at MIT, where Noyes had created research positions for recent recipients of the Ph.D., so too at Caltech: there were ample opportunities to prolong the freedom of graduate school and postpone the burdens of full-time instruction. Badger profited from five years of research fellowships after receiving his degree; Yost and Pauling each enjoyed similar freedom for three years. Some of the money came from Caltech, some from the national fellowship programs of the philanthropic foundations, and some from Noyes's own pocket.[50] But by one device or another, means were found to ensure that these young scientists, and others as well, had the time and opportunity to develop their full potential. As at the Research Laboratory of Physical Chemistry, those who showed signs of unusual originality were treated with patience and generosity exceptional in American universities.

Although the Gates Laboratory was much larger than the Research Laboratory of Physical Chemistry, there were strong continuities in the management style that Noyes brought to the two institutions. There were also strong continuities in research, although these are less easily discerned both because of the bewildering number and variety of the investigations conducted at Caltech in the 1920s and because of the many new methods by which these investigations were pursued. Work on the behavior of solutions, especially solutions of strong electrolytes, upon which Noyes had been engaged since the 1890s, was continued at Caltech, as were investigations of the free-energy changes accompanying chemical reactions. The study of crystal structure by means of X-ray diffraction, which Noyes had planned to undertake at MIT, was prosecuted vigorously at Caltech. And even Tolman's research, which dealt primarily with the rates and mechanisms of chemical reactions and the development of a relativistic thermodynamics, had antecedents at MIT. Noyes, during the 1890s, had published several papers on reaction kinetics, and Tolman, as we have seen, first published on relativity theory while in Boston.

X-ray crystallography was the most important of these lines of work. The great majority of publications coming from the Gates Laboratory in the early twenties were studies of crystal structure, and for the next three decades Caltech would be known as the leading American center for crystallographic study. For a significant fraction of that time, it was the only important American center for such work. At MIT's Research Laboratory of Physical Chemistry, where Burdick had also built a spectrometer, research involving the technique was discontinued after the departure of Noyes.[51] During 1916 and 1917 a promising beginning had been made on X-ray crystal studies at General Electric Research Laboratory by the physicist Albert W. Hull. In William D. Coolidge, General Electric had one of the world's leading authorities on the design of X-ray apparatus, and in Irving Langmuir it had an outstanding physical chemist who was becoming more and more intrigued by problems of molecular structure. But the work was distant from the corporation's interests and was discontinued after a few years.[52] Another start was made during the war years at Cornell, when a visiting crystallographer communicated his interest in the technique to a graduate student, Ralph W. G. Wyckoff. Wyckoff made the interpretation of X-ray diffraction patterns his specialty, but ended up at the Geophysical Laboratory of the Carnegie Institution of Washington, where he had few opportunities to propagate his knowledge and enthusiasm among younger scientists.[53] The upshot was, as Wyckoff has noted, that

> of the three original studies of structure in America only one had a continuing existence in a university. In this way the California Institute assumed from the outset a dominant position both as a center of crystal structure research and as a source of trained personnel, able, when opportunities later arose, to initiate new centers in other institutions.[54]

In retrospect, it may seem remarkable that a technique so powerful as X-ray crystallography should have excited so few imaginations in the United States. After all, scores of American scientists had rigged up primitive X-ray machines in the 1890s, only months after Röntgen's announcement of the discovery of the penetrating rays, and the first European studies of crystal structure, especially those of the Braggs, attracted nearly as much attention in the American scientific community as Röntgen's results of twenty years earlier. Nevertheless, the comparison is deceptive. The generation of X rays involved little special knowledge and few experimental tricks; indeed, scientists found Röntgen's discovery astonishing largely because they unknowingly had been producing X rays for years with their cathode-ray equipment.[55] It was far more difficult to produce a meaningful diffraction pattern by exposing a crystal to a beam of X rays and to extract the structural information that was inherent in such patterns.

The theory, largely worked out by the Braggs between 1913 and 1916, was straightforward enough. A beam of X rays fired at a crystal would be diffracted by the planes of the crystal in much the same way that ordinary light is diffracted by a mechanically ruled diffraction grating. The major difference is that, in the case of the crystal, the incident radiation passes through not one but many gratings. If each plane of the crystal reflects a small portion, the reflected radiation will have appreciable intensity only if it is in phase, that is, if the scattered wavelets reinforce one another. This will occur if the difference in the distances traversed by the reflected X rays are whole-number multiples of the wavelength λ of the radiation. A little geometry reveals that the difference between the distances traveled by rays reflected from successive planes is $2d \sin \theta$, where d is the distance between the planes and θ is the angle between the incident ray and the reflecting plane. Hence, a simple equation may be framed to express the conditions for constructive interference of reflected X rays:

$$n\lambda - 2d \sin \theta,$$

where n is an integral number. With this equation, named after W. L. Bragg, it became possible to calculate the distance d between planes of atoms in a crystal.[56]

Simple in theory, the study of crystal structure could be exceedingly complex in practice. The X-ray spectrometer, which measured the intensity of reflected X-rays over a range of angles, permitted measurement of the angle θ, but it was a cranky device, as were all apparatus in the 1910s involving X-ray tubes. And even with the simplest specimens, the Braggs' method could be tedious, since it revealed only one parameter of crystal structure at a time; to determine the complete set of interplanar distances in a three-dimensional lattice required many time-consuming experiments. More troubling than the shortcomings of apparatus and experimental technique, however, were the irregularities of crystals themselves. The Braggs' method demanded the use of large, well-formed crystals, and these were available for only a small number of substances. And in crystals in which successive planes of atoms were not equidistant, or in which the planes had different scattering and reflecting power (as in substances containing atoms of disparate size), the Bragg equation did not hold, making it necessary to develop an array of auxiliary techniques for interpreting data.

Many of these problems were resolved or ameliorated in the late teens and twenties. X-ray tubes were made more powerful and reliable, a method was developed for obtaining diffraction patterns from powders (randomly ordered microcrystals), improvements were made in the analysis of data, and studies were made of the scattering power of different atoms. There were, in other words, steady and impressive improvements in the range and power of X-ray techniques.[57] Nevertheless, the study of crystal structure remained time-consuming and, with the introduction of ever more sophisticated techniques for analyzing data, more mathematically demanding as well. Few chemists were attracted to the new field, for while they had a strong and growing interest in molecular structure, few had the requisite combination of patience, mathematical cunning, and physical sense to make good X-ray crystallographers. Physicists had experience with X rays and stronger foundations in mathematics, but they generally lacked a strong interest in experimental work on molecular structure. Not only were molecules, by custom, in the domain of the chemist, physicists could hardly find their study attractive while the atom itself still presented so many intriguing puzzles. And so, X-ray crystallography grew up something of an orphan, finding a place in laboratories where local circumstances proved especially favorable: the Braggs' laboratory, where the father had a physicist's interest in X rays and the son a chemist's interest in molecular structure; the Geophysical Laboratory, where disciplinary boundaries were less important than in universities; and at Caltech, a new institution where a conscious effort was being made to integrate chemistry and physics.

The emphasis at the Gates laboratory was decidedly on the chemical uses of X-ray data. Caltech's crystallographers became known not so much for their contributions to the hardware of X-ray crystallography as for their rapid exploitation of others' innovations. They did not invent a better X-ray tube, but were among the first to use the high-powered tube developed by William D. Coolidge at General Electric. They did not devise the powder or rotating-crystal methods of obtaining Bragg reflections—the two major advances in crystallographic technique during the late teens and early twenties—but were quick to try these and other new methods. While other laboratories found X rays and their behavior as exciting as the geometry of crystals, the Gates Laboratory was single-minded in its pursuit of structural information.[58]

Initially, the targets were simple inorganic salts, such as potassium cyanide (KCN) and coordination compounds like ammonium hexafluorosilicate, $(NH_4)_2SiF_6$. These crystals were selected largely for their simplicity. So long as the aim was an exhaustive analysis of structure, an analysis entailing no assumptions extraneous to the X-ray data themselves, simplicity was essential. Despite the serious constraints on choice of targets, much of the work proved valuable. X-ray analysis, for example, provided powerful evidence in support of the configurations of coordination compounds that the Swiss chemist Alfred Werner had hypothesized in the 1890s. By the mid-twenties, however, the simple crystalline substances had been largely worked over. To undertake the analysis of more complex crystals, it was necessary to sacrifice some rigor, to filter out by other than crystallographic means the most probable structures from the unmanageably large set of possible structures. A powerful set of rules for doing this was published in 1929 by a postdoctoral fellow at Caltech, Linus Pauling.[59]

LINUS PAULING: FROM STUDENT TO TEACHER

Pauling was Noyes's greatest discovery. A graduate of Oregon State Agricultural College, Pauling had arrived in Pasadena in 1922 on one of the graduate fellowships that Noyes had helped create. His selection was a risk. Oregon State did not have a tradition of research and offered no advanced courses in physical chemistry; even its elementary course, in Noyes's eyes, appeared substandard. Nor was Pauling even a chemistry major. Urged by his family to acquire a marketable skill, ignorant of the opportunities open to chemists, and limited by the options available at Oregon State, Pauling had taken his degree in chemical engineering. Nevertheless, his teachers thought well of him—he was made a teaching assistant after his junior year and later was nominated for a Rhodes Scholarship. More important, he displayed enthusiasm and initiative. Independent study of back issues of the *Journal of the American Chemical Society* had given him a taste for theoretical chemistry, especially the work of Lewis and Langmuir on valence. His interest in molecular structure, a topic high on the agenda at the Gates Laboratory, must have caught Noyes's eye. After awarding him a fellowship, Noyes sent Pauling proof sheets of the new edition of his textbook, *Chemical Principles*, and told Pauling to work his way through its numerous problems before the start of classes. He also advised Pauling to consider X-ray crystallography as an area for thesis work.[60]

After arriving in Pasadena, Pauling quickly proved himself more than an ordinary student. He began research immediately on the determination of crystal structures and completed his first published work under the direction of Dickinson during his first year.[61] By the time he was awarded his Ph.D. three years later, he had published the results of six additional structural studies. Not only had he learned the art of crystal analysis, he also had followed a punishing regimen of studies in mathematics, physics, and physical chemistry: seminars on complex numbers, potential theory, and number theory with Caltech's mathematicians Harry Bateman and E. T. Bell; on statistical thermodynamics and relativity theory with Tolman; and on quantum theory with two of Europe's leading theoreticians, Paul Ehrenfest and Arnold Sommerfeld, both of whom visited Caltech during Pauling's student years. When another distinguished European physicist, Peter Debye, visited Pasadena in 1925, Pauling collaborated with him on a paper on the theory of solutions. The green student of 1922 had turned out to be a prodigy.[62]

Noyes had seen talent before, but none equal to Pauling's. After Pauling received his Ph.D., Noyes went to exceptional lengths to ensure that he would remain at Caltech. The principal threat was G. N. Lewis, whose nose for talent was as discriminating as Noyes's. When, in 1925, Lewis told Noyes that he wished to offer Pauling a job, Noyes neglected to convey the message. When Pauling won a National Research Council fellowship that same year, Noyes insisted that he stay in Pasadena rather than use it to travel to Berkeley. A few months later Noyes encouraged Pauling to resign his NRC fellowship and instead to apply for a Guggenheim fellowship to visit Europe. To expedite matters, Noyes offered to pay Pauling's passage and support him until the Guggenheim fellowship would take effect. As the historian Judith Goodstein has observed, Noyes was surely anxious to put as much distance as possible between Pauling and Berkeley, to which Noyes of course had a history of losing talented young chemists.[63]

But there were other reasons for Pauling to go to Europe. He was, in fact, only one among a score or more of able young American physical chemists to study there in the mid-twenties (see Table 6.2). This was the largest overseas migration of American physical chemists since the heyday of Ostwald's insti-

TABLE 6.2
Postdoctoral Study by American Physical Chemists in Europe, 1920–1929

Name	Ph.D.-degree-granting Institution	Dates in Europe	Sites of Postdoctoral Study
David C. Jones	Johns Hopkins	1921–22	London
E. O. Kraemer	Wisconsin	1921–22	Uppsala
J. A. Beattie	MIT	1922–23	Leiden
Harold C. Urey	California	1923–24	Copenhagen
Roscoe G. Dickinson	Caltech	1924–25	Cavendish
Donald H. Andrews	California	1925–26	Leiden
Rudolph L. Hasche	Johns Hopkins	1925–26	Berlin; Vienna
Robert S. Livingston	California	1925–26	Copenhagen
T. J. Webb	Princeton	1925–26	Zürich
Louis Harris	MIT	1925–27	Zürich
Martin Kilpatrick	NYU	1926–27	Copenhagen
Linus Pauling	Caltech	1926–27	Munich; Copenhagen; Zürich
Donald S. Villars	Ohio State	1926–27	Göttingen
Thorfin Hogness	California	1927–28	Göttingen
Bernard Lewis	Cambridge	1927–28	Berlin
John B. Taylor	Illinois	1927–28	Hamburg
John W. Williams	Wisconsin	1927–28	Copenhagen; Leipzig
Eli Lurie	MIT	1928	Berlin
Richard M. Badger	Caltech	1928–29	Göttingen; Bonn
John R. Bates	Princeton	1928–29	Berlin
Joseph H. Simons	California	1928–29	Cambridge
Don M. Yost	Caltech	1928–29	Uppsala; Berlin
Hubert N. Alyea	Princeton	1929–30	Berlin
Henry Eyring	California	1929–30	Berlin
Joseph E. Mayer	California	1929–30	Göttingen
Oscar K. Rice	California	1929–30	Leipzig

Source: National Research Council, *National Research Fellowships, 1919–1938* (Washington, D.C., 1938), pp. 28–39, supplemented by biographical information from *American Men of Science*.

tute at Leipzig. The students in this new wave differed from their predecessors in several respects. Like Pauling, they were far better trained: they knew more mathematics and physics, the great majority already possessed Ph.D.s, and many had begun to contribute to the journal literature. They were primarily interested in learning theory rather than experimental techniques. And many were going to institutes of theoretical physics rather than to the laboratories of physical chemists: Niels Bohr's institute in Copenhagen, Arnold Sommerfeld's in Munich, Erwin Schrödinger's in Zürich, and Max Born's in Göttingen. These young Americans were not seeking to become physical chemists but to become better and more complete physical chemists. And this meant obtaining an education in theoretical physics that few physics departments at home were yet able to offer.[64]

Pauling and his compatriots had strong reasons to seek such training. Physics was in turmoil. Each month in 1925 and 1926 seemed to bring news of another revolutionary idea, and the ideas were coming from these institutes of theoretical physics, mostly from young men with unfamiliar names like Pauli, Heisenberg, and Dirac. When Pauling left for Europe in March 1926, it was too early to predict exactly how this new work would affect chemistry, but alert physical chemists sensed that a new quantum physics was emerging that might revolutionize understanding of the atom. And a new theory of the atom was sure to carry implications for chemists interested in chemical bonds and molecular structure.

Given both their long-standing interest in these problems and their conviction that physical chemists needed to stay in touch with contemporary trends in physics, it is hardly surprising that A. A. Noyes and G. N. Lewis were especially insistent in encouraging their students to make the trip to Europe. Lewis could hardly have challenged the wisdom of Noyes's advice to Pauling when he himself was sending his best students abroad.[65] But the special concern that Noyes and Lewis had in promoting physical study among their students was more than simply a product of habit or general orientation. Both Noyes and Lewis were acutely conscious of seams that recently had opened up between chemistry and physics, and both were anxious to see those seams repaired. Paradoxically, Lewis, whose enthusiasm for physics was surpassed only by his love of chemistry, played a role in opening the most important of these breaches. It related to the topic that was closest to Pauling's heart, the structure of the atom and molecule.

The basic problem was that for much of the preceding decade chemists and physicists had been working with different and largely incompatible models of atomic structure. Physicists generally found that the model of the atom described by Niels Bohr in 1913 offered the most promising account of the available evidence, especially that gleaned by spectroscopy. Bohr's atom was a dynamical system in which electrons revolved about a positive nucleus in coaxial rings. Within their orbits electrons behaved in accordance with the laws of classical mechanics, much like planets revolving about a sun. But the orbits calculated by Bohr were restricted by quantum rules that had no basis in classical mechanics, and when electrons moved from one orbit to another, as when excited by heat, they did so in a decidedly nonclassical manner—that is, by leaps. Nonclassical too was their interaction with radiation. Electrodynamics demanded that a charged particle subject to acceleration continuously radiate energy; Bohr's electrons did not. Instead they emitted or absorbed a specific quantum of radiation with each leap between orbits. The frequency of this radiation was related to its energy by Planck's constant; its discreteness accounted for the spectral lines characteristic of the elements. It was the existence of line spectra that both stimulated Bohr to invent his atom and provided him with a means of testing it. And in its tests, it performed surprising well. As elaborated by Bohr and others, especially Arnold Sommerfeld, the Bohr model was shown to account quantitatively for much of the spectral data of the hydrogen atom and one-electron ions.[66]

Despite its successes, few if any scientists found the Bohr atom satisfying. Physicists' sensibilities were offended by its mixed classical and quantum parentage and even more by the arbitrary choices sometimes made by Bohr and Sommerfeld in calculating the positions and intensities of line spectra. They were also disturbed by the model's persistent failure to account for the spectra of multielectron systems.[67] Many physicists had come to believe that spectra held the key to the problem of atomic structure. The frequencies and intensities of spectral emissions, they felt, must contain clues to the distribution and energies of an atom's electrons. For them, the success of Bohr's model in predicting the optical properties of the simplest atom was its greatest asset; its failure when applied to larger atoms was its greatest liability.

Chemists did not share physicists' affection for spectra; indeed, they generally were bewildered by their colleagues' preoccupation with such data. The Bohr atom's achievements in that arena mattered less to them than its ability to account for such basic chemical phenomena as the formation of molecules. And here Bohr's model seemed incompatible with ideas that were so well supported as to seem facts of laboratory experience. The isomers of organic chemistry, for instance, offered overwhelming evidence of the persistence of molecules that differed from one another only in the orientation of the carbon atom's bonds. How, chemists wondered, could a system whose electrons were in constant motion explain the directedness and stability of those chemical bonds?[68]

More satisfactory, according to the chemists' criteria, was the model of the atom proposed by G. N. Lewis in 1916. Like Bohr, Lewis assumed that an atom's central positive charge was surrounded by electrons equal in number (in the case of neutral atoms) to the element's atomic number. Unlike Bohr, Lewis postulated that the electrons occupied concentric shells about the nucleus, the first containing a maximum of two electrons, the next two a maximum of eight each. These electrons, Lewis believed, were "held in position by more or less rigid constraints," whose positions and magnitudes were determined by the nature of the atom and of its partners in combination.[69] As depicted in Lewis's doodles of 1902 and his paper of 1916, the electrons, except for those of hydrogen and helium, were usually stationed at the corners of concentric cubes surrounding the positive nucleus (see Figure 6.1).

Lewis's model of the atom was physically farfetched. It did not explain how negatively charged electrons could maintain static positions in the vicinity of a large positive charge, it did not take notice of spectroscopic evidence, and it was not even internally consistent. Lewis, for example, had to suppose that the electrons at the corners of the carbon atom's outermost cube somehow were drawn together at the center of the cube's edges to explain the tetrahedral orientation of carbon's valences.[70]

Yet Lewis's principal concern was not the atom, but the molecule; not spectra, but such chemical phenomena as valence and ionization. And here the cubical portrait of the atom led Lewis to several penetrating and powerful insights. The picture of two atoms, each lacking one electron in its outermost cube, suggested that both atoms could achieve the stability that came with full complements of electrons by sharing a pair (see Figure 6.2). And this led, in turn, to Lewis's suggestion that the chemical bond itself typically could be identified with the sharing of two electrons.[71]

The concept of the bond as a pair of electrons held physically intermediate between two atomic kernels proved enormously rich. It pointed toward a resolution of the century-old problem: that of explaining how the ionic species like NaCl and molecules composed of identical atoms like H_2 could be understood within a single framework. In the former case, Lewis argued, the shared electron pair was closer to one atom than to another; in the latter, the shared electron pair was equidistant from the two atomic kernels. The position of the electrons, Lewis proposed, would determine whether a compound exhibited polar or nonpolar properties. Molecules in which the shared electrons were much more closely bound to one kernel than another would be readily ionized in solution, have high dielectric constants, form molecular complexes, and show other behavior typical of strong electrolytes. Molecules in which the shared electrons were held equidistant between two kernels would be difficult to ionize, have low dielectric constants, form no molecular complexes, and show other behavior typical of such very poorly conducting solutes as the hydrocarbons. And since polar and nonpolar molecules were not really different in kind but only in degree, one might expect to encounter a variety of molecules intermediate in their properties between the two extremes.[72]

But Lewis's concept of the shared-pair bond did much more than supply a unified account of the disparate electrical properties of molecules. It also accounted for why "odd molecules"—those with an odd number of electrons— were rare and highly reactive, related the valences of elements to their posi-

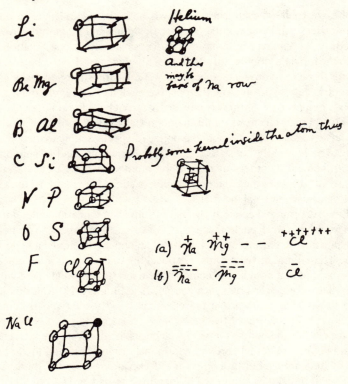

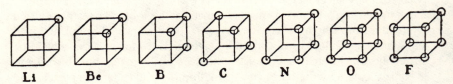

FIGURE 6.1. G. N. Lewis's memorandum of 1902 and his 1916 depiction of the outer electronic structure of the elements from lithium to fluorine. Reproduced from G. N. Lewis, *Valence and the Structure of Atoms and Molecules* (New York, 1923), p. 29 and ''The Atom and the Molecule,'' *Journal of the American Chemical Society 38* (1916): 767.

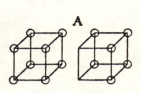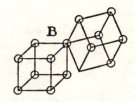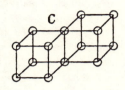

FIGURE 6.2. Different forms of the molecule I_2 according to G. N. Lewis (1916): (A) the molecule completely ionized, (C) joined by an electron pair bond, and (B) in an intermediate condition. Reproduced from G. N. Lewis, "The Atom and the Molecule," *Journal of the American Chemical Society 38* (1916): 775.

tions in the periodic table, suggested structures for hitherto puzzling molecules like the ammonium salts, and much more.[73] Like Bohr's model, Lewis's could both account for a wealth of data and generate testable predictions. It was not a mathematical model, but within the range of chemists' experience it was at least as effective as Bohr's was within the domain of spectroscopy.

History presents other instances in which specialists of different stripes held contradictory ideas about phenomena of common interest. Memories of the decades-long conflict between geologists and physicists over the age of the earth were still fresh in the 1910s.[74] In the case of the Bohr and Lewis theories, however, the resolution of conflict was much speedier. By 1923, when Pauling was a second-year student at Caltech, most of the contradictions between the two models had been resolved. Bohr, during the early twenties, gave greater attention to systems with several electrons and abandoned or modified several of the assumptions that had caused Lewis great difficulty. In particular, he ceased to envision electrons as sharing orbital rings and instead adopted the idea that electronic orbits could be arranged about the nucleus in three-dimensional shells. Like Lewis, Bohr suggested that the rows of the periodic table reflected the arrangement of electrons in these shells. Going beyond Lewis, he suggested that the filling of those shells was an orderly process governed by certain quantum rules. Lewis, for his part, de-emphasized the importance of cubes and conceded that the positions of electrons in his static atom might represent the average positions of more or less mobile electrons. Indeed, by 1923, Lewis pronounced himself willing to accept Bohr's model of the hydrogen atom in its entirety.[75]

Despite the convergence of Bohr's and Lewis's positions on atomic structure, there were still disturbing discontinuities. Lewis's empirical rule—that bonds were formed by shared electron pairs—afforded chemists useful guidance to the question of how atoms interacted with one another to form molecules. But the physical basis of both pairing and sharing was murky. Overwhelming chemical evidence supported his idea that shared electron pairs played a fundamental role in bonding, but precisely how and why two electrons could constitute a bond was a mystery.[76] Lewis suspected that magnetic forces held the key. Electron orbits, he suggested, act as magnets; when orbits were coupled their magnetic moments were neutralized, thus enhancing the stability of the system. But the notion that magnetic forces could overcome the powerful repulsive forces between two electrons seemed implausible to physicists, and Lewis could supply no quantitative evidence to support his idea.[77] Bohr and his fellow physicists, however, could offer no better explanation. They seemed able to specify the behavior of a single electron with some precision, but their theory broke down when it came to forecasting the behavior of systems containing two or more electrons. And unlike Lewis's theory, Bohr's could offer chemists little practical assistance when it came to assigning structures to molecules.

So, while the blatant contradictions between the theories of Bohr and Lewis may have been eliminated, one theory had not been reduced to the other. And neither was fully competent to explain the nature of the chemical bond or the structure of atoms and molecules. Here was an exciting agenda for research, especially for a chemist like Pauling, whose earliest interest had been in the structure of molecules, whose research had been in the determination of crystal structures, and whose training included generous doses of physics and mathematics.

The prospect was made all the more enticing by the revolution that occurred in physics in the months after Pauling received his Ph.D. During the summer of 1925, Werner Heisenberg, a young German physicist on holiday from his work at Bohr's institute in Copenhagen, developed a new mechanics that was competent to deal with the discontinuous phenomena of the quantum realm. This mechanics, based upon use of the esoteric algebra of matrices, obviated the need to shuttle back and forth between Newtonian and quantum techniques when interpreting spectra and proved capable of explaining a far wider range of spectral data than had the methods of Bohr and Sommerfeld.

A few months later, an Austrian physicist, Erwin Schrödinger, published an account of another form of quantum mechanics, based not upon matrix algebra but rather upon the more familiar differential equations of wave phenomena. Following a suggestion made by a French physicist, Louis de Broglie, Schrödinger treated electrons as having the properties of waves. Adapting the classical equation for wave propagation, Schrödinger derived his celebrated equation for the electron wave of the hydrogen atom:

$$\partial^2\psi/\partial x^2 + \partial^2\psi/\partial y^2 + \partial^2\psi/\partial z^2 + (8\pi^2 m/h^2)(E - V)\psi = 0,$$

where m is the mass of the electron, h is Planck's constant, $V(x, y, z)$ is the electron's potential energy in a field of force, E is a constant representing the total energy of the particle, and the wave function ψ seemed to be equivalent to the amplitude of the electron's matter wave. When Schrödinger solved the equation for ψ he found that it yielded physically meaningful results only when certain values of E were employed. These values corresponded to the energies that Bohr had assigned to the stationary or nonradiating states of hydrogen's electron. Schrödinger's new mechanics, like Heisenberg's, proved capable of achieving all that Bohr's older methods had and much more. When, in 1926, Schrödinger, Born, Heisenberg, and Pauli demonstrated in a variety of ways that matrix mechanics and wave mechanics were formally equivalent, confidence in the essential reliability of the new methods swelled.[78]

Although Pasadena was a continent and an ocean away from the sites of this revolution in physics, it was hardly isolated. While a student, Pauling had learned some of the uses and limitations of the old quantum theory from his seminars with Tolman and the lectures of Sommerfeld and Ehrenfest. In two papers published in 1926, Pauling explored ways to determine the stability of certain classes of molecules by using the best features of the Bohr model of the atom and the Lewis model of the chemical bond. And just before leaving for Europe he had attended lectures on the matrix mechanics delivered by another visitor to Pasadena, Max Born, Heisenberg's teacher and collaborator. At no other university, in America or Europe, was a chemist likely to learn as much about quantum physics as at Caltech. Noyes and Millikan had done their work well.[79]

After arriving in Munich in April 1926, Pauling plunged into the study of the new quantum mechanics. Attending Sommerfeld's first lectures on Schrödinger's wave mechanics, Pauling discovered that unlike Heisenberg's austere matrix mechanics, Schrödinger's formulation of quantum mechanics afforded him "a simple and satisfactory atomic model, more closely related to the chemist's atom than to that of the old quantum theory."[80] In particular, the wave function provided information about electron density about the nucleus. Just as in optics the square of the amplitude is proportional to the intensity of light, in wave mechanics the square of the Schrödinger wave function constituted a measure of the electron density. In the case of a hydrogen atom, the wave equation yielded a picture

of a nucleus embedded in a ball of negative electricity—the electron distributed through space. The atom is spherically symmetrical. The electron density is greatest at the nucleus, and decreases exponentially as r, the distance from the nucleus increases. It remains finite, however, for all finite values of r, so that the atom extends to infinity; the greater part of the atom, however, is near the nucleus—within 1 or 2 Å.[81]

The nature of this ball of negative electricity was not clear. Some physicists, following Schrödinger, interpreted it to be a stationary distribution of the electron through space—a smear, as it were; others, following Born, interpreted it as a time average of all the positions assumed by a point electron. But this issue mattered less to Pauling than the information that Schrödinger's equation yielded about the shape and size of the atom.

Pauling had already formed the habit of building physical models of atoms and molecules, and he had thorough knowledge of such empirical data on the sizes of atoms as could be obtained by X-ray diffraction studies.[82] Now he possessed a theoretically sound technique for calculating the size and shape of an atom. Taken separately, each method had serious liabilities. X-ray crystallography could provide reliable information on the sizes of atoms or ions only if those atoms were in crystals sufficiently simple to be amenable to rigorous analysis. Schrödinger's equation could be solved only for atoms and ions with a single electron. Systems including a nucleus and two or more electrons raised much the same problem that astronomers faced in treating systems of three or more bodies: exact solutions were impossible and approximate solutions could only be obtained by laborious calculations involving more or less plausible assumptions about the magnitudes of interactions. Neither approach as yet yielded much useful information about the nature of chemical bonds. But together the two methods might be made to complement one another.

During his eighteen months in Europe, Pauling threw himself into the task of integrating these traditions of research. His attack was two-pronged. One entailed the development of methods whereby the Schrödinger equation could be made to provide approximate solutions for multielectron atoms and ions. Here, the data of X-ray crystallography, patchy as it was, could furnish independent checks on the accuracy of his solutions. The other prong entailed the use of theoretically derived conclusions about the sizes and shapes of ions to simplify the task of analyzing new and more complex crystals by X-ray techniques.

His results were impressive. In a path-breaking paper published in the *Proceedings of the Royal Society* in 1927, Pauling used an atomic model based upon the wave mechanics of Schrödinger to calculate the radii of multielectron ions; these values were in tolerably good agreement with the interatomic distances obtained by X-ray analysis of crystals.[83] In that same year, he published a paper in the *Journal of the American Chemical Society* in which he used theoretically derived values of the ionic radii to draw conclusions regarding the structures of comparatively simple crystals. This paper was soon followed by others in which Pauling developed rules governing the structure of complex crystals.[84] These rules greatly simplified crystallographic analysis by restricting the set of plausible crystal structures to a small fraction of the universe of possible structures. In the eyes of the younger Bragg, Pauling's work was "a landmark" revealing "the fundamental principles underlying all inorganic crystals."[85]

Pauling's exceptional knowledge of the results of X-ray crystallography had helped him prove that Schrödinger's model of the hydrogen atom could be extended; his knowledge of wave mechanics and his techniques for using wave mechanics to estimate ionic sizes helped him to overcome an impasse in the analysis of crystal structures by X-ray diffraction. This movement back and forth between theory and empirical data, and between the realms of physicist and chemist, became characteristic of Pauling's work. So too was the emphasis upon visualizable structures. Pauling was quite prepared to accept the stricture laid upon science by Heisenberg's uncertainty principle, formulated while Pauling was still in Europe. It might be impossible to specify simultaneously the position and momentum of an electron. But it was possible, using the wave mechanics, to discuss the sizes and configurations of the clouds of negative electricity that surrounded nuclei. And with such knowledge, Pauling could build models of such larger entities as crystal lattices—models that could be both tested against and used as guides in the interpretation of empirical data.[86]

Pauling's interests, however, were not confined to ions and crystals; he was also vitally interested in molecules whose bonds were of a non-ionic character—those in which, according to Lewis, both nuclei had significant claim on shared electrons. Here, as in the case of the predominantly ionic bonds of inorganic crystals, the new wave mechanics proved capable of development, but only in conjunction with the now-famous exclusion principle of Wolfgang Pauli. This principle, first advanced in January 1925, stated that no two electrons could occupy the same atomic state, where atomic state is defined in terms of four quantum numbers.

Three of Pauli's four quantum numbers were already familiar to physicists. The first or principal quantum number appeared in Bohr's initial treatment of the hydrogen atom and was used in specifying the energy level of the atom. Additional quantum numbers were introduced as new layers of complexity were added to Bohr's relatively simple vision: an azimuthal quantum number to specify the angular momentum of electrons conceived to be moving elliptically at relativistic speeds, and an inner or magnetic quantum number to express the orientation of the plane of the electron's orbit in a magnetic field. In the Bohr-Sommerfeld model of the atom, each of these three numbers had a role in accounting for features of line spectra, each quantized some aspect of an electron's periodic motions, and each was restricted to certain values.[87]

During the early 1920s, Bohr and several of Europe's other leading physicists had turned their efforts to describing the electronic structure of the heavy elements by using the degrees of freedom afforded by these quantum numbers. The most successful of these schemes, proposed by E. C. Stoner in 1924, built up the periodic table by distributing electrons among energy levels according to their principal quantum number, and among sublevels according to their azimuthal quantum number (see Figure 6.3). The number of possible orbits in each sublevel was set equal to twice the value of the inner quantum number. Electrons, Stoner proposed, could enter a group until all the possible orbits were occupied. At that point the structure would be symmetrical and have the stability of the noble gases. Stoner's scheme coordinated a great deal of spectroscopic evidence relating to electronic structure, and it also made sense chemically, since the number of electrons he assigned to the outer subgroups

Element.	Atomic Number.	Level (n).	Sub-Level. (k,j.)						
			I	II	III	IV	V	VI	VII
			1,1	2,1	2,2	3,2	3,3	4,3	4,4
He...	2	K (1)	2						
Ne...	10	L (2)	2	2	4				
A...	18	M (3)	2	2	4	(4	6)		
Kr...	36	N (4)	2	2	4	(4	6)	(6	8)
Xe...	54	O (5)	2	2	4	(4	6)		
Nt...	86	P (6)	2	2	4				

FIGURE 6.3. E. C. Stoner's suggested distribution of electrons in the inert elements (1924). The distribution of the electrons in atoms is given by the part of the table above and to the left of the thick lines; *n* represents the principal quantum number; *k*, the azimuthal quantum number; and *j*, the inner quantum number. After the table in Edmund C. Stoner, "The Distribution of Electrons among Atomic Levels," *Philosophical Magazine 48* (1924): 722.

of elements generally corresponded to known valences.[88] But his rules for distributing electrons among subshells were essentially empirical. Moreover, his scheme did not account for certain features of the fine structure of the spectrum; in particular, spectroscopic data suggested that electrons assigned identical energy levels in Stoner's table could differ in their responses to strong magnetic fields.

By proposing a fourth quantum number, Pauli successfully accounted for these spectroscopic data and at the same time suggested an unambiguous rule for assigning energy levels to electrons. The fourth number, according to Pauli, could have two possible values, thus explaining both why seemingly identical electrons could behave differently and why Stoner had found the number of electronic orbits in closed shells to be twice the inner quantum number. With this extra degree of freedom, each electron in the atom could be assigned a unique energy state corresponding to a unique quartet of quantum numbers. And by postulating that no two electrons in an atom may occupy the same state, Pauli devised a simple rule for assigning electronic configurations to all the elements.

Pauli did not describe his quantum number in the language of geometry or classical physics. But by the end of 1925 two Dutch physicists, G. Uhlenbeck and Samuel Goudsmit, had proposed that the two-valuedness of Pauli's quantum number reflected a fourth periodicity in an electron's motions. It moved in elliptical orbits, the ellipses precessed around the nucleus, the orbital planes precessed in an imposed magnetic field, and the electron rotated on its own axis. This spin could be either parallel or antiparallel to the direction of its orbital motion.[89]

Formulated within the framework of the atomic models of Bohr and Sommerfeld, the idea of electron spin could be said to be the last significant contribution to the old quantum theory of the atom. When Uhlenbeck and Goudsmit's paper appeared in November 1925, Heisenberg had already published his first paper on matrix mechanics, and Schrödinger was writing his on wave mechanics. It soon became clear that pictorial representations of the motions of individual electrons were impossible. Nevertheless, the quantum numbers of earlier days did find a place in the new quantum mechanics: the first three were reinterpreted to describe the size, shape, and orientation of the electron clouds or "orbitals" described by Schrödinger wave functions, and the fourth as an inherent property of the electron which, although unmechanical, was universally referred to as "spin."

Chemists conversant with recent developments in physics greeted the hypotheses of Pauli and Uhlenbeck and Goudsmit with delight. Whereas Stoner's periodic system had told chemists little if anything about the chemical properties of atoms that they did not already know, the notion of a spinning electron seemed to support Lewis's long-standing belief that pairing of electrons played an important role in atomic and molecular stability. According to this new conception, a pair of electrons having identical values of the first three quantum numbers would have spin axes oriented in opposite directions. Paired in single orbitals they would neutralize each other magnetically. "It will be recognized by the chemist," wrote Lewis's student, Worth Rodebush,

that Pauli's rule is only a short hand way of saying what Lewis has assumed for many years as the basis of his magnetochemical theory of valence. If the electrons are paired in the atom magnetically, it is easy to see how two unpaired electrons in different atoms may be coupled magnetically and form the nonpolar bond.[90]

As it turned out, it was not quite so simple as Rodebush suggested. The magnetic forces between oppositely spinning electrons, for example, were minuscule by comparison with the electrical forces in a molecule—something physicists already understood by the time Rodebush wrote.[91] If a shared pair of electrons constituted the chemical bond, more than magnetic coupling was involved. More generally, the Lewis theory, even when reinterpreted in light of the exclusion principle and spinning electrons, did not constitute a quantitative account of chemical combination and offered no hope of becoming such. Such a treatment of chemical bonds would require more than the rules of Lewis and Pauli.

Pauling had the good fortune to be in Europe while solutions to these problems were beginning to take form. Among the friends he made in Munich were two young German physicists who shared his interest in the chemical bond: Fritz London and W. Heitler. After completing studies under Sommerfeld, Heitler and London went on to Zürich to work with Schrödinger. There, in the summer of 1927, they wrote a seminal paper on the hydrogen molecule.[92]

Their point of departure was Schrödinger's wave equation. It was no secret that wave equations could be written for molecules as well as for atoms simply by adding terms. Other physicists had also begun to explore ways of applying Schrödinger's equation to multinuclear systems, and with some success.[93] In principle, such equations could yield complete information about interatomic distances and molecular stability. The problems of solving these equations, however, were formidable. One was mathematical. For any system with three or more charged bodies, which is to say all molecules, exact solutions were out of the question. In the case of the simplest non-ionic molecule, H_2, the total energy of the molecule was dependent on the energy of repulsion of the two nuclei, the energies of each electron in the fields of the two nuclei, and the energy of repulsion of the two electrons. The best that could be achieved in the face of such complexities were approximate solutions. Here, however, mathematicians and physicists had amassed a rich store of techniques upon which Heitler and London were able to draw.

In setting up these equations, however, it was also necessary to make physical assumptions. In Heitler and London's treatment of the hydrogen molecule, one was crucial. Noting that according to Heisenberg the electrons of the two hydrogen atoms were in no way distinguishable from one another, London and Heitler proposed that they could exchange places without altering the system in any way. The very rapid exchange of places would significantly reduce the energy of the system and shorten internuclear distance. It would, that is, so long as the spins of the two electrons were antiparallel. If parallel, the interaction of the two electrons would increase the energy of the system and result in a repulsion rather than an attraction. In short, they suggested that in addition to the well-known forces of electrostatic attraction and repulsion between the parts of two hydrogen atoms, there were also hitherto neglected exchange forces. The opposed spin of hydrogen's two electrons made chemical combination possible, but the exchange forces that supplied most of the energy of the bond itself were electrical rather than magnetic in origin. Expressed in yet another way, the wave equation of the hydrogen molecule was not simply the product of the wave functions of the two original atoms A and B,

$$\psi_{AB}(1,2) = \psi_A(1)\,\psi_B(2),$$

where the integers 1 and 2 represent the electrons initially attached to atoms A and B respectively, but rather the wave equation was

$$\psi_{AB}(1,2) = \psi_A(1)\,\psi_B(2) \pm \psi_A(2)\,\psi_B(1),$$

where $\psi_A(1)\,\psi_B(2)$ and $\psi_A(2)\,\psi_B(1)$ are equally probable resonance structures of the molecule. When the spins of electrons 1 and 2 are antiparallel, the operation in the above equation is a sum; when the spins are parallel, it is a difference. In a large collection of hydrogen atoms, the two outcomes are equally likely.[94]

The method of Heitler and London resulted then in a wave equation for the hydrogen molecule with which it was possible to calculate approximate values of the molecule's ionization potential, heat of dissociation, and other constants. These predictions proved reasonably consistent with empirical values obtained by spectroscopic and chemical means. After Heitler and London's method was refined by other scientists, the fit between predictions and experimental data became astonishingly close. "We have here," wrote a young American physicist,

an example of the computation of a chemical affinity by unambiguous methods directly from the quantum postulates, without any undetermined constants or other arbitrary features. Is it too optimistic to hazard the opinion that this is perhaps the beginnings of a science of "mathematical chemistry" in which chemical heats of reaction are calculated by quantum mechanics just as are the spectroscopic frequencies of the physicist?[95]

But effective as Heitler and London's method proved in treating the hydrogen molecule, it was subject to limitations. The exact treatment of larger molecules, for instance, entailed mathematical difficulties that were insuperable (and remain so even with the aid of digital computers), and even approximate treatments demanded the use of implausible physical assumptions. Nevertheless, where their methods could not be readily extended, their theory of valence could. In a paper published in January 1928, London drew out the general principles of non-ionic valence bonds that had been implicit in his earlier work with Heitler. In essence, his theory was a quantum-mechanical version of Lewis's shared-pair theory—a version whose credibility was bolstered by the successful quantitative treatment of the simplest molecule. Like Lewis, London assumed that the atoms participating in the formation of a molecule retained their integrity. Like Lewis, he postulated that the pairing of two lone electrons with opposite spins is characteristic of valence bonds. And like Lewis, he equated the valence of an atom with its number of unpaired electrons; electrons already coupled with partners of opposite spin were chemically inert.[96]

By the time Pauling returned to the United States in the spring of 1928, he had become an enthusiastic advocate for the ideas of London and Heitler. In a review essay, written while he was still in Europe, Pauling gave American chemists an introduction to their treatment of the hydrogen molecule.[97] In a second and briefer paper published in the *Proceedings of the National Academy of Sciences*, Pauling presented a succinct review of London's theory of valence, cast in a nonmathematical language that would appeal to chemists. Moving freely back and forth between the electron dot formulas of Lewis and the exchange energies of Heitler and London, Pauling emphasized that, in his view, the two theories were essentially one. But while pointing out that the new quantum-mechanical theory of valence was "in simple cases entirely equivalent to G. N. Lewis's successful theory of the shared-electron pair," he also cited examples of how it was "more detailed and correspondingly more powerful than the old picture."[98]

One of Pauling's illustrations merits special attention, both because it marks the appearance of an idea that would prove enormously fruitful in Pauling's later work and because it effectively resolved the last major contradiction between physical and chemical versions of the atom. It involved the valences of the carbon atom. According to Stoner's and subsequent physical theories of electronic structure, the outermost electron shell of the unexcited carbon atom consisted of two discrete subshells, each containing two electrons. One of these subshells, by 1928 labeled the $2s$ orbital, was filled. Its electrons differed only in the value of their spin quantum numbers. The other subshell, labeled $2p$, consisted of three orbitals, corresponding to the three permissible values of the magnetic quantum number. In carbon's ground state, it was presumed that two of these orbitals contained an unpaired electron.[99]

This arrangement was consistent with the rules governing the distribution of electrons formulated by Bohr, Stoner, and Pauli, and it reflected the known spectroscopic properties of the carbon atom. It also accounted for the existence of molecules like carbon monoxide, in which carbon was bivalent. But, as every chemist knew, carbon could, and usually did, assume a quadrivalent state in which its four bonds were of equal strength. Lewis, in recognition of this fact, had placed four unpaired electrons in the outer shell of his carbon atom, accepting the embarrassment of molecules like carbon monoxide in exchange for a theory that accounted for the valence that carbon typically assumed. Neither view accounted for all the evidence, but the Lewis version of carbon's electronic structure came nearer to meeting the need of chemists, while the physical version of carbon came much closer to satisfying the principles of physics. Here was the greatest remaining discrepancy between the atomic models of the physicist and chemist.[100]

And here, according to Pauling, was where London's theory of valence showed its power. Noting that the energy separating the $2s$ and $2p$ subshells was known to be unusually small in the carbon atom, Pauling suggested that the energy released by the formation of four shared-pair bonds was sufficient to destroy the distinction between $2s$ and $2p$ sublevels. In later papers, he would refer to this as a hybridization of the bond orbitals. The result was that the carbon atom possessed not two unpaired electrons, but four; these occupied four orbitals that could be considered equivalent to one another, each directed to one corner of a tetrahedron. This hybridization of carbon's $2s$ and $2p$ orbitals, Pauling suggested, was not only possible, but the preferred outcome, thus explaining the unusual stability of saturated carbon compounds. The exchange or resonance energy released by the formation of four shared-pair bonds more than compensated for the energy consumed in the hybridization of orbitals.[101]

Pauling presented this suggestion in a single paragraph in his 1928 paper; it would be three years before he returned to the subject, later explaining that it took him that long "to work out hybridization of bond orbitals in a simple enough way so that I could get somewhere in a finite length of time in making calculations."[102] When he overcame this obstacle, results flowed smoothly and rapidly. Between 1931 and 1933, Pauling published a series of papers sharing the general title, "The Nature of the Chemical Bond."[103] In these, Pauling greatly extended the valence theory of London and his own notion of hybridization. The molecules he considered were far too complex for rigorous mathematical treatment; instead, he employed ingenious approximations to derive from quantum mechanics a few simple rules for the electron-pair bond, some being restatements of the principles of Lewis and London, others of his own formulation. He then demonstrated how these rules, in conjunction with a wide variety of experimental data, could be used to predict the structures, bond strengths, rotational motions, magnetic moments, and other properties of a wide variety of molecules and complex ions (see Figure 6.4).

It was in these papers, too, that Pauling developed the notion of resonance, showing, in his celebrated paper on benzene, that the energy and other properties of the molecule could not be explained in terms of the wave function of a single valence-bond structure, but instead were the product of a linear combination of the wave functions of several structures, each making a specifiable contribution to the properties of the whole.[104] Like the elephant in the tale of the blind men, the benzene molecule did not correspond to any one of its many descriptions, but rather was captured by all of them.

Pauling's papers, far richer than these few sentences can indicate, were not the only important articles on the chemical bond to appear in the late 1920s and early 1930s. Pauling owed much, for example, to a young American physicist, John C. Slater, whose technique for formulating approximate wave functions was critical for the treatment of complex molecules like benzene.[105] And the work of both Pauling and Slater rested on the contributions of Heitler and London, whose treatment of the hydrogen molecule was a prototype for their treatment of more complex systems.

Nor was Heitler and London's the only method for achieving approximate solutions of the wave equation for molecules. Between 1926 and 1935, other

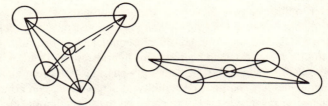

Tetrahedral, sp³ (2.000). C in diamond and compounds. Zn in [Zn(CN)₄]⁼, etc.

Square dsp² (2.694). Ni in [Ni(CN)₄]⁼, Pd in [PdCl₄]⁼, Pt in [PtCl₄]⁼, Au in [AuCl₄]⁻, etc.

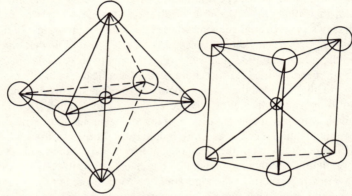

Octahedral d²sp³ (2.923) Co in [Co(CN)₆]≡, Pd in [PdCl₆]⁼, Pt in [PtCl₆]⁼, etc.

Trigonal–prismatic (2.983) Mo in MoS₂ and perhaps in [Mo(CN)₆]⁼.

FIGURE 6.4. Linus Pauling's depiction of the relative orientations of various electron-pair bonds (1932). Taken from Linus Pauling, "The Nature of the Chemical Bond. III. The Transition from One Extreme Bond Type to Another," *Journal of the American Chemical Society 54* (1932): 993.

physicists, Friedrich Hund and Robert S. Mulliken most prominent among them, invented an entirely distinct technique which treated the molecule not as a composite of two atoms held together by valence bonds but rather as an individual in which the constituent atoms lost their integrity.[106] Although starting with physical assumptions different from those of Heitler, London, Slater, and Pauling, the method of Hund and Mulliken proved equally competent in most cases and superior in some and was adopted by many physicists in the 1930s.[107]

Among the scientists who applied quantum mechanics to the molecule in the interwar era, however, Pauling was exceptional in at least one sense: he was the only chemist to hold a secure place in the first rank.[108] Wielding a technique that seemed a natural extension of the intuitively satisfying Lewis theory of valence, located at an institution that was already among the leading centers of research and graduate study in chemistry, and gifted with an exceptional talent for exposition, Pauling exercised enormous influence on the next generation of chemists. As Mulliken later acknowledged, Pauling "persuaded chemists all over the world to think of typical molecular structures in terms of the valence bond method."[109]

In many ways, Pauling's work may be seen as both the culmination of classical physical chemistry and the beginning of a new chemical physics. In the broadest sense, it realized Ostwald's dream for physical chemistry, for in Pauling's papers the principles of chemistry could be seen flowing logically from those of physics. "I remember clearly," Pauling later wrote,

> how much different my own thinking about molecular structure and the chemical bond was in 1935 from what it had been ten years earlier. In 1925 I had accepted the idea that the covalent bond consists of a pair of electrons shared jointly by two atoms. I had no sound theoretical basis, however, for any detailed consideration of chemical bonding . . . no way of distinguishing between the good ideas and the poor ideas about the electronic structure of molecules. By 1935, however, I felt that I had an essentially complete understanding of the nature of the chemical bond. This understanding had been developed in large part through the direct application of quantum mechanical principles to the problem of the electronic structure of molecules, and also in large part through the formulation of new empirical principles, based upon the observed properties of substances (especially thermodynamic properties and bond lengths, bond angles, and other details of molecular configuration), and usually suggested by quantum mechanical considerations.[110]

Chemistry, to be sure, was not made fully mathematical. Nor could its rules be deduced solely from the laws of physics. The complexity of chemical phenomena still demanded the use of approximations, many of a largely empirical character. Pauling's results, to which he and others often applied the word "semiquantitative," could only be as precise and rigorous as his refractory subject permitted. Even so, they were far more definite than those given by Lewis's entirely qualitative theory of valence. And, despite the frequent use of simplifying assumptions and empirical data, they were far more intimately related to physics. Pauling's treatment of the electron-pair bond and his concepts of hybridization and resonance were moored to quantum mechanics. His chemistry, if not fully determined by the new physics, was consistent with and dependent upon it.

Pauling's work was a culmination of the old physical chemistry in another sense. Ostwald had not only sought to link chemistry with physics, he had also sought to unify chemistry—to build a structure in which all branches of the science could find a place. The first generations of physical chemists had attained only part of this goal. Through the application of thermodynamics, especially to solutions, many hitherto disparate phenomena of analytical chemistry, inorganic chemistry, mineral chemistry, and electrochemistry were linked within a single explanatory framework. But ample as was the mansion of thermodynamics, it was too cramped to house all of chemistry. Despite Ostwald's efforts, room could not be found for much of organic chemistry, especially phenomena explained through reference to structure. The theory of the tetrahedral carbon atom, although developed by Ostwald's close associate, van't Hoff, found no place in the physical chemistry of the 1890s.

Ostwald had believed that an *allgemeine Chemie* would be a chemistry purged of structures and concerned only with energy transformations. His successors, among them A. A. Noyes and G. N. Lewis, gradually lost this faith—not so much because structural theory remained indispensable to organic chemists as because of the developing needs of their own discipline. Progress in the study of solutions seemed to demand that attention be given to bonds and the structures they held together. Yet even Lewis, who pursued this line farther than his colleagues, treated his work on valence and molecular structure as a diversion from his main interest, thermodynamics.

Pauling, while learning much from Lewis, Noyes, and others of their generation, reversed their priorities. For Pauling, the chemistry of solutions became a mine of data from which he could occasionally extract a valuable nugget; thermodynamics, a useful—but essentially unexciting—tool. The structure of atoms, molecules, and crystals were central to his work. Lewis's diversion had become Pauling's preoccupation.

Through a series of gradual changes physical chemistry had become a form of structural chemistry. How did this outcome fulfill Ostwald's ambitions? Certainly not in outward form. Pauling's taste for the *anschaulich*, his penchant for models, were contrary to all that Ostwald had preached in the 1890s.

But in a deeper sense, there was continuity. Ostwald's focus had been the nature of chemical affinities; a theory of affinity, he believed, would be the cornerstone of a truly general chemistry. In analyzing chemical bonds in terms of the forces acting between subatomic particles, Heitler, London, Pauling, and their colleagues developed such a theory of affinity. The quantum-mechanical treatment of chemical bonds offered guidance in the assignment of structures to molecules. But it also was able to indicate why some atoms combine to form stable molecules while others do not, to yield semiquantitative values of energies of formation and dissociation, and to explain optical and other physical properties of compounds. Pauling's chemistry was not just a structural chemistry.

Nor was it a chemistry of only certain types of molecules. The quantum-mechanical treatment of bonds eventually found application in treatments of the ionic crystals of geochemistry, the intermetallic compounds of metallurgy, the coordination compounds of inorganic chemistry, and the polymers of biochemistry. Nor, of course, should organic chemistry be omitted here. Most of the distinctions between inorganic and organic chemistry fell away when viewed from Pauling's perspective. His use of the concepts of hybridization and resonance to explain both the tetrahedral carbon atom of van't Hoff and the stable benzene rings of Kekulé were triumphs of the new quantum chemistry. But those same concepts found application in his treatment of the inorganic compounds of nitrogen and other elements. Far more than Ostwald's and van't Hoff's, Pauling's chemistry merited the name *allgemeine Chemie*.

Pauling's work, and that of those who were associated with or competing with him in the development of quantum chemistry, fulfilled Ostwald's grand ambitions for physical chemistry. But it also fulfilled many of Noyes's more modest dreams. Among Noyes's first publications was a review article on the problem of benzene.[111] Written while Noyes was an aspiring organic chemist, in the days before he met Ostwald, this article reviewed the various structures that Kekulé and others had proposed for the benzene molecule. It is unlikely that Pauling knew of this very minor contribution to the enormous literature on benzene when he developed his solution to the problem, but Noyes must have recalled his own uncertain effort when he read his protégé's treatment and smiled to see the circle closing.

Likewise, Noyes must have taken satisfaction in Pauling's clever treatment of the relation between ionic and covalent bonds. One of the mysteries of classical physical chemistry had been the question of why certain substances ionize in solution while others do not, or do so only to insignificant degrees. The reader may recall Paul Walden's characterization of this problem as the skeleton in the closet of physical chemists. This puzzle had led Noyes to postulate the existence of electrical and chemical molecules in 1907, it had led him and many of his colleagues at the Research Laboratory of Physical Chemistry to interest themselves in problems of molecular structure, and it had been among the issues that Lewis considered as he formulated his theory of the shared-electron-pair bond.

Pauling was a member of a generation for whom such a puzzle was beginning to seem as old-fashioned as Noyes's high collars. Debye and Hückel had given the problem of strong electrolytes an elegant solution while Pauling was a graduate student, and the remaining theoretical problems consisted largely of finding ways to treat concentrated and nonaqueous solutions. The quantum chemistry of the 1920s and 1930s, largely concerned with matter in the solid and gaseous states, made no direct contributions to the solution of these puzzles. Nor did Pauling. After a brief flirtation with solution theory in 1925, he left the field, never to return. Yet, Pauling's work did furnish chemists with a much clearer understanding of the physical differences between ionic and covalent bonds and of the nature of the molecules in which such bonds occurred. As Lewis had first suggested, nonpolar and polar species could be placed at two ends of a spectrum; the middle of the spectrum was populated with a variety of molecules whose bonds were of a mixed ionic and covalent character. Ease of electrolytic dissociation was not entirely dependent on bond type, but in general, the closer a bond was to the ionic end of the continuum, the more susceptible it was to disruption in polar solvents like water.[112]

Moreover, with the development of spectroscopy and of an understanding of the electronic structure of atoms, much of the mystery was stripped away from the process of ionization itself. Thermodynamics dictated that a salt such as sodium chloride would ionize in water only if the energy required to remove the sodium atom's outermost electron (the ionization energy) was less than the sum of the energy released by the union of the electron and the neutral chlorine atom (the electron affinity) and the energy of electrostatic attraction between the two resulting ions. During the 1920s, it became possible to calculate ionization energies and electron affinities from spectroscopic data and to correlate these properties with atomic number. Noyes himself dedicated two of his final contributions to theoretical chemistry to this topic.[113]

More important, as Pauling and others demonstrated, variations among the elements in such physical properties as ionization energy and electron affinity could readily be explained through reference to electronic structure.[114] The outermost electron in the atom of sodium, for instance, could be shown to be effectively screened from the nuclear charge by the spherically symmetric charge distribution of its ten inner electrons. The affinity of this atom for its outermost electron is, in fact, somewhat weaker than that of a single proton— hence sodium's very low ionization energy. Wave mechanics revealed, however, that a neutral atom of chlorine lacked a spherically symmetric cloud of electrons. An approaching electron is not well shielded from the nucleus, and hence is subject to a powerful force of attraction. Consequently, the electron affinity of chlorine is very high. In solution, where the stability of sodium and chlorine ions is enhanced by their electrostatic attraction to one another and to the ambient polar molecules of water, ionization will occur. If sodium chloride is vaporized, however, it will dissociate into neutral atoms rather than ions, the reason being that the atoms are separated by distances sufficiently large to make electrostatic interaction between ions negligible. Here, among other places, the new quantum chemistry was able to account for phenomena that classical physical chemistry could describe but not explain.

Small wonder that, when speaking of Pauling, Noyes sometimes sounded like John the Baptist describing the Messiah. In 1923 Noyes was referring to him as "the exceptional fellow from Oregon"; by 1927 Pauling had become "his understudy"; by 1931, "the most able chemist he had ever seen"; and by 1933 Noyes was telling foundation officials that "[w]ere all the rest of the Chemistry Dept. wiped away except P., it would still be one of the most important departments of chemistry in the world."[115] But others were hardly less enthusiastic. When, in 1939, Pauling drew together his earlier work into a book and dedicated it to G. N. Lewis, the older chemist responded:

I have returned from a short vacation for which the only books I took were a half a dozen detective stories and your "Chemical Bond." I found yours the most exciting of the lot. I cannot tell you how much I appreciate having a book dedicated to me which is such a very important contribution. I think your treatment comes nearer to my own views than that of any other authors I know and there are very few places where I could possibly disagree with you and those perhaps because I have not thought about the thing sufficiently.

Between 1929 and 1932 Pauling received visiting appointments at Berkeley, Manchester, and MIT, was awarded the first A. C. Langmuir Prize of the American Chemical Society, and was issued a call from Harvard.[116] Critics sometimes grumbled about "his advocacy of the doctrine of the infallibility of Pasadenean research" and "his somewhat pontifical style." But none could dispute his genius for making the esoteric concepts of quantum physics comprehensible to chemists, his encyclopedic command of the data of chemistry, or his exceptional skill at moving back and forth between the realms of experiment and theory.[117]

During the early 1930s, as Noyes's health deteriorated and Tolman became more and more preoccupied with issues in cosmology, Pauling became the linchpin of Caltech's program in chemistry. His consuming interest in molecular structure was reflected in the continuing emphasis on X-ray diffraction at Caltech, but also in a variety of new initiatives undertaken in the early 1930s. While on a visit to Germany in 1930, Pauling was shown Herman Mark and R. Wierl's apparatus for obtaining diffraction patterns of substances in the gaseous state with a beam of electrons. Their technique, which made it possible to determine interatomic distances and bond angles of many molecules that could not be studied by X-ray techniques, immediately excited Pauling's enthusiasm. He quickly arranged for the construction of similar apparatus in Pasadena, thus commencing a program of research that would become as important at Caltech in the 1930s as X-ray crystallography had been in the previous decade. About the same time, Richard M. Badger began his decades-long investigations in molecular spectroscopy, focusing on the infrared spectra of polyatomic gases. Among the early results of this research was a simple rule for determining interatomic distances from spectroscopic data alone, thus affording Pauling yet another means of assessing bond lengths. And while this work was underway, Howard J. Lucas, for many years Caltech's lone and neglected organic chemist, was engaged in studies of organic compounds that made free use of the idea of resonance and other concepts drawn from the writings of his colleague, Pauling. Important in its own right, Lucas's work became one of the principal inspirations for the development of physical organic chemistry in the United States. Finally, with the aid of the Rockefeller Foundation and the encouragement of Noyes, Pauling himself began to lead graduate students into research on the structure of biologically important molecules—work that would culminate in 1950 with the publication of a paper on helical structures in proteins.[118]

Colleagues interested in other topics sometimes felt overshadowed by Pauling's inordinate success, and Pauling, never very tolerant of obstacles, became more and more demanding as he learned his own worth. These demands sometimes threatened to fracture the harmony of the laboratory.[119] But like most good managers, Noyes recognized the need to adapt his plans to the materials at hand. A star of Pauling's magnitude came along once in a generation, and Noyes was willing to tolerate some grumbling in order to keep him. Caltech never conformed exactly to the hopes Noyes entertained for it in 1919; it did not become that utopia he had imagined. But the Gates Laboratory, despite occasional bickering, was enormously productive and influential—stronger than any other laboratory of comparable size in the world.

The development of a chemistry that was consistent with quantum mechanics stands among the most important accomplishments of twentieth-century science. Noyes lacked the mathematical skill, physical understanding, and youth to contribute to this achievement or even fully to appreciate it. Nevertheless, he was largely responsible for putting Caltech in the vanguard of those institutions in which the new quantum chemistry developed. Without Noyes, there would have been no Gates Laboratory. It was his reputation and connections that had brought chemists like Tolman and Dickinson to Pasadena and that had enabled Caltech to secure funds for their work. It was Noyes who was the talent-hunter who spotted Pauling and other green but promising prospects and who created the graduate fellowships that brought them to Pasadena. It was Noyes who chose to make X-ray crystallography an integral part of the Chemistry Division's research. And it was Noyes who designed a program that encouraged Caltech's chemists to view their science as intimately connected with physics. He was, like Liebig and Ostwald in earlier generations, a research director par excellence—a scientist capable of identifying and nurturing exceptional talent, of matching that talent with important and timely opportunities in his science, and of securing the resources necessary to realize those opportunities. His shortcomings as a scientist and original thinker did not prevent him from shaping his discipline and the institutions in which it evolved.

CHAPTER 7

A Dissenter's Decline

ARTHUR A. NOYES made skillful use of the abundant resources available to physical chemists in the postwar era. Not only was he a master of grantsmanship, he also possessed a sharp eye for talent and a superb sense of proportion. He could not solve a crystal structure or a wave equation. But he had the capacity to discern, well ahead of most of his generation, the chemical significance of X-ray crystallography, quantum mechanics, and other new ideas and techniques. Like an explorer following a river to its source, he doggedly pursued a few fundamental questions that he had first encountered as a young man—questions about differences between strong and weak electrolytes and the nature of ionization in solution. These questions ultimately led him into terrain that he was ill-equipped to explore, but into which he could despatch younger and better-prepared chemists.

Noyes's counterpart at Cornell, Wilder D. Bancroft, proved less adaptable. Like Noyes, Bancroft was able to harness new resources in the postwar years. Like Noyes, he had the ability to attract able young chemists to his laboratory and to infuse them with enthusiasm for their science. And like Noyes, Bancroft pursued a few topics energetically: already before the war he had made himself an authority on heterogeneous equilibria. Lacking in Bancroft, however, was that openness to the ideas and techniques of physics which was so characteristic of Noyes and his students. While Noyes embraced the physics of the twentieth century, Bancroft drew back. While Noyes encouraged his students to master as much mathematics as they could, Bancroft fostered among his students a suspicion of mathematical reasoning and any science that depended upon it. While Noyes was heir to the Ionists' ambition of building a dynamic *physikalische Chemie*, Bancroft cast himself as defender of a static and increasingly obsolescent *allgemeinen Chemie*—a chemistry that eschewed both molecular-kinetic and quantum theory, and that relied instead upon an essentially qualitative version of thermodynamics to tie together empirical data.

Ironically, it was the efforts of Noyes's research school that ultimately led to a twentieth-century version of Ostwald's general chemistry. By relating the properties and behaviors of molecules to their electronic structures, quantum chemists like Pauling constructed a new mental map of their science far more detailed and inclusive than any dreamed of by Ostwald. Bancroft, however, could take no satisfaction in this achievement. His mind was closed to the physics upon which it depended; his imagination was captivated by the chimera of a chemistry that would account for the phenomena of industry, biology, and daily life in terms comprehensible to nonmathematicians.

Bancroft's inflexibility proved a prescription for disaster. During the 1920s and early 1930s, as Noyes and his students marched from triumph to triumph, Bancroft and his dwindling band of disciples stumbled from one debacle to another. By 1932, when Noyes was basking in the reflected glory of Pauling's accomplishments, Bancroft was fighting to preserve his reputation and to save the institution to which he had devoted much of his career, the *Journal of Physical Chemistry*.

FROM PHASE RULER TO COLLOID CHEMIST

Bancroft's tune changed remarkably little over the decades, but there were some changes in his lyrics. His concern with the industrial and qualitative side of chemistry found new expression as his passion for the phase rule gradually was displaced by an equally intense concern with the chemistry of colloids.

Colloid chemistry was nothing new. Its roots went back well into the nineteenth century. The term "colloid" was coined in 1861 by the British chemist Thomas Graham to describe substances that would not pass readily with a solvent through a parchment membrane that was permeable to salt solutions. When isolated, these substances (gelatin, albumin, and a variety of other materials both organic and inorganic) did not appear as crystals but rather had an amorphous form—hence the name colloid, meaning "glue-like." Although practically inert to chemical reaction, colloids were easily displaced from solution by electrolytes. They also appeared to have molecular weights much higher than ordinary molecules, leading Graham and others to suppose that they might be combinations of smaller molecules. Whether these complexes were held together by valence bonds or by mechanical means was an open question.[1]

During the next four decades, the colloidal state and substances that could enter it received sporadic attention from a variety of scientists: biochemists, physical chemists, and physicists. This work resulted in a great deal of empirical data, a proliferation of special terms, and some valuable generalizations.[2] But until the turn of the century, colloids remained an esoteric topic, typically treated in footnotes rather than textbooks. This attitude began to change, however, during the decade or so preceding the world war.

The new interest in this old subject had several sources. In part it was due to advances in technique. New filtration methods, for example, facilitated the isolation of colloidal particles of uniform size; improvements in methods of measuring osmotic pressure and freezing-point depression afforded more reliable means of measuring molecular weights; and with the invention of the ultramicroscope in 1903, it even became possible to watch colloidal particles dancing in solution.[3] Important, too, was the growth of biochemistry and industrial chemistry, which provided an ever-larger number of scientists with opportunity and motive to study colloidal systems. Industry already produced scores of such colloidal products as paints and resins, and was eager to produce others, such as artificial rubber. Many biological substances were also of colloidal dimensions—too small to be observed under an ordinary microscope and too large to diffuse easily across organic membranes. Since Graham's day there had been chemists who suspected that the peculiar properties of living matter might find a basis in the chemistry and physics of colloids. During the years around the turn of the century, a growing number of chemically-minded biologists and biochemists, impressed by similarities between the *in vitro* behavior of colloidal systems and the *in vivo* behavior of protoplasmic constituents, came to agree with them.[4] By the eve of World War I, the ingredients were present for an explosion of interest in colloids, much as had happened in the case of the chemistry of solutions thirty years earlier. And just as an Ostwald had struck the match in the 1880s, so another Ostwald, Wilhelm's son Wolfgang, provided the spark in the 1910s.

Parallels between the roles of the Ostwalds, father and son, in the history of these two fields were not entirely fortuitous. Described by his sister as a young man who longed to found and provide for a "scientific family," Wolfgang Ostwald modeled his career on his father's.[5] At first the young Ostwald was attracted by the prospects in experimental biology, taking his doctorate at the age of twenty-one at Leipzig's zoological institute. But during two postdoctoral years in Berkeley, at the laboratory of Jacques Loeb, the young Ostwald's interests shifted more and more toward the physico-chemical basis of vital phenomena—a basis that he believed could be found in the chemistry of colloids. After returning to Leipzig in 1906, Ostwald soon secured a position in his father's institute, founded a journal, the *Kolloid-zeitschrift*, and commenced work on a textbook.[6]

Like the proverbial prophet, Ostwald was little honored at home. Despite his father's best efforts, it was not until 1935—nearly thirty years after he began teaching—that Wolfgang Ostwald attained the rank of ordinary professor at Leipzig. Although endowed with his father's energy and ambition, the younger Ostwald lacked his iron discipline. "Wolf did not possess the Apollonian composure of my father," wrote his sister; "rather his was a Dionysian temperament." Fond of the opera and a composer of *Lieder*, Ostwald had a penchant for the dramatic, but little of his father's skill in the laboratory, either as a researcher or director of research.[7] Nor did he have his father's good fortune: while colloid chemistry had its Ostwald, it had no Arrhenius or van't Hoff.

Despite the neglect of academic authorities at home, Ostwald was lionized on his many excursions abroad. His enthusiasm, broad knowledge of colloids, literary skill, and showmanship made him a superb lecturer. Perhaps the most effective of these tours was a visit he made to the United States in 1913–1914. Giving 56 lectures in 74 days at universities from New York to Nebraska, Ostwald stunned many listeners with his stirring account of the wonders and prospects of the new science of colloids. "We have," he wrote in the published version of these lectures,

> entirely overlooked the fact that between matter in mass and matter in molecular form there exists a realm in which a whole world of remarkable phenomena occur, governed neither by the laws controlling the behavior of matter in mass nor yet those which govern materials possessed of molecular dimensions. We did not know that this middle country existed, how large a number of natural phenomena belong to it, nor how greatly the degree of dispersion determined their behavior. We have only recently come to learn that every structure assumes special properties and special behavior when its particles are so small that they can no longer be recognized microscopically while they are still too large to be called molecules. Only now has the true significance of this region of the colloid dimensions—The World of Neglected Dimensions—become manifest to us.[8]

The special properties he referred to were essentially those which Graham and Graham's successors had discovered in the nineteenth century. Possessing surface areas large relative to their mass, colloids could be incredibly powerful adsorbing agents, capable of binding to themselves other substances many times their own weight. By concentrating other chemicals on their surfaces, they could function as powerful catalysts, either in protoplasm or in industrial vessels. Colloids could also assume strikingly different forms depending upon their degree of dispersion. As sols—that is, when in liquid suspension—some differed from ordinary solutions only in their poor diffusivity; as gels—that is, when removed from suspension by mechanical means or by the addition of salts—they typically became viscous, jelly-like substances. Left alone, these gels displayed a tendency to separate into two phases by throwing off a dilute colloidal solution, and this, Ostwald suggested, was clearly analogous to the biological process of secretion. If immersed in liquids, the gels swelled, and here, Ostwald suggested, was a colloidal analogue to edema and inflammation. Parading these and other phenomena before his audience, Ostwald insisted that "since the birth of the so-called classical physical chemistry of the molecular solutions some thirty years ago, no branch of physics or chemistry has arisen which can be compared in importance . . . with that of colloid chemistry."[9]

Few of the details of Ostwald's treatment were novel. What was new was Ostwald's strong emphasis on the omnipresence of colloids, on the possibility that their behavior was governed by as yet undiscovered laws, and on the urgent need of a new science that would uncover and explicate those laws. Colloids, he insisted, were not a peculiar class of compounds, but rather embraced all matter that was divided into particles intermediate in size between molecular and macroscopic dimensions. At a minimum, colloid chemistry deserved recognition as a well-formed and vital branch of physical chemistry—but better that it be granted even greater *Lebensraum*: a right ''to existence as a separate and independent science.''[10] ''Perhaps you think . . . that I am possessed of a colloid mania because I see colloids everywhere,'' he admitted in an aside that was typical of his style:

> Let me admit that I do see colloids everywhere, only I do not believe that because of this I must be adjudged insane. It is simply a fact that colloids constitute the most universal and the commonest of all the things we know. We need only look at the sky, at the earth, or at ourselves to discover colloids or substances closely allied to them. We begin the day with a colloid practice—that of washing—and we may end it with one in a bedtime drink of colloid tea or coffee. Even if you make it beer, you still consume a colloid. I make these remarks in full earnest and with the request that if I do not prove my assertions to your satisfaction, you challenge me in the matter.[11]

These lectures were not the products of a highly disciplined or self-critical mind. But they did find a wide audience and warm reception in the United States.[12] Some of those who attended no doubt were intrigued by passing acquaintance with one or more of the oddities of colloids, others simply out of curiosity to see and hear this new Ostwald. Whatever their reasons for attending, many went away impressed with Ostwald's prospectus. Those of a biological bent were intrigued by the hope that colloid chemistry would supply tools better suited to the task of explaining life than had so far been afforded by organic or physical chemistry. Physical chemists were excited by the discovery of an unanticipated area of ignorance adjacent to their own field. Colloids carried electrical charge, but did not appear to obey Faraday's law. When dispersed in solvents, they seemed to exist at or near the boundary between true solutions and mechanical suspensions, and it was by no means clear whether they were best treated by the laws of homogeneous or heterogeneous equilibrium. The peculiar adsorptive powers of colloidal surfaces could be described by empirical equations and classical thermodynamics, but the nature of the forces at work—whether chemical or physical—was unclear. Biochemists were especially eager to understand colloids, but physical chemists, too, had motives for studying these peculiar substances and considered themselves especially well-equipped to do so. Indeed, physical chemists so dominated research in the field that it generally came to be seen as a branch of physical chemistry.[13]

American entry into World War I cooled American chemists' infatuation with Ostwald, but did little to dampen their growing passion for colloids. Quite to the contrary: the war strengthened the movement. Colloid chemists could make a powerful case for the centrality of their subject to the war effort. The foods that soldiers ate, the charcoal in their gas masks, the smoke they used to screen their movements, and the rubber their vehicles moved on were all colloids, as were the catalysts used in the high-pressure fixation of nitrogen. Research on these and other materials brought a significant fraction of America's chemists into touch with the new subject during the war, and their exposure led to an explosion in the literature on colloids. Between 1909–1913 and 1914–1918, articles on colloid and surface chemistry in the *Journal of the American Chemical Society* quadrupled; during the subsequent five years the number doubled yet again.[14]

Wilder D. Bancroft was among the first American physical chemists to develop an interest in colloids. By 1910 he was advising students that colloidal phenomena might prove important in the study of biological problems; soon thereafter he began systematic study of the literature on emulsification in connection with work on the chemistry of the photographic plate.[15] This flirtation with colloid chemistry became an engagement during the winter of 1915–1916, when a fire destroyed his apparatus and very nearly consumed the entire chemical laboratory. Until the new Baker Chemical Laboratory was completed in 1923, Bancroft and his students had to work, he later wrote, "in the patched-up ruins of a building of which the whole top story had been burned off, . . . [O]ur equipment consisted chiefly of an inadequate supply of burettes and beakers. That meant that we had to do colloid chemistry whether we wanted to or not."[16]

"This was very likely a blessing in disguise," Bancroft added, and not without reason. His laboratory's research on alloys had made little progress after Carnegie funds were cut off in 1912, and subsequent work on electrochemical and photochemical problems had led to few concrete results. In all of these areas, other laboratories, better equipped and led by scientists with stronger credentials in physics and mathematics, had seized leadership. In colloid chemistry, however, Bancroft found a field that seemed well suited to his interests and talents. It was, for one thing, a subject of great and growing practical significance. Echoing Ostwald, Bancroft described it as "the chemistry of everyday life," since it dealt with matter in the form of bubbles, drops, grains, filaments, and films.[17] A knowledge of colloid chemistry was essential, Bancroft thought, to understanding the behavior of such disparate substances as cement, oils, paints, plastics, inks, wine, dairy products, smoke, and fog, since all consisted of matter in a finely divided state. Secondly, since intensive study of the chemistry of colloids was just beginning, the subject was at a stage of development, in Bancroft's view, when mathematical ingenuity was not essential; it could benefit, he believed, from treatment with broad, qualitative strokes. Bancroft, in other words, was drawn to colloid chemistry for very much the same reasons he had been attracted to the phase rule two decades earlier.

After service during the war as chief of the editorial section of the Chemical Warfare Service, Bancroft attempted to consolidate a position of leadership among American colloid chemists. Drawing on knowledge of the industrial uses of colloids that he had acquired during the war, Bancroft published a textbook in which he emphasized the abundant practical uses to which an understanding of colloids could be put. Relying on a few thermodynamic principles and eschewing all but the most elementary mathematics, the book was intended, Bancroft wrote, for "those who are interested in colloid chemistry as chemistry rather than mathematical physics." In addition to writing this textbook, which won sufficient readers to go through three editions, Bancroft also turned over more and more of his *Journal* to articles in the field. By the mid-1920s nearly half of each volume of the *Journal of Physical Chemistry* dealt with colloids. He also initiated an undergraduate lecture course on colloid chemistry at Cornell and repeatedly sought funds for an Institute of Colloid Chemistry. The aim of the latter was to be the investigation of phenomena fundamental to those industries whose processes or products involved colloidal solutions. Although this institute remained a dream, Bancroft and his students did conduct a series of studies on the preparation and properties of emulsions, on the theory of emulsification, and later did attention-grabbing work on the biological effects of drugs as interpreted from the standpoint of colloid chemistry.[18]

Bancroft gained a following among colloid chemists in the 1920s not only because of his textbook, his editorial work, and his own contributions to the literature, but also because he championed a point of view. He was, together with Wolfgang Ostwald, a leading proponent of what has been called the "isolationist" position among colloid chemists.[19] For Bancroft, as for Ostwald, the chemistry of colloids was essentially distinct from the chemistry of ordinary molecules. This is best exemplified by his treatment of proteins, substances which, according to Bancroft and his allies, were colloidal aggregates—complexes of many smaller molecules—and not chemical compounds. These aggregates, according to Bancroft, did not behave at all like simple molecules of definite proportions. For one thing, they did not form salts. So, for example, when a protein such as wool was colored with a dye or acquired a positive charge through exposure to an acid, it did not participate in a chemical reaction, but rather in an adsorption. The dye or hydrogen ion was bound to the surface of the protein by electrostatic forces, forces of physical cohesion, or by residual chemical affinities, rather than by valence bonds. The result was not a new chemical compound of definite proportions, but a new and larger colloidal aggregate whose properties were modified by the adhesion of additional molecules or ions to its surface. The model he applied to these processes, in other words, was not that of the chemical reaction but rather that of the essentially physical binding of gases to activated charcoal.[20]

Nor did Bancroft believe that proteins were capable of entering the state of solution. In water, particles of protein remained suspended because they were constantly agitated by collisions with the far smaller molecules of the solvent, but unlike true solutions these suspensions consisted of two phases rather than one. The particles were visible under the ultramicroscope and were separable from the solvent by mechanical means, such as dialysis or ultrafiltration. By definition, Bancroft insisted, such heterogeneous mixtures did not obey the laws of true solution.[21]

By no means did all students of colloids share Bancroft's isolationism. In Zürich, the organic chemist Hermann Staudinger argued forcefully for a unionist position; many colloidal particles, he insisted, were nothing more than giant molecules which could be synthesized by classical methods from simpler units. In Paris, the dean of French physical chemists, Jean Perrin, believed that he had proved that dilute colloidal solutions obeyed the laws of solution, as did The Svedberg in Uppsala. And no less a physicist than Albert Einstein had reached the same conclusion by theoretical reasoning.[22]

Closer to home, Albert P. Mathews, former head of the physiology department at the University of Chicago and professor of biochemistry at the Medical College of Cincinnati, roundly condemned Bancroft, Ostwald, and other colloid chemists for confusing descriptive and explanatory terms. ''Adsorption,'' Mathews wrote, ''is a name descriptive of a physical (or chemical) phenomenon. It says . . . nothing as to the cause or causes which produce the phenomenon described by it.''[23] Bancroft, by treating practically all processes in which colloids participated as adsorptions and their products as adsorption complexes, simply confessed his own ignorance and confused his readers:

> The exact mechanism of this surface condensation or adsorption is not yet clear. The fact that the energy relations are satisfied gives us no picture at all of the mechanism of the process, and before we really understand it we must have such a picture.[24]

In fact, Mathews asserted, such a picture was beginning to emerge in the form of evidence that suggested that proteins were nothing more than larger versions of the definite chemical compounds of organic chemistry. Containing both free amino and carboxyl groups, they were capable of forming salts with both acids and bases. The protein did not physically adsorb dyes or hydrogen ions, as Bancroft suggested, but rather reacted with them to form compounds no different in kind from sodium chloride. Even if a protein could not be isolated as a definite chemical compound, ''it leads only to confusion to treat it as if it were a compound of some other nature,'' for in that case, ''the extreme specificity of the reactions, the clear-cut substitutions of one base or acid for another, pointing clearly to a chemical union, remain wholly unexplained.'' ''[T]he whole subject [colloid chemistry] at the present time,'' Mathews concluded, ''is a perfect morass and those who wander in this field with physical adsorption for their lantern climb out of one mud hole only to fall into another.''[25]

During the early 1920s, however, the most outspoken of unionists was Jacques Loeb. Since hosting the young Wolfgang Ostwald at Berkeley, Loeb had grown increasingly dubious about the colloid movement and its works.

Now ensconced in the splendor of the Rockefeller Institute, Loeb lost all patience with those who claimed that the colloid dimension was subject to special laws. In his *Proteins and the Theory of Colloidal Behavior*, published a year after Bancroft's book, Loeb deployed all his considerable knowledge and ingenuity to demolish Bancroft's position. Using a relatively new technique for measuring the concentration of hydrogen ions in solution, Loeb prepared proteins in what he understood to be a non-ionized form and described, in quantitative terms, their reactions with acids and bases. Contrary to the adsorption theory, he found that a protein such as gelatin in a non-ionized state was essentially inert in the presence of anions or cations, although on either side of this isoelectric point it would readily combine with ions of opposite charge in stoichiometric proportions. The clear implication was that gelatin was an amphoteric electrolyte of definite composition which could behave like an acid or a base depending upon the acidity of its solution. Having established to his satisfaction that proteins were chemical compounds, Loeb proceeded to demonstrate that the laws of solution could be applied to predict such physical properties of protein solutions as their osmotic pressure. Loeb's book summarized and significantly strengthened the case for proteins as macromolecules.[26]

Beset on all sides by critics, many with impeccable credentials, Bancroft gave little ground. Bancroft dismissed the earlier work of Einstein and Perrin, which demonstrated that, from the standpoint of kinetic theory, no distinction could be made between molecules and colloidal particles. This result Bancroft called "an over-hasty conclusion" that conflicted with chemical evidence and would eventually be disproved.[27] And while praising Loeb's experimental technique, Bancroft denied the force of his conclusions. Peppering Loeb with small but troublesome questions about the breadth of his evidence and the quality of his reasoning, Bancroft insisted that all of Loeb's results could also find explanation within the framework of adsorption theory.[28] Loeb, for his part, was frustrated by Bancroft's obtuseness. "In the protein industries in the United States," he wrote in 1923,

> the new work on proteins [Loeb's own research] has been accepted. Of course, the medical quacks probably will continue to stick to the old fashioned colloid chemistry which they find remunerative. One occasionally finds also a physical chemist like Wilder Bancroft sticking to that confusion, but I do not think Bancroft has ever done an experiment in his whole life, and since he is neither a mathematician nor capable of rationalistic [sic] thinking, I think he can be ignored.[29]

Despite Loeb's self-confident tone, the controversy was by no means over. Loeb's evidence, though powerful, was circumstantial. He could not specify the composition and structure of a single protein molecule, nor adduce an unambiguous example of a protein that had been isolated in pure form. Furthermore, calculations of the molecular weights of proteins, based on Loeb's assumptions, yielded values so high as to make many chemists cringe: 33,800 for egg albumin, 45,000 for serum albumin, and 50,000 for hemoglobin. That molecules so much larger than the ordinary molecules of organic chemistry could exist seemed inherently implausible to many chemists.[30] So, although Loeb's position gradually won adherents, the isolationist viewpoint retained the allegiance of a significant number of reputable American and European scientists. Only after 1930, as evidence drawn from the study of polymeriza-tion reactions, the electrolytic behavior of proteins in solution, and their sedi-mentation in ultracentrifuges became overwhelming, did the molecular nature of proteins become generally acknowledged to be a fact.[31]

This, of course, did not destroy colloid chemistry. It was still possible and useful to study the colloidal properties of proteins and other large molecules. It is just that after 1930 few chemists did so with the romantic notion that those properties were inexplicable under the laws governing matter in a molecular condition. Bancroft, however, continued to think of proteins as colloidal ag-glomerates into the mid-1930s. His stubborn refusal to accept the clear impli-cations of evidence developed through use of these new techniques cost him much of the reputation he had earned through his earlier research on hetero-geneous equilibria.

BANCROFT'S *JOURNAL* AND AMERICAN PHYSICAL CHEMISTS

Bancroft, champion of the isolationist position in colloid chemistry, was also an isolationist in a larger sense. Through the 1920s he fought a prolonged rearguard action against those who he believed would make physical chemis-try a department of physics. His skepticism about Einstein's proof that col-loidal particles and molecules are essentially indistinguishable was typical of his reaction to twentieth-century physics; his impatience with Loeb's quanti-tative reasoning was symptomatic of a deeper aversion to the "tyranny" of numbers.

These attitudes were reflected in his editorship of the *Journal of Physical Chemistry*. During the 1920s, its pages only hinted at the enormous progress being made in the study of the free energies and entropies of chemical reac-tions or in the understanding of the chemistry of dilute solutions. Nearly ab-sent were articles on the problem of valence, on the use of X-ray diffraction techniques, and on the applications of quantum mechanics to the problem of molecular structure. A reader of Bancroft's journal could hardly discern in its pages evidence of the revolutionary changes in physical chemistry then under way.

It might be thought that a journal with such a decidedly anti-progressive stance would soon expire. In fact, Bancroft had no easy task in sustaining the *Journal of Physical Chemistry* during the 1920s. Its continued existence de-pended upon his money, his skill in fund-raising, his papers, and papers of friends and students. That it survived testifies to Bancroft's dedication and ingenuity. It also suggests that his message found a sympathetic hearing among at least some of his colleagues. Although out of step with the most eminent of his contemporaries, Bancroft was not simply a crank or a fool, and his prejudices were shared by not a few chemists. Ultimately, however, these resources proved inadequate. In 1933 Bancroft was forced to retire as editor of the journal he had founded nearly four decades earlier. By then the *Journal* itself had become a point of contention between those who would maintain and strengthen their discipline's ties with physics and those who would chart an independent course. The story of the end of Bancroft's editorship may be read, then, as having more than simply biographical interest; it also reflects the nearly total victory of the tradition represented by Noyes, Lewis, and Pauling in the competition to define the aims and agenda of physical chemistry.

Before the war, the *Journal of Physical Chemistry* had been Bancroft's voice: an outlet for his own publications and those of his students and associates. The *Journal* lacked broad support among physical chemists and operated at a deficit. Nevertheless, it had a nucleus of about five hundred subscribers, and its annual deficit was never too large for Bancroft's purse. The war changed this equation. Postwar inflation forced up costs of production and distribution while it eroded the purchasing power of the *Journal*'s subscribers: professors and university libraries. For the *Journal* to survive, Bancroft felt, it was necessary to maintain a stable subscription price. Each year this was achieved at greater personal cost until finally, during the recession of 1921, even Bancroft flinched.[32]

Bancroft responded to the crisis in two ways: by seeking to broaden the *Journal*'s subscriber base and by seeking a patron who might supply cash without threatening his editorial autonomy. He met early success in both aims. By appealing to the widespread concern that Germany might reestablish a position of supremacy in chemical publications, Bancroft obtained formal votes of confidence from the American Chemical Society, the Chemical Society of London, and the British Faraday Society.[33] Although Bancroft received no financial support from these arrangements, he was able to broaden and strengthen his board of associate editors. The new editors would lend their prestige and, it was hoped, their papers to the *Journal*; Bancroft would continue to enjoy the final say as to the contents and format of publication.

While these negotiations were under way, Bancroft also obtained the patron he sought—the Chemical Foundation. Organized to administer royalties from the licensing of German patents seized during the war, the Chemical Foundation, under the leadership of its president, Francis Garvan, awarded the *Journal of Physical Chemistry* a $3,000 annual subsidy in October 1921. After Bancroft successfully concluded negotiations with the British and American chemical societies, this subvention was increased. By 1931 the Chemical Foundation was pouring over $17,000 a year into the *Journal of Physical*

Chemistry. These funds allowed Bancroft to hold his subscription rates down and, at the same time, to enlarge the *Journal*. The artificially low price also stimulated sales: by 1932 its circulation had doubled.[34]

An enlarged editorial board and an annual subvention relieved Bancroft of many of his immediate problems. But there were costs attached. For one thing, his appeal to the American Chemical Society for recognition alienated those who were closely identified with the *Journal of the American Chemical Society*: its former editor and chairman of the department of chemistry at the University of Illinois, William Albert Noyes; its new editor, Arthur B. Lamb of Harvard; and many of those physical chemists who habitually published in its pages and served on its editorial committees, among them G. N. Lewis and Arthur A. Noyes.[35] These chemists had long been scornful of Bancroft's ideas and editorial policies. Now they were being asked to give Bancroft and his journal a vote of confidence, a vote that some feared might be misunderstood as an endorsement of his views.

Predictably, the proposal precipitated a clash. Bancroft's critics argued that the *Journal* had outlived whatever usefulness it had once possessed. Edward Washburn and G. N. Lewis, both members of the committee organized to consider the question, went so far as to charge that the very name *Journal of Physical Chemistry* now perpetuated a fiction. "The fact is," wrote Lewis,

> that physical chemistry no longer exists. The men who have been called physical chemists have developed a large number of useful methods by which the concrete problems of inorganic chemistry, organic chemistry, biochemistry, and technical chemistry may be attacked, and as the applications of these methods grow more numerous, it becomes increasingly difficult to adhere to our older classification. Nowadays a journal of physical chemistry can hardly differ from a journal of general chemistry.[36]

And how, Lewis added, could the *Journal of Physical Chemistry* compete with the larger, far richer, and better-known *Journal of the American Chemical Society* in that realm?

As Lewis saw it, Bancroft's arguments were self-serving. The goal of Ostwald had been attained. Physical chemistry had become synonymous with general chemistry. There was no longer a need for a special journal to defend the interests of physical chemists before a hostile establishment; physical chemists now were the establishment. Bancroft was not representing the interests of physical chemistry but rather was speaking for a small party of malcontents—chemists unable or unwilling to enter the twentieth century.

From Bancroft's perspective, Ostwald's goals seemed very far from realization. The highly touted advances in chemical thermodynamics and solution theory, Bancroft asserted, were largely illusory. The quantitative chemistry of solutions, the chemistry of Noyes and Lewis, Debye and Hückel, was still a chemistry of dilute aqueous solutions. Concentrated aqueous solutions and nonaqueous solutions still resisted treatment, and Lewis's distinction between concentrations and activities was simply a fig leaf concealing physical chemists' ignorance. Moreover, there was need for a concentration of effort if the tools of physical chemists were to be successfully applied:

> While the organic chemist does make conductivity determinations, freezing-point and boiling-point measurements, and does study reaction velocity occasionally, it would be a bold man who would claim that anything more than a beginning has been made in the application of physical chemistry to organic chemistry. We are only just working out the theory of colloid chemistry and it will be quite a while before we can say that we really understand about dyeing, leather, paints, rubber, pottery, cement, etc. The development of pure and applied colloid chemistry will be hastened and the impression produced will be greater if the articles along these lines can be published together.[37]

Nor was it only his discipline that needed *The Journal of Physical Chemistry*. So did the nation. It would be a tragedy, Bancroft suggested, if American chemists, having seized international leadership, should now forfeit their gains. And such, Bancroft suggested, would be the result if the American Chemical Society turned its back on specialized journals such as his own. The demise of the only English-language journal devoted specifically to physical chemistry would only redound to the benefit of Germany. This was a potent argument for chemists with fresh memories of the war. "We have heard much of dyes as a 'key-industry,' " wrote Eugene C. Bingham, the chair of the committee to consider the future of the *Journal of Physical Chemistry*,

> but *the publication of knowledge is a true key-industry*. It is creditably reported that important articles in America are already seeking publication in Germany for lack of opportunity here. How stupid it would be for the Anglo-Saxon, slowly awakening to the value of research, to hand over the results of his own studies to his clever Teutonic rival . . .[38]

As it turned out, Bancroft won his recognition. A majority of the members of the committee were sympathetic to this former president of the society who was now seeking their help. But it was a victory empty of much meaning. G. N. Lewis, T. W. Richards, John Johnston of Yale, Edward W. Washburn of Illinois, and others rejected invitations to join Bancroft's new editorial board. While achieving his immediate aim, Bancroft had done nothing to win the good will of those whose support would have mattered most.[39]

Bancroft's securing of support from the Chemical Foundation likewise brought immediate benefits but long-term costs. Alarmed by the chaotic condition of the *Journal*'s accounts, Francis P. Garvan, the lawyer who presided over the Chemical Foundation, insisted that his associate, William Buffum, be appointed business manager of the *Journal*. Intent on making the *Journal* a self-supporting enterprise, Garvan and Buffum maintained constant pressure on Bancroft to show results that would justify their investment. Telling his patrons what they wanted to hear, Bancroft described the enormous practical consequences that might ensue from putting the largely empirical practices of the soap, paint, dyeing, and tanning industries on a scientific footing. The *Journal*, he suggested, would become the outlet for the papers of the great new institute of applied colloid chemistry that he hoped to build at Cornell. When that dream foundered for lack of patrons, Bancroft turned to medicine, suggesting that research on colloids would ultimately lead to the discovery of powerful new drugs that might even cure cancer. Once again, Bancroft offered sizzle but no steak. A ten-thousand-dollar grant from Cornell's Heckscher Research Fund and a smaller award from the Cancer Research Fund of the University of Pennsylvania led to no results. Instead of reaping kudos for the Chemical Foundation, their investment in Bancroft and his *Journal* was only yielding failures and criticism.[40]

By the end of 1929, Bancroft was growing desperate to achieve a spectacular coup—a scientific success that would at once vindicate his commitment to colloid chemistry and capture the headlines (and subscriptions) that Garvan and Buffum seemed to covet. Nor did he crave fame simply to satisfy his patrons. Now in his sixties, Bancroft realized that opportunities to match the achievements of his grandfather and his European mentors were fast dwindling. "One cannot count on having somebody else exploit one's discoveries," he would soon be telling a graduating class at the University of Southern California.

> Consequently, [the scientist] must make up his mind to sell himself to the scientific world if he is not going to run the risk of being classified as a man whose ideas, thought excellent, came when the time was not ripe for them. . . . Since the greatest discoveries are likely to be ones for which the world is least ready, we see that the greatest scientific men should really be super-salesmen.[41]

The answer to Bancroft's needs seemed to appear that summer in the form of a paper by a National Research Fellow working in his laboratory, George H. Richter. Richter had been sent to Cornell by his dissertation advisor, Harry B. Weiser, a Bancroft protégé at the Rice Institute. His interest was the chemistry of anesthetics. "I knew nothing about anaesthetics or the nervous system," Bancroft later wrote, "but I never let a promising research man get away from me. I asked him to write a critical summary of the theories of anaesthesia . . ."[42] Among the theories that Richter discussed was one developed by the great French physiologist, Claude Bernard, in 1875. Anesthetics, according to Bernard, induced drowsiness and unconsciousness by effecting a "semi-coagulation of the substance of the nerve cells."[43] Bancroft, reading Richter's paper, was struck by Bernard's idea and quickly translated his terms into the language of colloid chemistry. Anesthetics, Bancroft suggested, acted much like salts added to a sol; they were nothing more than agents that caused a coagulation of the colloids in the protoplasm of sensory nerves. By causing a cell's colloids—chiefly albumin-like proteins—to flocculate or agglomerate, an anesthetic reduced their surface area and thus slowed all the cell's many catalytic reactions.

It was known that under certain conditions the coagulation of an albumin sol could be reversed by the addition of peptizing agents—chemicals that increased the dispersion of colloids. So, too, Bancroft supposed that the effects of an anesthetic wore off as the result of the gradual displacement of the foreign agglomerating agent by the normal electrolytes of the cell. As the cell's proteins returned to their normal state of dispersion, their catalytic activity increased and alertness was restored. If, however, coagulation went too far, granulation of colloidal proteins ensued, resulting in the death of the organism. "From my knowledge of colloid chemistry," Bancroft wrote, "it was evident that the objections against Claude Bernard's theory were unsound. We therefore proceeded to show that the theory was right."[44]

Convinced that after decades of work he had now stumbled upon a truly important discovery, Bancroft hastened to share it with the world. During the next three years, Bancroft, Richter, John E. Rutzler (one of Bancroft's graduate students), and a pair of physicians, R. S. Gutsell and H. E. Merriam, published more than a score of papers on this colloid theory of anesthesia, some in Bancroft's *Journal* and others in the prestigious *Proceedings of the National Academy of Sciences*. Theirs, however, was not just a theory of anesthesia. It very soon became a theory of poisoning, drug addiction, alcoholism, and insanity as well. In each instance, Bancroft and his co-workers claimed that changes in the dispersion of the colloids of nerve cells produced dysfunction. So, for example, in the case of morphine addiction and alcoholism, prolonged exposure to agglomerating agents induced a coagulation of colloidal proteins that was, to some degree, irreversible. In cases of depression, the normal balance between dispersion and agglomeration was displaced and the colloids of the brain were abnormally coagulated; in cases of schizophrenia, the brain colloids were over-peptized.[45]

Reasoning that if foreign agents could disturb the natural state of the proteins of nerve cells, Bancroft and his associates concluded that it should be possible to counter their action by the administration of other substances with antagonistic effects. So, for example, the addition of a powerful peptizing agent ought to reverse the effects of coagulating agents. Experiments with both egg albumin and anesthetized rabbits and dogs convinced Bancroft that sodium thiocyanate, also known as sodium rhodanate, was the most effective such substance that could be tolerated by a living organism in therapeutic dosages. Announcing that sodium rhodanate was a veritable elixir that "alleviates all troubles due to reversible coagulation of proteins," Bancroft and his colleagues plunged into a program of clinical trials, using his colleagues' private patients—morphine addicts, alcoholics, and manic depressives—as subjects.[46]

The course of this research need not be followed in detail. Suffice it to say that Bancroft's results, while unfailingly optimistic, resisted duplication elsewhere. Indeed, it would have been surprising had they been duplicated, for his procedures violated just about every standard of clinical research. His trials involved a handful of subjects who suffered from ill-defined maladies, little effort was made to establish controls, and follow-up was almost nonexistent. It took time, however, for his methods to receive critical scrutiny, and for a year or so Bancroft rode high on a wave of publicity and public acclaim. Written up in all the major New York newspapers and even in *Time*, Bancroft was touted as a scientist who had found a cure for alcoholism, insanity, and the "drug habit."[47]

Nor was it only journalists who were impressed. Bancroft's theory was seductively simple, and while his evidence was scanty and would soon be exposed as hopelessly flawed, it was coming from a reputable source. Bancroft, after all was a chair-holder at Cornell, a member of the National Academy of Sciences, and the holder of an honorary degree from Cambridge. His work was supported in part with funds from the Eli Lilly Company, and his associates included younger chemists and physicians with spotless credentials. Bancroft's fellow chemists in the New York Section of the American Chemical Society were enthusiastic. Meeting in February 1933, their awards committee voted to bestow on Bancroft the prestigious William H. Nichols Medal in recognition of his work on the colloid chemistry of the nervous system.[48]

The announcement of this award precipitated an avalanche of congratulatory letters and newspaper stories. It also triggered a sharp rebuke from the *Journal of the American Medical Association*, whose editors had already expressed skepticism about Bancroft's claims and worry about his infringements on the prerogatives of physicians. Chemists, they charged, were casting "doubt on the whole system of rewards and prizes in the field of scientific research and discovery," by awarding Bancroft a medal "for his extraordinary views on the effects of sodium thiocyanate and for his theory of agglomeration—or maybe it is conglomeration . . ."[49] Chauncey D. Leake, professor of pharmacology at the University of California Medical School, was more direct:

> There is not objection to Professor Bancroft amusing himself in biologic speculation. But one may justifiably object when he claims scientific validity for what is certainly speculative on his part, even though he may try to disguise it by plausible argument, superficial experimentation, and selected reference to the scientific literature.[50]

Noting that potassium thiocyanate was known to be toxic to human beings and that he and other pharmacologists had been unable to confirm Bancroft's results, Leake concluded that "it is reprehensible for him [Bancroft] to claim scientific validity for the application of his notions to medical fields."[51]

Appalled to discover that their would-be medalist was being charged with quackery, the Nichols Award Committee hurriedly sought to dissociate themselves from the controversy. Three weeks before the medal was to be presented, the chairman of the committee asked Bancroft to accept the award for his work on applications of the phase rule rather than for his "agglomeration theory." Bancroft, nettled by their fickleness, told the committee's chairman that he would refuse the medal before he would accept an alteration in the announced terms of the award. Taking Bancroft at his word, the awards committee announced that Bancroft had declined to accept the honor and that no award would be made in 1933.[52]

Bancroft, ever ready to cast himself in the role of righteous dissenter, never abandoned his belief in his colloid theory of nerve function. Indeed, as criticism mounted, Bancroft's claims for sodium rhodanate became ever more extravagant. Disease might be a result of an excessive degree of coagulation or of dispersal of bodily colloids and might be cured by the administration of agents that restored tissues to their natural state. The process of aging itself might be a process of coagulation of cellular colloids which might be reversed simply by the regular administration of a suitable dispersing agent. Citing personal experience, Bancroft asserted that daily doses of sodium rhodanate would increase resistance to infection, improve sleep, and prolong life by hindering the aging of protoplasmic colloids. By 1935, however, when these claims were advanced, only a handful of scientists were still listening.[53]

Like the Wall Street speculators of 1929, Bancroft had gambled on a flyer. It had briefly carried him upward but eventually proved worthless. In the crash, it was not money that Bancroft lost, but his reputation—and his journal.

THE END OF BANCROFT'S EDITORSHIP AND THE BIRTH OF THE *JOURNAL OF CHEMICAL PHYSICS*

The end of Bancroft's editorial control of the *Journal of Physical Chemistry* came in 1932, two years after his first paper on sodium rhodanate and the year before the Nichols Medal fiasco. Bancroft did not relinquish his editor's chair voluntarily; it was taken from him through the joint actions of the Chemical Foundation, the newly organized American Institute of Physics, the American Chemical Society, and a group of young physical chemists led by Harold C. Urey of Columbia.

Each of these parties had its own motive for wishing to see Bancroft's retirement. Physical chemists who stressed the intimate links between chemistry and physics were dissatisfied with Bancroft's editorial policies and, in some cases, hostile to Bancroft personally. Officers of the American Chemical Society, many of whom could be counted among Bancroft's critics, were interested to see greater cooperation between the editors of American chemical journals, and this, in their view, meant that Bancroft had to go. The leaders of the American Institute of Physics sought to gain a new member for their family of journals that might cover the borderland between physics and chemistry more effectively than did Bancroft's. At the same time, they wished to establish a record of aggressive leadership that might benefit their new institution. The precarious position of Bancroft's journal presented a perfect opportunity to achieve both of these aims. Finally, the officers of the Chemical Foundation had come to see Bancroft's enterprise as a bottomless pit for subsidies. They could no longer console themselves with the thought that their subvention was performing a valuable scientific service now that Bancroft was embroiled not only with physically-minded chemists but also with protein chemists and pharmacologists. Anxious to economize during the Depression, the Chemical Foundation was eager to extricate itself from an increasingly embarrassing commitment to Bancroft.[54]

The crisis was precipitated in February 1932 when a conference was held in New York among Bancroft, Henry Barton, the director of the American Institute of Physics, and William Buffum, who not only was an officer in the Chemical Foundation and business manager of Bancroft's journal, but had recently been appointed treasurer of the American Institute of Physics. Barton and Buffum had already discussed the creation of a journal that would be managed by the American Institute of Physics, subsidized by the Chemical Foundation, and cover topics on the border between chemistry and physics. Thinking it better to begin with an established journal than start one afresh, they had called the conference to try to secure Bancroft's cooperation and subscription list. Buffum, speaking for the Chemical Foundation, told Bancroft that the *Journal of Physical Chemistry* was unlikely to receive additional subsidies in future years. Barton in turn expressed interest on behalf of the American Institute of Physics in acquiring a periodical to cover the region where chemistry and physics overlapped, one that might appeal to the concerns of scientists like A. A. Noyes, G. N. Lewis, Irving Langmuir, and Linus Pauling. Together they invited Bancroft to cede ownership of his journal to the American Institute of Physics so that it might become such a periodical.[55]

Bancroft could hardly have been surprised by the suggestion. The American Institute of Physics itself had evolved out of conversations between Francis Garvan and the physicists Karl Compton and George Pegram. In December 1930, Compton and Pegram had approached Garvan seeking help for the financially strapped *Physical Review* similar to that which was being given Bancroft's journal. Garvan, although a notoriously poor administrator, had a fondness for schemes that promised to foster cooperation among scientists and greater efficiency in their work. Instead of granting them their subsidy, he proposed that they bring him a comprehensive plan "to coordinate and strengthen the activities of the great groups of physicists who are now in several organizations or perhaps not now members of any organization." The outcome was the American Institute of Physics, an organization that centralized the editorial functions of five of the societies representing American physicists. Until the new Institute could find its legs, the Chemical Foundation housed it and paid the salaries of Barton and its other officers.[56]

Bancroft knew about the special relationship between the Chemical Foundation and the American Institute of Physics and also knew that Garvan and Buffum were sorely disappointed in him and his journal. Nevertheless, he would not consider the proposal that Barton and Buffum had laid before him. The *Journal of Physical Chemistry*, he argued, was directed toward an audience of chemists, not physicists. Instead of absorbing his journal, Bancroft suggested that the American Institute of Physics create a section on chemical physics in the *Physical Review*, which might draw articles of a more mathematical character from the *Journal of the American Chemical Society*. Over the course of a year or two, this section might evolve into a separate periodical. Once the physicalists had withdrawn from the American Chemical Society, Bancroft seemed to believe that he would be left in a favorable position to obtain funding for continued publication of his journal. The meeting concluded with a tentative agreement that this would be the best course of action for all concerned. This understanding did not last long.[57]

Two months later Barton sent letters to approximately fifty American physicists and chemists working on topics that fell between chemistry and physics, telling them that the American Institute of Physics was formulating plans to publish a new journal of chemical physics that would bring together contributions currently appearing in *Physical Review*, the *Journal of the American Chemical Society*, and the *Journal of Physical Chemistry*.[58] The letter brought a flurry of replies. Some, such as R. T. Birge, head of the physics department at the University of California, were less than enthusiastic about the new journal: "I think there is prevalent now a very dangerous tendency toward specialization, which, if carried to its logical conclusion, will destroy science entirely. The purpose of science is to *correlate* facts, and the more the specialization the less the opportunity to make the correlation." More typical was the response of Harold C. Urey, who applauded the initiative of the American Institute of Physics and castigated Bancroft's journal. Publication of physical papers in the *Journal of Physical Chemistry*, he wrote, "was burial without a tombstone," since so few physicists read it.[59]

Encouraged by letters such as Urey's, Barton organized a meeting at the convention of the American Physical Society in Washington at which he announced that the Chemical Foundation planned to withdraw support from Bancroft's journal at the end of the year and that the American Institute of Physics was going ahead with plans to start a new journal of chemical physics. Upon returning to New York, Barton sent out letters to interested chemists and physicists reiterating these announcements and soliciting advice regarding the organization of the new journal.[60]

Bancroft was enraged. Not only had he not received official notification of the decision of the Chemical Foundation to withdraw its support, but Barton had stated in his letter that the new journal was being launched with Bancroft's complete cooperation.[61] Bancroft fired off a letter to Francis Garvan, president of the Chemical Foundation, imploring him not to shift money from the *Journal of Physical Chemistry* to the proposed journal of chemical physics:

> We have just developed a method for treating drug addicts. We have made it probable that the medical men will soon be able to control practically all cases of functional insanity; we have cured one case of dementia praecox paranoid and one acute case of depression psychosis, not to mention several milder ones. . . . Our methods are now being tested in this country, Germany, and Japan. We have started a new period in the application of chemistry to medicine. It would seem extremely fitting that the further developments should be published in the Journal of Physical Chemistry backed by the Chemical Foundation. In view of your interest in the application of chemistry to medicine, it seems almost incredible that we should be sacrificed to help out the physicists.[62]

Nor was Bancroft's the only protest. Ross A. Gortner, a prominent colloid chemist at the University of Minnesota, wrote to Barton requesting an explanation of the apparent duplicity of the American Institute of Physics and vowed in another letter to the secretary of the American Chemical Society that he would send his articles to Germany before giving them to a journal of chemical physics. Gortner added:

> I think there is a bunch of politics in this thing myself, and if it is a question of squeezing the chemists out of the physical chemistry field, as I have already indicated, you can count on me for any kind of support you want, and I think you will find the same reaction from a large group of people who are interested in the borderline between physics and chemistry, and who publish in the Journal of Physical Chemistry.[63]

These letters provoked some alarm at the American Institute of Physics. Barton and his colleagues had no desire to publish papers on colloid chemistry, so Gortner's threat to withhold articles posed little danger; but the threat that the American Chemical Society might perceive the new journal as an encroachment on matters within its jurisdiction was a concern, as was the possibility that Bancroft would persuade others that the behavior of the American Institute of Physics and the Chemical Foundation had been less than proper.[64] If a journal of chemical physics was to become a self-supporting venture, it would need the articles and subscriptions of physical chemists, but such support was clearly endangered by the growing suspicion in some quarters that the physicists were seeking "to swipe the Journal of Physical Chemistry away from the chemists."[65]

The leaders of the American Institute of Physics responded to this suspicion by embarking on a campaign to assuage chemists' concerns. Barton assured Bancroft and Gortner that it was never his intention to do harm to the *Journal of Physical Chemistry* or to enter into a rivalry over publications with the American Chemical Society. Karl Compton, chairman of the Institute's governing board, twice invited Bancroft to join the honorary board of editors planned for the new journal. In addition, Barton held conferences with A. A. Noyes and Arthur B. Lamb to ensure that the journal of chemical physics would enjoy their complete support and cooperation and asked Farrington Daniels, professor of physical chemistry at the University of Wisconsin, to work on behalf of the Institute at an upcoming meeting of the American Chemical Society in Denver.[66]

These efforts to mend fences met with mixed success. After extensive discussions, Lamb agreed to divert to the new journal contributions on the "physical or mathematical side" that he received as editor of the *Journal of the American Chemical Society*. Moreover, such prominent physical chemists as A. A. Noyes, G. N. Lewis, William Albert Noyes, Jr., and Linus Pauling agreed to lend their names to the masthead of the journal. To these scientists the title and sponsorship of the journal mattered little; its subject coverage and editorial quality were of primary importance. Harold C. Urey, the Columbia physical chemist who had been selected to edit the new journal, held their confidence. Although he was not yet forty, his credentials were impeccable: a Ph.D. in Lewis's department at Berkeley, a year of postgraduate work in physics under Bohr at Copenhagen, and a record of important work on atomic and molecular structure. His plan, which was to emphasize such topics as the mathematical theory of solutions, reaction kinetics, molecular spectroscopy, crystallography, equations of state, and radiochemistry, met with their approval, as did his decision to avoid articles dealing with the applications of the phase rule and colloid chemistry, fields Urey considered too technical or qualitative.[67]

Less successful was Daniels's mission to Denver. Daniels introduced four resolutions at the business session of the Division of Physical and Inorganic Chemistry, all of them directed toward promoting closer ties between chemists and physicists and encouraging subscriptions to the new journal of chemical physics. "To these innocent resolutions," Daniels wrote,

> there was violent opposition on the part of many, including some who will (I am sure) subscribe to the new journal and contribute articles to it. . . . The objections seemed to be that these resolutions would white-wash the under-cover attack on the Journal of Physical Chemistry. Rather than have the resolutions voted down or subjected to a closely divided vote, I withdrew them.

The resistance that met his resolutions was in marked contrast to the treatment Ross Gortner received when he introduced a motion urging physical chemists to support the *Journal of Physical Chemistry*. Gortner's motion was passed unanimously.[68]

Opposition to Daniels's resolutions came from several quarters. Several older chemists, including Charles L. Parsons, the long-time secretary of the American Chemical Society, and Harrison E. Howe, the editor of *Industrial and Engineering Chemistry*, failed to see any need for cooperation between chemists and physicists. Colloid chemists viewed the physicists' initiative with distaste, at least in part because they feared, with some justification, that if the *Journal of Physical Chemistry* were to be displaced by a journal of chemical physics, they would have no American outlet for their papers.[69] But as the fate of Daniels's resolutions suggests, opposition was broader than this. The reasons for this reluctance to support the new enterprise were complex, involving personal loyalties, professional jealousies, and institutional politics. Behind all else, however, lay an abiding suspicion shared by many chemists that the integrity of their subject was threatened by the expansion of physics. The proposed journal of chemical physics had come to symbolize that expansion to many rank-and-file chemists.

Compton's overtures had done little to mollify Bancroft, but there was little he could do to hinder the progress of Barton and his associates. His appeal to the Chemical Foundation fell on deaf ears, and Buffum, after finally giving Bancroft notice that subsidies were to be terminated at the end of 1932, was pressing Bancroft for a decision on how to treat orders for new subscriptions and renewals. Knowing that it would be impossible to pay upwards of $15,000 a year from his own pocket to meet operating deficits, and knowing that there was little chance of persuading others to do so, Bancroft had no recourse but to offer the name and subscription list of his journal to either the American Institute of Physics or the American Chemical Society. In mid-August 1932, he decided to surrender the journal unconditionally to the latter. To Karl Compton, Bancroft wrote: "It is very polite of your people to ask me to be the Honorary Chairman of the Advisory Council and I appreciate the courtesy. I am not and never shall be a chemical physicist and it is not desirable to masquerade so I have to decline. I shall watch with much interest the career of the Journal of Chemical Physics."[70]

The *Journal of Physical Chemistry* did not founder after Bancroft ceded ownership to the American Chemical Society. Under a new editor, Samuel C. Lind, a Leipzig graduate and professor of physical chemistry at Minnesota, the number of pages published annually was reduced sharply and more critical standards were applied to submissions. Colloid chemistry, albeit in a somewhat humbler form than that of Bancroft, continued to be the *Journal*'s main emphasis; pointedly excluded were papers on clinical medicine, including those by its former editor.[71] The *Journal* did not become a glamorous showcase, but with support from the American Chemical Society it did find a secure place on library shelves.

Far more illustrious was the future of the *Journal of Chemical Physics*. With its first issues of 1933, it established itself as the outstanding periodical treating the borderland between chemistry and physics. Not only did the first volume include the final three installments of Linus Pauling's multipart series on the nature of the chemical bond, it also included contributions from Farrington Daniels and John Van Vleck of Wisconsin, Henry Eyring and Hugh S. Taylor of Princeton, William D. Harkins and Robert S. Mulliken of Chicago, Wendell Latimer and G. N. Lewis of California, G. B. Kistiakowsky of Harvard, Irving Langmuir of General Electric, and John C. Slater of MIT. In short, the roster of contributors read like a directory of scientists active in regions in which chemistry was most intimately related to new physical theories and techniques. Concentrated in one place, the work of these scientists, most of them educated as chemists, offered overwhelming testimony to the importance of physics, especially quantum mechanics, to chemistry.[72] It also offered tangible evidence that leadership in this new chemical physics, born like the old physical chemistry in Europe, had passed to the United States.

BANCROFT AND THE TRADITIONS OF PHYSICAL CHEMISTRY

Stripped of his journal, Bancroft continued to teach at Cornell until his retirement in 1937 at the age of seventy. Although he kept busy by consulting for industrial firms and publishing occasional articles and reviews, he was a disappointed man, embittered by the loss of his journal, by the hostility of the medical community to his research, and by a younger generation's repudiation of his agenda for physical chemistry. When he died in 1953, he was forgotten except by former students and a few surviving friends, for whom he remained a "gentleman-scholar" of somewhat eccentric but always stimulating ideas. His obituary in the *Journal of the American Chemical Society* commenced with the no-longer rhetorical question, "Who was Wilder Dwight Bancroft?"[73]

His life may be read in several ways. Bancroft may be seen as his adversaries viewed him: a man who had lost touch with his changing subject, a physical chemist uncomfortable with modern physics and with little taste or talent for mathematics, a scientist who appealed to the prejudices and resentments of his colleagues in an effort to fight a hopeless rearguard action against those who considered physics to be the fundamental science. Or, adopting the viewpoint of his friends and students, he may be seen more sympathetically as a romantic defender of the traditional domains of chemistry, an acute and independent-minded critic of conventional wisdom, and a talented mediator between basic and industrial science whose greatest weakness was his excessive enthusiasm.

Bancroft may be recognized in both portraits, but neither is especially illuminating. Rather, both Bancroft and his adversaries may be viewed as representatives of two traditions in physical chemistry that were present from its inception as a distinct specialty in the 1880s. As practitioners of a hybrid specialty, physical chemists were from the outset torn between the values and attitudes of physics and chemistry. The discipline did not so much resolve these differences as internalize them. The same Ostwald who portrayed himself as one who would illuminate the dark recesses of chemistry with the torch of physics also fought a protracted war with his colleagues in physics over the value of molecular-kinetic theory. An outspoken champion of the unity of science, he could also organize his history of electrochemistry around the century-long conflict between physical and chemical theories of galvanic action, maintaining throughout the superiority of the chemists' view. A proponent of stronger mathematical training for chemists, he freely confessed his own limitations as a mathematician. While encouraging his students to study physics, he also set great store on the traditional skills of the laboratory chemist and sometimes used dissertation defenses as opportunities to criticize the views of colleagues in physics. His ambivalence was even reflected in the names he applied to his new specialty, in some place describing it as *allgemeine Chemie* and in others as *physikalische Chemie*. Questions were concealed in his use of these terms. Did he envision a new discipline that, while borrowing from physics, would be essentially independent of it? Or did he seek to create a specialty that, while taking its problems from chemistry, would merge insensibly into physics in its methods? Was the principal goal of the new discipline to unify and strengthen chemistry or to show how chemistry could be reduced to physics?

Ostwald provided no clear answers for these questions, instead leaving them to his successors to puzzle over. Like citizens of a buffer state between two great powers, physical chemists could turn their faces in two directions. Some, among them Nernst, Noyes, and Pauling, kept their eyes focused on physics, adopting not only its concepts and tools, but also its values. Although not oblivious to the possibility of eventual practical application, their primary goal was understanding. This understanding would be achieved by advancing steadily from relatively simple systems to more complex: from dilute solutions to concentrated, from weak electrolytes to strong, from aqueous solutions to nonaqueous, from homogeneous systems to heterogeneous, from crystals of sodium chloride to those of coordination compounds, from the hydrogen molecule to molecules of hemoglobin. Understanding, for these scientists, meant the ability to fit phenomena to equations, equations linked with those of physics. Their faith in the unity of science ultimately made it inconsequential whether the nexus was thermodynamics, kinetic theory, or quantum mechanics.

Others, such as Bancroft, kept their eyes focused on chemistry. For them, physical chemistry was not simply to be a department of physics. It was to afford chemists principles around which they might reorganize their sprawling and chaotic subject. Physical chemists would show how the innumerable and diverse phenomena of industrial and agricultural chemistry, geochemistry and biochemistry, analytical and organic chemistry could all be ordered by a few powerful concepts and general techniques. Their learning would be the common denominator of all those interested in matter and its transformations. As such, physical chemistry would be a boon to teachers and students of chemistry. It would also be an instrument of power—a tool for obtaining greater control over the reactions of industry and biology.

To be effective, however, the physical chemist could not throw over the traditions and values of chemistry for those of physics. "Of course a man must specialize," Bancroft told his students; "He must be a chemist rather than a physicist."[74] For Bancroft this meant speaking the language of the laboratory rather than that of the mathematics seminar and confronting the problems of industry and medicine head-on rather than obliquely. As an undergraduate teacher, he aimed to convey to nonspecialists an understanding of the manifold ways in which chemistry shaped modern life and to majors an understanding of a few simple but widely applicable principles, like the phase rule and Le Châtelier's theorem. As a researcher, he consistently focused his work and that of his students upon topics of immediate practical value. Instead of starting with simple systems that lent themselves to mathematical treatment, Bancroft preferred to begin with the complex substances with which engineers and physicians had to contend: alloys and albumins, dyes, paints, and anesthetics. As an editor he spurned papers that, to his mind, concerned only the positions of decimal points or the manipulation of equations. Asked in 1926 to review recent work on the determination of free energies, Bancroft refused, saying, "I can't do it. I abominate exact measurements myself and I don't know what has been done."[75]

Tensions between these rival traditions intensified during the early twentieth century as Ostwald's students came to better appreciate their differences. Conflicts arose over the worth of such techniques and ideas as activity coefficients and the Debye-Hückel theory, but the greatest antagonisms resulted from differences over priorities in research: Is science better served by research on dilute solutions or on applications of the phase rule? Is the more important research frontier to be found in the study of colloids or quantum mechanics? There was, of course, no way to predict answers to such questions; the rival programs had to compete for resources, the allegiance of young chemists, and standing in the larger community of scientists.

Bancroft was at a disadvantage in this competition from the outset. He found himself a dissenter not because he was a step ahead of his time, like his grandfather, but because he was a step behind. His vision of a unified and autonomous science of chemistry belonged to the nineteenth century, as did his distaste for mathematics and his skepticism about theoretical physics. No less obsolescent were the methods he chose to promote his program. His journal was a proprietary podium, much like Liebig's *Annalen der Chemie* or Ostwald's *Zeitschrift für physikalische Chemie*. But the day when a single editor could direct the growth of physical chemistry had long since passed. Instead of an asset, the journal became a liability, draining Bancroft of time, energy, and money. After World War I, when many scientific periodicals flourished, subscriptions to the *Journal of Physical Chemistry* grew slowly and deficits mounted—symptoms of the extent to which Bancroft had, in his own words, become a "back number."[76]

As Bancroft's program and journal together moved toward bankruptcy, his adversaries prospered. New physical theories and techniques gave physical chemists fresh insights into the structure of matter and the nature of chemical bonds. At the same time, new and richer sources of financial support placed the equipment, laboratories, and fellowships required to exploit these tools within the grasp of scientists. It was the physicalists such as Noyes and Lewis who adopted the new theories and applied the new techniques, gained access to the new funds and created the institutions necessary to manage the expanded scientific enterprise. Ultimately, their students, scientists such as Pauling and Urey, would demonstrate that the shortest route to a truly general chemistry was through physics. Their conception of physical chemistry proved flexible, capable of growth, and congenial to cooperative endeavors. Bancroft, tied to a nineteenth-century image of his science and a proprietary attitude toward his laboratory and journal, was unable to exploit these opportunities.

Bancroft was not bereft of support, even as late as 1932. But more and more he found himself appealing to former students and friends bound to him by personal loyalties rather than intellectual convictions and to chemists who responded to physics with the same emotional mixture of envy and disdain that infantrymen had for the "glory boys" flying above the trenches in World War I. This currency was hard to trade on. The physical chemists whose opinions mattered most viewed Bancroft as less a leader than an oddity. Whether a journal covering the borderland between chemistry and physics was sponsored by a chemical or physical society was secondary in their view to the issues of editorial standards and coverage. Bancroft's journal did not satisfy their criteria; the *Journal of Chemical Physics* promised to. The resistance offered by Bancroft and his allies was sufficient to cause concern among the partisans of the new journal but fell far short of stopping their plans. Physical chemists, by and large, reaffirmed their links with physics during the 1920s and 1930s. Those who did not were passed by. The foundation of the *Journal of Chemical Physics* marked a victory of the physicalist tradition. It was both a manifestation and a symbol of physical chemists' enhanced interest in and identification with physics during the interwar years.

Notes

CHAPTER 1
Modern Chemistry Is in Need of Reform

1. Ostwald, *Volumchemische Studien über Affinität* (Dorpat, 1877); *OKEW*, no. 250, Leipzig, 1966, p. 32.

2. On Ostwald see Wilhelm Ostwald, *Lebenslinien, Eine Selbstbiographie*, 3 vols. (Berlin, 1926–1927); Erwin Hiebert and Hans Günther Körber, "Friedrich Wilhelm Ostwald," *DSB, Supplement I*, pp. 455–469; F. G. Donnan, "Ostwald Memorial Lecture," *JCS 136* (1933): 316–332; and Wilder D. Bancroft, "Wilhelm Ostwald, The Great Protagonist," *JCE 10* (1933): 539–542 and 609–613.

3. Ostwald, "Die Aufgaben der physikalischen Chemie," *Humboldt. Monatschrift für die gesamten Naturwissenschaften 6* (1887): 249–252; reprinted in *Forschen und Nutzen. Wilhelm Ostwald zur wissenschaftlichen Arbeit*, ed. G. Lotz, L. Dunsch, and U. Kring (Berlin, 1978), pp. 192–198.

4. Ibid. See also Ostwald, "An die Leser," *ZPC 1* (1887): 1–4.

5. These were J. H. van't Hoff (1901), Svante Arrhenius (1903), Wilhelm Ostwald (1909), and William Ramsay (1904).

6. On the changing boundaries of our categories of natural knowledge see Thomas S. Kuhn, "Mathematical vs. Experimental Traditions in the Development of Physical Science," *Journal of Interdisciplinary History 7* (1976): 1–31; Michael J. Mulkay, "Three Models of Scientific Development," *Sociological Review 23* (1975): 509–526 and 535–537; David O. Edge and Michael J. Mulkay, *Astronomy Transformed: The Emergence of Radio Astronomy in Britain* (New York, 1976), esp. pp. 350–398; and Robert E. Kohler, *From Medical Chemistry to Biochemistry: The Making of a Biomedical Discipline* (Cambridge, 1982), esp. pp. 1–8.

7. This view is argued most strongly by Maurice P. Crosland in "The Development of Chemistry in the Eighteenth Century," *Studies on Voltaire and the Eighteenth Century 24* (1963): 369–441. See also Charles C. Gillispie, "The *Encyclopédie* and the Jacobin Philosophy of Science: A Study in Ideas and Consequences," in *Critical Problems in the History of Science*, ed. Marshall Clagett (Madison, 1959), pp. 255–290; Arnold Thackray, *Atoms and Powers: An Essay on Newtonian Matter Theory and the Development of Chemistry* (Cambridge, Mass., 1970); Robert E. Schofield, *Mechanism and Materialism: British Natural Philosophy in an Age of Reform* (Princeton 1970); and J. K. Bonner, "Amadeo Avogadro: A Reassessment of His Research and Its Place in Early Nineteenth-Century Science," Ph.D. dissertation, The Johns Hopkins University, 1974, chapter 1.

8. William Nicholson, for example, devoted most of the second volume of his *Introduction to Natural Philosophy* (London, 1790) to the discussion of chemistry, whereas Thomas Rutherford avoided the subject in *A System of Natural Philosophy*, 2 vols. (Cambridge, 1748).

9. Nicholson, *A Dictionary of Chemistry*, vol. 1 (London, 1795), pp. 254–255.

10. On chemical education in the eighteenth century see Karl Hufbauer, *The Formation of the German Chemical Community (1720–1795)* (Berkeley, Calif., 1982); and Arthur L. Donovan, *Philosophical Chemistry in the Scottish Enlightenment* (Edinburgh, 1975).

11. Thackray, *Atoms and Powers*, pp. 269–278.

12. Henry C. Bolton, *A Select Bibliography of Chemistry, 1492–1892* (Washington, D.C., 1893), pp. 1068–1158.

13. Karl Hufbauer, "Social Support for Chemistry in Germany during the Eighteenth Century: How and Why Did It Change?" *HSPS 3* (1971): 207.

14. Joseph Ben-David, *The Scientist's Role in Society: A Comparative Study* (Englewood Cliffs, N.J., 1971), pp. 108–138.

15. Jack B. Morrell, "The Chemist Breeders: The Research Schools of Liebig and Thomas Thomson," *Ambix 19* (1972): 1–46.

16. Not, however, without resistance; see R. Steven Turner, "Justus von Liebig versus Prussian Chemistry: Reflections on Early Institute Building in Germany," *HSPS 13* (1982): 129–162.

17. On foreign responses to German innovations in chemical training, see Gerrylynn K. Roberts, "The Establishment of the Royal College of Chemistry: An Investigation of the Social Context of Early Victorian Chemistry," *HSPS 7* (1976): 437–485; D.S.L. Cardwell, *The Organization of Science in England* (London, 1957); Harry W. Paul, *The Sorcerer's Apprentice: The French Scientist's Image of German Science, 1840–1919* (Gainesville, Florida, 1972); Margaret W. Rossiter, *The Emergence of Agricultural Science: Justus Liebig and the Americans* (New Haven, 1975); Charles Rosenberg, *No Other Gods: On Science and American Social Thought* (Baltimore, 1976), esp. pp. 133–210; and Owen Hannaway, "The German Model of Chemical Education in America: Ira Remsen at Johns Hopkins," *Ambix 23* (1976): 145–164.

18. J. Jacob Berzelius, *Lehrbuch der Chemie*, 4th German ed., trans. F. Wöhler (Dresden, 1835), p. 1. See also Andrew Ure, *Dictionary of Chemistry and Mineralogy*, 4th ed. (London, 1835), p. 317.

19. James C. Booth and Campbell Morfit, *The Encyclopedia of Chemistry* (Philadelphia, 1850), p. 434.

20. Russell McCormmach, "Editor's Foreword," *HSPS 3* (1971): xi.

21. Lothar Meyer, *Modern Theories of Chemistry*, 5th ed., trans. Phillips Bedson and W. Carleton Williams (London, 1888), p. xxv (Introduction to the 1st ed. of 1864). On chemists and mathematics, see Keith J. Laidler, "Chemical Kinetics and the Origins of Physical Chemistry," *AHES 32* (1985): 46.

22. Susan G. Schacher, "Robert Wilhelm Bunsen," *DSB*, vol. 2, pp. 586–590; Henry Roscoe, "Bunsen Memorial Lecture," *JCS 77* (1900): 513–554.

23. Roscoe, "Bunsen Memorial Lecture," pp. 550–551.

24. Kopp discussed the history of his experimental work in "Über die Molekularvolume von Flüssigkeiten," *AC 250* (1889): 1–117. See also J. E. Thorpe, "Kopp Memorial Lecture," *JCS 63* (1893): 775–815.

25. J. H. van't Hoff, "Landolt Memorial Lecture," *JCS 99* (1911): 1653–1660.

26. Ostwald, *R. W. Bunsen* (Leipzig, 1905).

27. See Svante Arrhenius, *Theories of Solution* (New Haven, 1912), esp. p. 72, where Arrhenius asserts that physical chemists should stress their ties with earlier

workers, "for it is the most convincing proof of their soundness," that the principles of physical chemistry "should have developed quite continuously and organically from all the results of chemical experience."

28. On Berthollet, see Michelle Sadoun-Goupil, *Le chimiste Claude-Louis Berthollet, 1748–1822: Sa vie, son oeuvre* (Paris, 1977), and M. P. Crosland, *The Society of Arcueil: A View of French Science at the Time of Napoleon I* (Cambridge, Mass., 1967).

29. Thackray, *Atoms and Powers*, pp. 213–214.

30. Berthollet, "Suite des recherches sur les lois de l'affinité," *Annales de chimie* 37 (1801): 151–181; *Recherches sur les lois de l'affinité* (Paris, 1801); *Essai de statique chimique* (Paris, 1803).

31. Frederick L. Holmes provides a detailed analysis of Berthollet's ideas on affinity and their subsequent history in "From Elective Affinities to Chemical Equilibria: Berthollet's Law of Mass Action," *Chymia 8* (1962): 105–146. See also Maurice W. Lindauer, "The Evolution of the Concept of Chemical Equilibrium from 1775 to 1923," in *Selected Readings in the History of Chemistry*, ed. Aaron J. Ihde and William F. Kieffer (Easton, Penn., 1965).

32. Satish Kapoor analyzes this controversy in "Berthollet, Proust and Proportions," *Chymia 10* (1965): 53–110.

33. See Bonner, "Amadeo Avogadro," pp. 130–134.

34. Stuart Pierson, "Heinrich Rose," *DSB*, vol. 11, pp. 540–542; Alexander Williamson, "Results of a Research on Etherification," and "Suggestions for the Dynamics of Chemistry Derived from the Theory of Etherification," both reprinted in *Papers on Etherification and on the Constitution of Salts*, Alembic Club Reprints, no. 16 (Edinburgh, 1902), pp. 5–17 and 18–24; Marcellin Berthelot and L. Pean de Saint-Gilles, "Recherches sur les Affinités," *Annales de chimie et physique*, iii 65 (1862): 382–389 and 414–422; 68 (1863): 225–369.

35. Discussed by Lothar Meyer in *Modern Theories of Chemistry*, pp. 449–452.

36. These papers are collected in *Untersuchungen über die chemischen Affinitäten, Abhandlungen aus den Jahren 1864, 1867, 1879, OKEW*, no. 104 (Leipzig, 1899). For biographical information, see George B. Kauffman, "Cato Maximillian Guldberg," and "Peter Waage," *DSB*, vol. 5, pp. 586–587 and vol. 14, pp. 108–109. On their research, see M. M. Pattison-Muir, "Chemical Affinity," *PM 8* (1879): 181–203; Holmes, "From Elective Affinities to Chemical Equilibria," pp. 136–139; and J. R. Partington, *A History of Chemistry*, vol. 4 (London, 1964), pp. 588–593.

37. Guldberg and Waage, *Untersuchungen über die chemischen Affinitäten* (1867), *OKEW*, p. 16.

38. Guldberg and Waage, *Untersuchungen über die chemischen Affinitäten* (1864), *OKEW*, pp. 10–14.

39. Guldberg and Waage, *Untersuchungen über die chemischen Affinitäten* (1867), *OKEW*, pp. 20–22.

40. Ibid., p. 125.

41. Julius Thomsen, "Über die Berthollet'sche Affinitätstheorie," *Annalen der Physik 138* (1869): 65–102, esp. 94–102; August Horstmann, "Theorien der Dissociation" (1873) and "Über ein Dissociationsproblem" (1877), both reprinted in *Abhand-*

lungen zur Thermodynamik chemischer Vorgänge, OKEW, no. 137 (Leipzig, 1903), pp. 26–41 and 42–55.

42. Holmes, "From Elective Affinities to Chemical Equilibria," pp. 136–137.

43. Trevor H. Levere, *Affinity and Matter: Elements of Chemical Philosophy, 1800–1865* (Oxford, 1971); Helge Kragh, "Julius Thomsen and Classical Thermochemistry," *BJHS 17* (1984): 257–258 and 269–271; R.G.A. Dolby, "Thermochemistry versus Thermodynamics: The Nineteenth Century Controversy," *HS 22* (1984): 375–400. See also Part III of Meyer's *Modern Theories of Chemistry*.

44. Stig Viebel, "Hans Peter Jörgen Julius Thomsen," *DSB*, vol. 13, p. 359.

45. See Donald G. Miller, "Pierre Duhem, un oublie," *Revue des questions scientifiques 28* (1967): 454–458 and 463–464; Mary Jo Nye, "The Scientific Periphery in France: The Faculty of Sciences at Toulouse (1880–1930)," *Minerva 13* (1975): 390n.

46. J. Willard Gibbs, "On the Equilibrium of Heterogeneous Substances," *Transactions of the Connecticut Academy of Arts and Sciences 3* (1875–1878): 108–248 and 343–524; reprinted in *The Collected Works of J. Willard Gibbs*, vol. 1 (New York, 1928). On Gibbs, see Martin J. Klein, "Josiah Willard Gibbs," *DSB*, vol. 5, pp. 386–393; and Lynde P. Wheeler, *Josiah Willard Gibbs: The History of a Great Mind* (New Haven, 1951), reprint ed. (Hamden, Conn., 1970). On Gibbs and physical chemistry, see F.G. Donnan, "The Influence of J. Willard Gibbs on the Science of Physical Chemistry," *Journal of the Franklin Institute 199* (1925): 457–483; John Johnston, "Willard Gibbs, An Appreciation," *SM 26* (1928): 129–139; Duncan A. MacInnes, "The Contributions of Josiah Willard Gibbs to Electrochemistry," *TES 71* (1937): 65–72; and Henry Le Châtelier, "L'Oeuvre de J. Willard Gibbs," *Chemisch Weekblad 23* (1926): 406–409.

47. The reprint list is reproduced in Wheeler, *Josiah Willard Gibbs*, pp. 236–248.

48. On van't Hoff, see Ernst Cohen, *Jacobus Henricus van't Hoff: Sein Leben und Wirken* (Leipzig, 1912); James Walker, "Van't Hoff Memorial Lecture," *JCS 103* (1913): 1127–1143; and Robert Scott Root-Bernstein, "The Ionists: Founding Physical Chemistry, 1872–1890," Ph.D. dissertation, Princeton University, 1980, pp. 182–350.

49. On Arrhenius, consult Ernst H. Riesenfeld, *Svante Arrhenius* (Leipzig, 1931); James Walker, "Arrhenius Memorial Lecture," *JCS 118* (1928): 1380–1401; and Root-Bernstein, "The Ionists," pp. 16–181 and 366–415.

50. Ostwald, *Lebenslinien*, vol. 1, p. 148.

51. Ibid., pp. 94–99. See also Hiebert and Körber, "Ostwald."

52. Ostwald, *Lebenslinien*, vol. 1, p. 99.

53. Ibid., p. 102 and pp. 117–118. Ostwald, *Volumchemische Studien über Affinität*, p. 37.

54. Ostwald, *Volumchemische und optisch-chemische Studien* (Dorpat, 1878); reprint ed., *OKEW*, no. 250 (Leipzig, 1966), p. 99; Donnan, "Ostwald Memorial Lecture," p. 318. While the methods of Thomsen and Ostwald often could yield good approximations of affinity constants, subsequent work would reveal flaws in their procedures. Thomsen's thermochemical method ignored changes in entropy accompanying reactions; Ostwald neglected secondary effects. See Hiebert and Körber, "Ostwald," p. 458.

55. Paul Walden, *Wilhelm Ostwald* (Leipzig, 1904), pp. 47–48.

56. Pattison-Muir, "Chemical Affinity"; Meyer, *Die Modernen Theorien der Chemie*, 4th ed. (Breslau, 1883), pp. 437–440 and 491–514.

57. Arnolds Spekke, *History of Latvia: An Outline* (Stockholm, 1951), p. 309.

58. A. F. Holleman, "My Reminiscences of van't Hoff," *JCE 29* (1952): 380; H. S. van Klooster, "Van't Hoff in Retrospect," *JCE 29* (1952): 379.

59. Holleman, "Van't Hoff," pp. 380–381.

60. Cohen, *Van't Hoff*, p. 19; Root-Bernstein, "The Ionists," pp. 184–186.

61. Van't Hoff, "Sur les formules de structure dans l'espace," *Archives néerlandaises des sciences exactes et naturelles 9* (1874): 445–454; translated into English in *Classics in the Theory of Chemical Combination*, ed. O. T. Benfey (New York, 1963), pp. 151–160; O. B. Ramsay, ed., *Van't Hoff-Le Bel Centennial: A Symposium* (Washington, D.C., 1975); Root-Bernstein, "The Ionists," pp. 189–193 and 212–216.

62. Van't Hoff, *Ansichten über die organische Chemie*, 2 vols. (Braunschweig, 1878–1881). For an analysis of this work see Root-Bernstein, "The Ionists," pp. 217–235.

63. Van't Hoff, "Wie die Theorie der Lösungen entstand," *BDCG 27* (1894): 7.

64. Van't Hoff, *Études de dynamique Chimique* (Amsterdam, 1884), pp. 13–33. For a fuller treatment of van't Hoff's and subsequent work on reaction kinetics, see Laidler, "Chemical Kinetics and the Origins of Physical Chemistry."

65. Horstmann, "Theorien der Dissociation."

66. See Root-Bernstein, "The Ionists," pp. 256–273.

67. Van't Hoff, *Études*, pp. 4–7.

68. Ibid., pp. 114–118; Root-Bernstein, "The Ionists," pp. 256–281.

69. Van't Hoff, *Études*, pp. 161–176.

70. Ibid., pp. 177–195.

71. Ibid., pp. 202–209.

72. Van't Hoff, "Conditions électrique de l'équilibre chimique," *Kongliga Svenska Vetenskaps-Akademiens Handlingar 21* (1886), no. 17; translated into German and reprinted in *Die Gesetze des chemischen Gleichgewichtes für den verdünnten, gasförmigen oder gelösten Zustand*, *OKEW*, no. 110 (Leipzig, 1900), pp. 78–79.

73. Root-Bernstein, "The Ionists," pp. 282–283.

74. Van't Hoff, "L'équilibre chimique dans les systèmes gazeux ou dissous à l'état dilué," *Archives néerlandaises 20* (1886): 239–302; "Lois de l'équilibre chimique dans l'état dilué, gazeux ou dissous," *Kongliga Svenska Vetenskaps-Akademiens Handlingar 21* (1886), no. 17 (translated into German and reprinted in *Die Gesetze des chemischen Gleichgewichtes*, pp. 3–61).

75. Van't Hoff, "Wie die Theorie der Lösungen entstand," pp. 7–8; van't Hoff, "Ein Blick in das neue chemisch-physikalische Forschungsgebiet," *Deutsche Revue 20* (1895): 113–119.

76. These papers include the three cited in notes 72 and 74 above, "Une propriété générale de la matiere dilué," *Kongliga Svenska Vetenskaps-Akademiens Handlingar 21* (1886), no. 17 (translated into German and reprinted in *Die Gesetze des chemischen Gleichgewichtes*, pp. 62–73), and "Die Rolle des osmotischen Druckes in der Analogie zwischen Lösungen und Gasen," *ZPC 1* (1887): 481–508. Root-Bernstein attempts to reconstruct van't Hoff's thoughts during this period in "The Ionists," pp. 291–331.

77. Van't Hoff, "Die Rolle des osmotischen Druckes in der Analogie zwischen Lösungen und Gasen," p. 481.

78. Ibid., p. 483.

79. Ibid, pp. 500–502; see also his earlier paper, "Lois de l'équilibre chimique dans l'état dilué, gazeux ou dissous."

80. The letter is reproduced in Riesenfeld's biography of Arrhenius, between pp. 24 and 25, and in Cohen's biography of van't Hoff, pp. 239–242.

81. Riesenfeld, *Arrhenius*, p. 7.

82. Arrhenius, "The Development of the Theory of Electrolytic Dissociation," *PSM* 65 (1904): 390.

83. Arrhenius, "Electrolytic Dissociation," *JACS 34* (1912): 358; Walker, "Arrhenius Memorial Lecture," pp. 1382–1383.

84. Arrhenius, "Recherches sur la conductibilité galvanique des electrolytes," *Bihang till Kongliga Svenska Vetenskaps-Akademiens Handligar 8* (1884), nos. 13 and 14 (translated into German and reprinted in *Untersuchungen über die galvanische Leitfähigkeit der Elektrolyte*, OKEW, no. 160 [Leipzig, 1907]).

85. On Williamson, see note 35 above; Rudolph Clausius, "Über die Elektricitätsleitung in Elektrolyten," *Annalen der Physik 101* (1857): 338–360.

86. Arrhenius, *Untersuchungen über die galvanische Leitfähigkeit der Elektrolyte*, p. 61.

87. Ibid., pp. 68–75.

88. Ibid.; quoted by Root-Bernstein, "The Ionists," p. 103.

89. Arrhenius's letter to Ostwald is reproduced in *Aus dem wissenschaftlichen Briefwechsel Wilhelm Ostwalds*, ed. Hans-Günther Körber, vol. 2 (Berlin, 1969), p. 3; Ostwald describes his reaction in *Lebenslinien*, vol. 1, pp. 216–217.

90. Arrhenius, "Über die Dissociation der in Wasser gelösten Stoffe," *ZPC 1* (1887): 630.

91. Ibid., pp. 630–631.

92. Ibid., p. 633.

93. Ibid., pp. 642–643.

94. Ostwald, "Elektrochemische Studien. II. Abhandlung: Das Verdunnungsgesetz," *Journal für praktische Chemie 31* (1885): 433–462. See John H. Wolfenden, "The Anomaly of Strong Electrolytes," *Ambix 19* (1972): 176–177 for a discussion of Ostwald's first dilution law.

95. Ostwald, "Zur Theorie der Lösungen," *ZPC 2* (1888): 36–37.

96. Ostwald, "Über die Dissociationstheorie der Elektrolyte," *ZPC 2* (1888): 270–283.

97. Ostwald, *Elektrochemie: Ihre Geschichte und Lehre* (Leipzig, 1896), p. 1147.

98. Walker, "Van't Hoff Memorial Lecture," p. 1143.

99. This is among the primary themes of Root-Bernstein, "The Ionists."

100. Ostwald, *Elektrochemie: Ihre Geschichte und Lehre*.

101. See H. J. Hamburger, "Zur Geschichte und Entwicklung der physikalisch-chemischen Forschung in der Biologie," *Internationale Zeitschrift für physikalisch-chemische Biologie 1* (1914): 6–27; F. Y. Loewinson-Lessing, *A Historical Survey of Petrology*, trans. S. I. Tomkeieff (Edinburgh, 1954), esp. pp. 33–41.

102. R.G.A. Dolby, "Debates over the Theory of Solution," *HSPS 7* (1976): 346,

349, and 389; Erwin N. Hiebert, "The Energetics Controversy and the New Thermodynamics," in *Perspectives in the History of Science and Technology*, ed. Duane H.D. Roller (Norman, Oklahoma, 1971), pp. 67–86. Root-Bernstein observes that only 13 of 77 contributors to the *ZPC* during 1887–1888 had degrees in physics; "The Ionists," p. 362.

103. Jeffrey A. Johnson, "The Chemical Reichsanstalt Association: Big Science in Imperial Germany," Ph.D. dissertation, Princeton University, 1980, pp. 166–168.

104. Dolby, "Debates over the Theory of Solution," p. 349.

105. Morris W. Travers, *A Life of Sir William Ramsay* (London, 1956), p. 36 and pp. 89–90.

CHAPTER 2
Physical Chemistry from Europe to America

1. On these portions of van't Hoff's career, see H.A.M. Snelders, "J. H. van't Hoff's Research School in Amsterdam (1877–1895)," *Janus 71* (1984): 1–30; and E. J. Cohen, *Jacobus Henricus van't Hoff. Sein Leben und Wirken* (Leipzig, 1912).

2. See, for example, Svante Arrhenius, "On the Influence of Carbonic Acid in the Air upon the Temperature of the Ground," *PM* v *41* (1896): 237–276; *Worlds in the Making*, trans. H. Borns (New York, 1908); *Immunochemistry: The Application of the Principles of Physical Chemistry to the Study of the Biological Antibodies* (New York, 1907). Arrhenius's contributions to immunochemistry have received considerable historical attention in recent years. See, for example, Lewis P. Rubin, "Styles in Scientific Explanation: Paul Ehrlich and Svante Arrhenius on Immunochemistry," *Journal of the History of Medicine 35* (1980): 397–425.

3. On the *Zeitschrift für physikalische Chemie*, see Robert Scott Root-Bernstein, "The Ionists: Founding Physical Chemistry, 1872–1890," Ph.D. dissertation, Princeton University, 1980, pp. 351–365.

4. F.H.N., "Some Scientific Centres: The Laboratory of Wilhelm Ostwald," *Nature 64* (1901): 428.

5. Wilhelm Ostwald, *Das physikalisch-chemische Institut der Universität Leipzig* (Leipzig, 1898), p. 5.

6. James Walker, unpublished autobiographical notes, quoted by James Kendall in "Sir James Walker," *JCS 138* (1935): 1350.

7. Grete Ostwald, *Wilhelm Ostwald. Mein Vater* (Stuttgart, 1953), p. 60; Paul Walden, *Wilhelm Ostwald* (Leipzig, 1904), p. 74.

8. Walden, *Wilhelm Ostwald*, pp. 46–47.

9. George Jaffé, "Recollections of Three Great Laboratories," *JCE 29* (1952): 232.

10. W. Jost, "The First 45 Years of Physical Chemistry in Germany," *ARPC 17* (1966): 3.

11. Ostwald, "An die Leser," *ZPC 1* (1887): 1. Ostwald was quoting words written five years earlier by Emil Du Bois Reymond.

12. When Ostwald reminded Emil Fischer, the great sugar chemist, that organic chemists ought to thank physical chemists for developing new methods of measuring molecular weights, Fischer replied: "I don't need your methods." Many years after Ostwald died, Richard Willstätter, another of Germany's elite organic chemists, still felt bitterness. Ostwald, he wrote,

considered it his mission to battle passionately with the theories and methods of organic chemistry, which made little sense. For years he created discord and anger. . . . He criticized, especially in his discussions in the *Zeitschrift für physikalische Chemie*, all of the contemporary literature of structural chemistry. He had no knowledge or understanding of the development, content, or effect of our organic chemical views which had proved themselves to be uncommonly fruitful . . . Ostwald also lacked sufficient depth in the theoretical physical foundations of organic chemistry, and, more important, lacked the modesty of the natural scientist. . . . Ostwald's great influence caused us some bad moments.

See Jeffrey A. Johnson, "The Chemical Reichsanstalt Association: Big Science in Imperial Germany," Ph.D. dissertation, Princeton University, 1980, p. 166; and Richard Willstätter, *From My Life*, trans. Lilli S. Hornig (New York and Amsterdam, 1965), pp. 94–95.

13. Charles E. McClelland, *State, Society, and University in Germany, 1700–1914* (Cambridge, 1980), pp. 293–294.

14. Jeffrey A. Johnson, "The Chemical Reichsanstalt Association," pp. 168–171. For statistics on faculty and *Dozenten* in chemistry and physical chemistry at German universities and *technischen Hochschulen*, see Christian von Ferber, *Die Entwicklung des Lehrkörper der deutschen Universitäten und Hochschulen, 1864–1954* (Göttingen, 1956), pp. 197 and 209. Despite the obstacles to institutional development, German physical chemists held intellectual leadership in their field, at least until World War I. This pattern of intellectual strength and institutional weakness is similar to that in biochemistry. See Robert E. Kohler, *From Medical Chemistry to Biochemistry: The Making of a Biomedical Discipline* (Cambridge, 1982), pp. 9–39.

15. R.G.A. Dolby, "Debates over the Theory of Solution," *HSPS* 7 (1976): 297–404.

16. Among the British chemists who enjoyed the hospitality of Ramsay's laboratory after studying physical chemistry on the Continent were James Walker, James Wallace Walker, Frederick George Donnan, Alexander Findlay, and W. C. McC. Lewis. The Walkers, Donnan, and Findlay had studied under Ostwald at Leipzig; Lewis, under Ostwald's former assistant, Georg Bredig.

17. Harry W. Paul, *The Sorcerer's Apprentice: The French Scientist's Image of German Science, 1840–1919* (Gainesville, Florida, 1972), pp. 13–14.

18. The editor was Phillippe Guye (1862–1922). Guye had studied chemistry under Berthelot during the years 1889–1891 and then toured Germany, where he made the acquaintance of Ostwald and several of Ostwald's co-workers. A survey of contributors to volumes 1, 4, 7, and 10 of his *Journal de chimie physique* reveals that only about one quarter of those contributing articles were natives of France.

19. Although the Ionists had their differences, prior to World War I their debates were pursued in the manner of family squabbles. Neither van't Hoff nor Arrhenius, for example, could accept the extreme attacks that Ostwald made on the atomic theory during the 1890s, but they did not make a public issue of their differences. In France, by contrast, clashes over issues such as atomism resulted in invective and fragmentation, even among those whose research would seem to have made for natural alliances. The sharp polarization of French intellectuals over religious and political matters greatly intensified this factionalism.

20. Best illustrated by the melancholy tale of Pierre Duhem, whose career was

blighted because he challenged Berthelot's principle of maximum work too vigorously. See Stanley L. Jaki, *Uneasy Genius: The Life and Work of Pierre Duhem* (The Hague, 1984), esp. pp. 45–69 and 161–170.

21. The first extended discussion of Gibbs's work in French appeared in Georges Lemoine's "Études sur les équilibres chimiques," in *Encyclopédie chimique*, ed. E. Fremy, vol. 1 (Paris, 1882), pp. 69–372, esp. pp. 361–370. Lemoine taught engineering and chemistry at the Catholic University of Paris. Soon thereafter, Pierre Duhem, a student at the École Normale Supérieure, devoted a doctoral thesis to extending the uses of Gibbs's notion of thermodynamic potential in chemistry and physics, a thesis rejected by his examiners, at least in part because of Duhem's assault on Berthelot's principle of maximum work. In 1888, Henry Le Châtelier, newly appointed professor of chemistry at the Collège de France, made use of Gibbs's papers in the treatment of equilibrium conditions in industrial reactions. During the next two decades, Duhem and Le Châtelier, although on frosty terms with one another, became the principal advocates of Gibbs's ideas in France.

Le Châtelier, perhaps the most influential of these French physical chemists, shared the Ionists' sense that chemistry was moving from a descriptive to an analytical stage in its development. Yet according to Le Châtelier, physical chemistry, which he preferred to call *mechanique chimique*, had its roots in the work of Frenchmen and of the unique American, Gibbs. "During the century following the discoveries of Lavoisier," he wrote,

> chemists devoted their efforts exclusively to the determination of the weight relations according to which bodies entered into combination, that is to say, the formulae of compounds of definite proportions. But since the discoveries of H. Sainte-Claire Deville and of J. W. Gibbs, a new orientation has evolved; chemists have begun to lose interest in the discovery of new chemical combinations, of which there appears to be an adequate number. Rather, they have tended to direct their efforts toward determining the conditions of the production and of the destruction of known compounds.

In recapping the history of the concept of chemical equilibrium, Le Châtelier managed to omit reference not only to Ostwald, but also to Guldberg, Waage, and van't Hoff. "The concept," he wrote, "was glimpsed in certain special cases by Berthollet, studied systematically by Berthelot during his work on esterification, but was fully illuminated only through the works of Henri Sainte-Claire Deville. Sadly, the entirely inappropriate word, dissociation, as used by certain scientists, has for some time served only to obscure the nature of this phenomenon." See Henry Le Châtelier, *Leçons sur le carbone, la combustion, les lois chimique* (Paris, 1908), pp. 344–345 and 319–320. On Le Châtelier, see Cecil H. Desch, "Henry Le Châtelier, 1850–1936," in *Memorial Lectures Delivered before the Chemical Society, 1933–1942* (London, 1951), pp. 87–98.

22. Whereas their colleagues in Germany, Britain, and America tended to organize textbooks around the laws of solution and mass action and the theories of electrolytic dissociation and galvanic action, French physical chemists generally preferred to give students a far more abstract and formal treatment, with emphasis on the sure and certain principles of thermodynamics. Jean Perrin's *Traité de chimie physique. Les principes* (Paris, 1903), based upon his lectures at the Sorbonne, is an outstanding example of this genre. Perrin's text begins with an extended discussion of such basic terms as

mass, force, energy, ether, and equilibrium. He then develops the laws of thermodynamics, but more in the manner of a physicist or engineer than a chemist; his examples
and illustrations, for instance, are drawn primarily from mechanics rather than chemistry. Not until the final chapter, in which Perrin introduces Gibbs's concept of chemical potential and derives the phase rule, does *Les principes* begin to resemble a chemical textbook. Even then, however, he stopped short of giving readers a sense of how
these tools could be applied to real chemical systems. This he deferred to a future
volume, which in fact was never written. Also postponed were discussions of the laws
of solution and electrochemistry—the bread and butter of texts on physical chemistry
outside of France. *Les principes*, in short, was an introduction to thermodynamics.
What the Ionists had treated as a means to an end had here become the primary object
of study.

The best study of Perrin's career is Mary Jo Nye, *Molecular Reality: A Perspective
on the Scientific Work of Jean Perrin* (London and New York, 1972).

23. On France's provincial science faculties, see Mary Jo Nye, *Science in the Prov-
inces: Scientific Communities and Provincial Leadership in France, 1860–1930*
(Berkeley, 1986). Also useful are Harry W. Paul, "Apollo Courts the Vulcans: The
Applied Science Institutes in Nineteenth-Century French Science Faculties," and Terry
Shinn, "From 'Corps' to 'Profession': The Emergence and Definition of Industrial
Engineering in Modern France," both in *The Organization of Science and Technology
in France, 1808–1914*, ed. Robert Fox and George Weisz (Cambridge, 1980), pp.
155–181 and 183–208, respectively.

24. Charles Baskerville, *et al.*, "Report of the Census Committee," in *Twenty-fifth
Anniversary of the American Chemical Society* (Easton, Pennsylvania, 1902), Table I.

25. Stephen S. Visher, *Scientists Starred 1903–1943 in "American Men of Science"*
(Baltimore, 1947), p. 177. The others were Roger Adams, Ira Remsen, and Julius
Stieglitz.

26. There had been 309 articles (excluding reviews and notes of less than five pages)
published in the *JPC* by the end of 1905, and 84 were by Leipzig alumni; 49 of the 145
articles in the *JACS* were by students of Ostwald.

27. Henry Leffmann, "Chemistry in 1876," *The Catalyst 11* (1926): 4.

28. See, for example, Daniel J. Kevles' analysis of the social composition of the
American physics community in "The Study of Physics in America, 1865–1916,"
Ph.D. dissertation, Princeton University, 1964, Appendix V.

29. Noyes, "Autobiographical Notes," NAS. A bibliography of Noyes's published
work is appended to Linus Pauling's biographical sketch of Noyes in *BMNAS 31*
(1958): 322–346. T. L. Davis, "Samuel P. Mulliken," *PAAAS 70* (1935–1936): 562–
563.

30. Noyes to H. M. Goodwin, 5 January 1890, box 1, HMG.

31. Noyes, "Über die gegenseitige Beinflussung der Löslichkeit von dissociierten
Körpern," *ZPC 6* (1890): 241–267. Noyes to H. M. Goodwin, 20 February 1890, box
1, HMG.

32. Samuel Sheldon, review of *Solution and Electrolysis* (by W. C. Dampier
Whetham), in *PR 3* (1895–1896): 315.

33. William Ramsay to Henry Fyfe, November 1898, quoted by Morris W. Travers
in *A Life of Sir William Ramsay* (London, 1956), p. 184.

34. Noyes to Goodwin, 20 February 1890.

35. Jones, review of *Wilhelm Ostwald* (by Paul Walden), in *ACJ 32* (1904): 90–91. For a similar testimonial see Wilder D. Bancroft's review of the same book in *JPC 8* (1904): 505–506.

36. "Diary of Summer 1900 June Bicycle Trip Through Eastern Pennsylvania and Catskill Mountains"; Langmuir to Sadie Langmuir, 25 June 1903; 21 July 1903; Langmuir to Herbert and Arthur Langmuir, 26 July 1903; all in box 1, IL.

37. F. G. Cottrell had to wait a year for an opening after applying for admission to Ostwald's laboratory in 1900. See Frank Cameron, *Cottrell, Samaritan of Science* (Garden City, N.Y., 1952), pp. 69–70.

38. On Cooke, see Sheldon J. Kopperl, "The Scientific Work of Theodore William Richards," Ph.D. dissertation, University of Wisconsin, 1970, p. 15, and Cooke's textbook, *Elements of Chemical Physics* (Boston, 1860). G. N. Lewis, G. W. Heimrod, A. B. Lamb, and H. W. Morse all were Richards' Ph.D. students. Shortly after completing their dissertations each went to Leipzig.

39. On MIT, see Samuel C. Prescott, *When MIT Was 'Boston Tech': 1861–1916* (Cambridge, Mass., 1964), and John W. Servos, "The Industrial Relations of Science: Chemical Engineering at MIT, 1900–1939," *Isis 71* (1980): 531–549.

40. Davis, "Samuel P. Mulliken." Mulliken, a boyhood friend of Noyes from Newburyport and MIT, accompanied Talbot, Gill, and Noyes on this trip. Mulliken received his Ph.D. under Wislicenus, but did not join the MIT faculty until later in the 1890s.

41. See chemistry course listings in the MIT *Catalogue* during the 1880s and early 1890s.

42. Noyes, "Instruction in Theoretical Chemistry," *TQ 9* (1896): 324.

43. Servos, "Industrial Relations of Science," pp. 534–537.

44. Bancroft went so far as to write that many "believed what Ostwald said even when they knew that he was not right." His influence did suffer after 1906, however. A. A. Noyes, when solicited for suggestions regarding nominations for the 1909 Nobel Prize, declined to nominate Ostwald, saying: "now that the former [Ostwald] has practically given up scientific work for philosophy, I am less enthusiastic about him." See Bancroft, "Wilhelm Ostwald, the Great Protagonist," *JCE 10* (1933): 613; Noyes to T. W. Richards, 23 November 1908, 1907–1910 box, Personal Correspondence, TWR.

45. W. D. Bancroft, "The Relation of Physical Chemistry to Technical Chemistry," *JACS 21* (1899): 1102. See also J. E. Trevor, "The Achievements and Aims of Physical Chemistry," *JACS 16* (1894): 516.

46. Terry S. Reynolds, "Defining Professional Boundaries: Chemical Engineering in the Early Twentieth Century," *TC 27* (1986): 698–700.

47. Morris Loeb, "The Fundamental Ideas of Physical Chemistry," inaugural lecture given at Clark University in 1889, printed in *The Scientific Work of Morris Loeb*, ed. T. W. Richards (Cambridge, Mass., 1913), p. 5. Loeb was paraphrasing a passage taken from the coda of Ostwald's "Die Aufgaben der physikalischen Chemie," *Humboldt. Monatsschrift für die gesamten Naturwissenschaften 6* (1887): 249–252, reprinted in *Forschen und Nutzen. Wilhelm Ostwald zur wissenschaftlichen Arbeit*, ed. Günther Lotz, Lothar Dunsch, and Uta Kring (Berlin, 1978), pp. 192–198.

48. Bigelow, *Theoretical and Physical Chemistry* (New York, 1912), p. 5. Frederick H. Getman, a pupil of H. C. Jones at Johns Hopkins, used this same analogy in his popular text, *Outlines of Theoretical Chemistry* (New York, 1913), p. 1.

49. For an account of attempts to resolve the problem of strong electrolytes, see John H. Wolfenden, "The Anomaly of Strong Electrolytes," *Ambix 19* (1972): 175–196.

50. Bancroft, "Future Developments in Physical Chemistry," *JPC 9* (1905): 217; "Physical Chemistry," in *A Half Century of Chemistry in America, 1876–1926*, p. 94. On Kahlenberg, see Dolby, "Debates over the Theory of Solution," pp. 354–373.

51. For a discussion of the relationship between physical and analytical chemistry see Ferenc Szabadvary, *History of Analytical Chemistry*, trans. Gyula Svekla (New York, 1966), esp. pp. 353–374.

52. Ostwald, "The Historical Development of General Chemistry," *School of Mines Quarterly 27* (1905–1906): 398–399.

53. A. A. Noyes, *Notes on Qualitative Analysis* (Boston, 1892), title varies through nine subsequent editions; last edition, published posthumously, *A Course of Instruction in the Qualitative Chemical Analysis of Inorganic Substances* (New York, 1942), coauthored by Ernest H. Swift; A. A. Noyes and W. C. Bray, *A System of Qualitative Analysis for the Rare Elements* (New York, 1927); O. F. Tower, *A Course in Qualitative Analysis of Inorganic Substances* (Philadelphia, 1909).

54. Richards, "The Relation of the Tastes of Acids to Their Degrees of Dissociation," *ACJ 20* (1898): 121–126. Kahlenberg, "The Relation of the Taste of Acid Salts to Their Degree of Dissociation," *JPC 4* (1900): 33–37; see also A. A. Noyes's critical review of Kahlenberg's article in *RACR 6* (1900): 73; Richards' reply to Kahlenberg in *JPC 4* (1900): 207–211; and Kahlenberg's rejoinder in *JPC 4* (1900): 533–537. Kopperl discusses Richards' interest in physiology in "The Scientific Work of Theodore William Richards," pp. 23–24.

55. Kahlenberg and Rodney H. True, "On the Toxic Action of Dissolved Salts and Their Electrolytic Dissociation," *Botanical Gazette 22* (1896): 81–124; Kahlenberg, "The Relative Strength of Antiseptics," *Pharmaceutical Review 15* (1897): 68–70; and Kahlenberg and Rollan M. Austin, "Toxic Action of Acid Sodium Salts on Lupinus Albus," *JPC 4* (1900): 533–569; W. Lash Miller, "Toxicity and Chemical Potential," *JPC 24* (1920): 562–569.

56. Frank K. Cameron, "Physical Chemistry in the Service of Agriculture," *JPC 8* (1904): 643–644.

57. Cameron received his doctorate at Johns Hopkins in 1894 but served as Bancroft's research assistant at Cornell in 1897–98.

58. Among them were James M. Bell (Cornell Ph.D. under Bancroft, 1905), Harrison Eastman Patten (Wisconsin Ph.D. under Kahlenberg, 1902), and Atherton Seidell (Johns Hopkins Ph.D. under H. C. Jones, 1903).

59. Cameron, "Physical Chemistry in the Service of Agriculture;" pp. 640–641; Cameron recognized that osmotic pressure alone could not explain root function.

60. Loeb, "On the Nature of the Process of Fertilization," in Jacques Loeb, *The Mechanistic Conception of Life*, ed. Donald Fleming (Cambridge, Mass., 1964), pp. 105–115. On Loeb and his work, see Philip J. Pauly, *Controlling Life: Jacques Loeb and the Engineering Ideal in Biology* (New York, 1987); J. H. van't Hoff, *Physical Chemistry in the Service of the Sciences* (Chicago, 1903), pp. 73–93.

61. Henderson summarized much of his research in *Blood: A Study in General Physiology* (New Haven, 1928). On Henderson, see Walter B. Cannon, "Lawrence Joseph Henderson," *BMNAS 23* (1945): 31–58; and John Parascandola, "Organismic and Holistic Concepts in the Thought of L. J. Henderson," *Journal of the History of Biology 4* (1971): 63–114.

62. J. E. Trevor, review of *Electrophysiology* (by W. Biedermann), in *JPC 1* (1896–1897): 304.

63. See Wilder D. Bancroft, "The Relation of Physical Chemistry to Technical Chemistry," *JACS 21* (1899): 1101–1107; A. A. Noyes's review of Bancroft's article in *RACR 6* (1900): 71; Alexander Smith, "The Teaching of Physical Chemistry," *Electrochemical Industry 1* (1902–1903): 385–386; W. H. Walker, "Some Present Problems in Technical Chemistry," *PSM 66* (1905): 447; F. W. Frerichs, *et al.*, "Report of Committee on Chemical Engineering Education," *TAICE 3* (1910): 123–124; Walter F. Rittman, "The Application of Physical Chemistry to Industrial Processes," *TAICE 7* (1914): 48.

64. Kurt Mendelsohn, *The World of Walther Nernst: The Rise and Fall of German Science, 1864–1941* (Pittsburgh, 1973); Morris Goran, *The Story of Fritz Haber* (Norman, Oklahoma, 1967); Cohen, *Jacobus Henricus van't Hoff*.

65. George Wise, *Willis R. Whitney, General Electric, and the Origins of U.S. Industrial Research* (New York, 1985); Robert H. Dalton, "Eugene Cornelius Sullivan," in *American Chemists and Chemical Engineers*, ed. Wyndham D. Miles (Washington, D.C., 1976), pp. 463–464; George Dubpernell, "The Development of Chrome Plating," *Plating 47* (1960): 35–53; Cameron, *Cottrell: Samaritan of Science*, pp. 113ff.

66. H. S. van Klooster, "Metallography in America," *Chemical Age 31* (1923): 291–292; Bancroft, "Autobiographical Notes," WDB; F. G. Keyes, "Arthur Beckett Lamb," *BMNAS 29* (1956): 200–234; E. C. Sullivan, "George Augustus Hulett," *BMNAS 34* (1960): 82–105; A. A. Noyes, "Autobiographical Notes," NAS.

67. Trevor, "The Achievements and Aims of Physical Chemistry," *JACS 16* (1894): 516–523; Trevor, review of "On the Indispensability of the Atomistic in Natural Science" (by Ludwig Boltzmann), in *JPC 1* (1896–1897): 436; Noyes, *The General Principles of Physical Science* (New York, 1902), see esp. chapter 2. For background on Ostwald and energetics, see Erwin N. Hiebert, "The Energetics Controversy and the New Thermodynamics," in *Perspectives in the History of Science and Technology*, ed. Duane H. D. Roller (Norman, Oklahoma, 1971), pp. 67–87; Peter Clark, "Atomism versus Thermodynamics," in *Method and Appraisal in the Physical Sciences: The Critical Background to Modern Science, 1800–1905*, ed. Colin Howson (Cambridge, 1976), pp. 41–105; and Arie Leegwater, "The Development of Wilhelm Ostwald's Chemical Energetics," *Centaurus 29* (1986): 314–337.

68. Ostwald recognized that Perrin's work on Brownian motion and Thomson's work on gaseous ions offered convincing evidence for the atomic theory in his preface to the fourth edition of his *Grundriss der allgemeinen Chemie* (Leipzig, 1909). Morgan, however, was still unwilling to accept the reality of atoms in 1911. Jacques Loeb mentions Morgan in a letter to J. Stieglitz dated 27 March 1911: "This year he [Leonard Loeb, Jacques' son] is taking physical chemistry with Morgan, who tries to get along without atoms and molecules." Container no. 14, JL.

69. Whitney, review of *Physical Chemistry for Electrical Engineers*, in *JACS 28* (1906): 803.

70. The eleven areas were solution theory, including experimental studies of electrolytic dissociation, osmotic pressure, solubility effects, and acid-base equilibria; electrochemistry; chemical kinetics; phase-rule studies; thermochemistry and investigations of free-energy changes; theory of chemical thermodynamics; catalysis; radiochemistry; colloid and surface chemistry; photochemistry; and studies of molecular structure using X-ray-diffraction techniques.

71. Clarence J. West and Callie Hull, "The Fourth Census of Graduate Research Students in Chemistry," *JCE 5* (1928): 882–884. This article includes summary data from earlier censuses. West and Hull reported that in 1924, 460 students were studying physical chemistry, colloid chemistry, catalysis, subatomic chemistry and radiochemistry, electrochemistry, and photochemistry, out of a total of 1,700 students in all lines of chemical study (27%). Comparable statistics for 1925 were 544/1,763 (31%) and for 1926 were 523/1,882 (28%).

72. A. A. Noyes saw these regulations as posing a formidable obstacle to MIT graduates planning German study. See Noyes, "Institute Graduates at German Universities," *TR 8* (1906): 1–11; S. C. Lind discusses the effects of the regulations on his own course of study in "Autobiographical Notes," NAS.

73. Ostwald, review of the *JPC*, in *ZPC 21* (1896): 528; Duhem, "Une science nouvelle: la chimie physique," *Revue philomathique de Bordeaux et du Sud-Ouest* (1899): 270; Maberry, "Education of the Professional Chemist," *Science 25* (1907): 691.

74. On Jones, see E. Emmet Reid's introduction to H. C. Jones, *The Nature of Solution* (New York, 1917), pp. vii–xiii.

75. Arrhenius to Ostwald, 21 October 1893, in *Aus dem wissenschaftlichen Briefwechsel Wilhelm Ostwalds*, vol. II, ed. Hans Günther Körber (Berlin, 1969), p. 123.

76. Nef to T. W. Richards, 28 July 1899, Personal Correspondence, 1896–1902 box, TWR.

77. On Jones's hydrate theory see H. C. Jones, *Hydrates in Aqueous Solution* (Washington, D.C., 1907); *A New Era in Chemistry* (New York, 1913); and *The Nature of Solution*.

78. W. Boettger, review of "On the Nature of Concentrated Solutions of Electrolytes" (by H. C. Jones), in *RACR 9* (1905): 67–69. On the English school of hydrate chemists, see Dolby, "Debates over the Theory of Solution."

79. Louis Kahlenberg, review of *Hydrates in Aqueous Solution* (by H. C. Jones), in *Science 25* (1907): 963; see also J. J. van Laar's comments on Jones's theory in *Chemische Weekblad 2* (1905): 6–8.

80. Richards to Franklin, 6 May 1914, Professional and Personal Correspondence, 1902–1914 box, TWR; Menzies, review of *Introduction to Physical Chemistry* (by H. C. Jones), in *JACS 32* (1910): 722–724; Franklin, review of *A New Era in Chemistry* (by H. C. Jones), in *Science 40* (1914): 174.

81. Noyes to Richards, 1 June 1916, Professional and Personal Correspondence, 1916 box, TWR; Reid, *My First Hundred Years* (New York, 1972), p. 112; H. N. Morse to W. A. Noyes, 22 February 1913 [1914?] and W. A. Noyes to H. N. Morse,

26 February 1914, both in box 4, WAN; Mathews to T. W. Richards, 1 June 1916, Professional and Personal Correspondence, 1916 box, TWR.

82. Alexander Smith to W. A. Noyes, 1 February 1910, box 4, WAN; *Proceedings of the American Chemical Society* (1907): 16; Richards to W. D. Bancroft, 13 January 1914, Correspondence as President of the ACS 1914, TWR; Richards to Ira Remsen, 29 March 1916, Professional and Personal Correspondence, 1916 box, TWR.

83. Richards, "Progress in Physical Chemistry," *Science 8* (1898): 726–727.

84. Kopperl, "The Scientific Work of Theodore William Richards"; Aaron J. Ihde, "T. W. Richards and the Atomic Weight Problem," *Science 164* (1969): 647–651.

85. Richards to Charles W. Eliot, 27 July 1901, and Eliot to Richards, 29 July 1901, both in Nathan and Ida Reingold, *Science in America: A Documentary History, 1900–1939* (Chicago, 1981), pp. 76–78.

86. Olive Bell Daniels, "Farrington Daniels: Chemist and Prophet of the Solar Age. A Biography," unpublished manuscript (Madison, Wisconsin, 1978), pp. 60–61.

87. Richards to W. A. Noyes, 8 May 1916, Professional and Personal Correspondence, 1916 box, TWR.

88. Richards, "The Possible Significance of Changing Atomic Volume" and "The Significance of Changing Atomic Volume. II.," both in *PAAAS 37* (1902): 3–17 and 399–411 (also in *ZPC 40* [1902]: 169 and 597); Sir Harold Hartley, "Theodore William Richards Memorial Lecture," *JCS* (1930): 1952–1956; Kopperl, "The Scientific Work of Theodore William Richards," p. 18 and pp. 145–184.

89. Richards to Arrhenius, 18 January 1923, Professional and Personal Correspondence, 1919–1926 box, TWR.

90. Richards, "The Significance of Changing Atomic Volume. III.," *PAAAS 38* (1902–1903): esp. pp. 300–307 (also in *ZPC 42* [1902]: 129); Erwin N. Hiebert, "Walther Hermann Nernst," *DSB, Supplement I*, esp. pp. 436–441.

91. Interview with J. Robert Oppenheimer conducted by Thomas S. Kuhn, 18 November 1963, quoted in *Robert Oppenheimer: Letters and Recollections*, ed. Alice Kimball Smith and Charles Wiener (Cambridge, Mass., 1980), p. 68.

92. Bancroft, "Physical Chemistry," in *A Half Century of Chemistry in America*, p. 104; Richards to Joseph D. Davis, 10 February 1928, TWR (quoted by Kopperl, "The Scientific Work of Theodore William Richards").

93. Alexander Findlay, "Wilder Dwight Bancroft," *JCS 56* (1953): 2506–2514; H. W. Gillett, "Wilder D. Bancroft," *IECNE 24* (1932): 1200–1201.

94. Whitney to Henry S. Pritchett, 3 May 1917, 1916–1917 box, WRW.

95. Noyes to Henry S. Pritchett, 12 January 1903, copy in A. A. Noyes file, NAS.

96. Pauling, "Arthur Amos Noyes," p. 326.

97. Linus Pauling, "Fifty Years of Physical Chemistry in the California Institute of Technology," *ARPC 16* (1965): 1–14; C. A. Russell, *The History of Valency* (Oxford, 1971), pp. 302–309.

98. Joel Hildebrand, "Gilbert Newton Lewis," *BMNAS 31* (1958): 210–235; Arthur Lachman, *Borderland of the Unknown: The Life Story of Gilbert Newton Lewis* (New York, 1955); Melvin Calvin, "Gilbert Newton Lewis," in *Proceedings of the Robert A. Welch Foundation, Conferences on Chemical Research. XX. American Chemistry— Bicentennial*, ed. W. O. Milligan (Houston, 1977), pp. 116–150; see also papers on Lewis in *JCE 61* (Jan.–March, 1984).

99. Nef to Harper, 26 May 1892, 4 June 1892, both in box 14, WRH; Felix Lengfeld to Nef, 16 July 1892; Nef to Harper, 16 February 1894; Nef to George Ellery Hale, 9 August 1903, box 1, JUN.

100. H. N. Stokes to W. R. Harper, 11 November 1893, box 16, PPC; "Professors at Outs," undated newspaper clipping, box 1, JUN.

101. Bancroft, review of *Acht Vorträge über physikalische Chemie* (by J. H. van't Hoff), in *JPC 7* (1903): 33; Alexander Smith to Harper, 2 February 1903, 19 January 1904, 23 November 1905; Judson to Smith, 7 February 1906, box 16, PPC; John Ulric Nef, Jr., "John Ulric Nef," undated biographical MS, box 2, JUN.

102. "Committee of Tenured Professors in the Chemistry Department to R. M. Hutchins," 21 February 1934, PPC; Nef to Harper, 6 June 1892, box 14, WRH.

103. Stieglitz to Harper, 2 June 1892, box 15, WRH; "Outline of the History of the Kent Chemical Laboratory and the Department of Chemistry of the University of Chicago," January, 1920, box 16, PPC.

104. Lawrence Badash, *Radioactivity in America: Growth and Decay of a Science* (Baltimore, 1979), pp. 67–73, 78–79, and 94–98.

105. James Kendall, "Alexander Smith, 1865–1922," *Proceedings of the American Chemical Society 44* (1922): 113–117. Smith may well have developed his interest in physical chemistry as a result of his friendship with the English physical chemist James Walker. They had been students together at the University of Edinburgh during the 1880s. Both went to Germany for doctorates, Smith working under Adolph Baeyer at Munich and Walker under Ostwald. Later they returned together to Edinburgh where both worked as assistants in the laboratory of Alexander Crum-Brown. See James Walker, "Autobiographical Notes," quoted by James Kendall in "Sir James Walker," *JCS 138* (1935): 1349–1350.

106. On Harkins, see Robert S. Mulliken, "William D. Harkins," *BMNAS 47* (1975): 49–81, and autobiographical comments in his "Surface Structure and Atom Building," *Science 70* (1929): 434; George B. Kauffman, "William Draper Harkins (1873–1951): A Controversial and Neglected American Physical Chemist," *JCE 62* (1985): 758–761.

107. Daniel J. Kevles, "Genetics in the United States and Great Britain, 1890–1930: A Review with Speculations," *Isis 71* (1980); 441.

108. Alexander Gerschenkron, "Economic Backwardness in Historical Perspective," in *Economic Backwardness in Historical Perspective*, ed. Alexander Gerschenkron (Cambridge, Mass., 1966), pp. 5–30.

109. Cottrell, "Random Notes on the Salvaging of Ideas and Personalities," undated memorandum quoted in *Cottrell: Samaritan of Science*, pp. 66–67.

110. George Wise, "Ionists in Industry: Physical Chemistry at General Electric, 1900–1915," *Isis 74* (1983): 7–21; Martha Moore Trescott, *The Rise of the American Electrochemicals Industry: Studies in the American Technological Environment* (Westport, Conn., 1981), pp. 156–157, 180–181, and 284–298. See also Leonard S. Reich, *The Making of American Industrial Research: Science and Business at GE and Bell, 1876–1926* (Cambridge, 1985); Reese Jenkins, *Images and Enterprise: Technology and the American Photographic Industry* (Baltimore, 1975); and David Hounshell and John K. Smith, *Science and Corporate Strategy: Dupont R & D, 1902–1980* (New York, 1988).

111. The fullest account of the dislocations caused the chemical industry by World War I is Williams Haynes, *American Chemical Industry*, vol. 3 (New York, 1945); see also Frederick E. Wright, *The Manufacture of Optical Glass and Optical Systems: A War-Time Problem* (Washington, D.C., 1921); Robert M. Yerkes, *The New World of Science: Its Development during the War* (New York, 1920); Howard R. Bartlett, "The Development of Industrial Research in the United States," in *U.S. National Resources Planning Board, Research—A National Resource*, vol. 2, *Industrial Research* (Washington, D.C., 1941), pp. 19–77; L. F. Haber, *The Chemical Industry, 1900–1930: International Growth and Technological Change* (Oxford, 1971); and Daniel J. Kevles, *The Physicists: The History of a Scientific Community in Modern America* (New York, 1978), pp. 102–138.

112. Arnold Thackray, Jeffrey L. Sturchio, P. Thomas Carroll, and Robert F. Bud, *Chemistry in America, 1876–1976: Historical Indicators* (Dordrech, 1985), p. 111.

113. Ibid., p. 120.

114. Dolby, "Debates over the Theory of Solutions," p. 309, and "The Transmission of Two New Scientific Disciplines from Europe to North America in the Late Nineteenth Century," *Annals of Science 34* (1977): 287–310.

115. George Wise, "A New Role for Professional Scientists in Industry: Industrial Research at General Electric, 1900–1916," *TC 21* (1980): 408–429.

116. Henry Adams, "The Rule of Phase Applied to History," in *The Degradation of Democratic Dogma* (New York, 1920), pp. 267–311; Charles Rosenberg, "Martin Arrowsmith: The Scientist as Hero," in *No Other Gods: On Science and American Social Thought* (Baltimore, 1976), pp. 123–132.

117. Joseph Ben-David, *The Scientist's Role in Society: A Comparative Study* (Englewood Cliffs, N.J., 1971), pp. 139–168.

118. Remsen, review of *Lehrbuch der allgemeinen Chemie*, vol. 1 (by Wilhelm Ostwald), in *ACJ 7* (1885–1886): 281–282; review of *Grundlinien der anorganischen Chemie* (by Wilhelm Ostwald), in *ACJ 25* (1901): 85; "Some Changes in Chemistry in Fifty Years," *Science 27* (1908): 973–977.

119. Remsen, *The Principles of Theoretical Chemistry*, 5th ed. (Philadelphia, 1897); for more on Remsen see Owen Hannaway, "The German Model of Chemical Education in America: Ira Remsen at Johns Hopkins," *Ambix 23* (1976): 145–164.

120. See, for example, Smith's *Electro-chemical Analysis* (Philadelphia, 1890). On Smith and his work in electrochemistry see Lisa M. Robinson, "The Electrochemical School of Edgar Fahs Smith, 1878–1913," Ph.D. dissertation, University of Pennsylvania, 1986.

121. University of Chicago, *Register*, 1902–1903 and preceding years.

122. Wesleyan University, *Annual Catalogue*, 1908–1909 and preceding years.

123. Pritchett to Noyes, 16 February 1903, Roll 27, GEH.

124. Noyes to Pritchett, 12 January 1903; Kahlenberg, "Charles Kendall Adams," unpublished manuscript, box 12, LK; L. H. Duschak to Hulett, 14 July 1942, box 5, GAH; E. C. Sullivan, "George Augustus Hulett," *BMNAS 34* (1960): 86.

125. Massachusetts Institute of Technology, *President's Report*, 1912, pp. 51 and 55.

126. See the U.S. Department of the Interior, Bureau of Education, *Biennial Survey of Education* for information on new laboratories and their cost.

127. Ibid., Bulletin 1924, no. 14, 1920–1922, vol. 2 (Washington, D.C., 1925), pp. 297–298.

128. Thackray, *et al.*, *Chemistry in America, 1876–1976*, p. 257.

129. Robert E. Kohler, *From Medical Chemistry to Biochemistry*, pp. 121–193.

CHAPTER 3
King Arthur's Court: Arthur A. Noyes and the Research Laboratory of Physical Chemistry

1. Richards, "Progress in Physical Chemistry," *Science 8* (1898): 723.

2. Samuel C. Prescott, *When MIT Was "Boston Tech," 1861–1916* (Cambridge, Mass., 1954), p. 133; Maurice Caullery, *Universities and Scientific Life in the United States*, trans. J. H. Woods and E. Russell (Cambridge, Mass., 1922), p. 269.

3. Prescott, *When MIT Was "Boston Tech,"* p. 185.

4. James Phinney Munroe, "The Massachusetts Institute of Technology," *New England Magazine 33* (1902–1903): 157.

5. Prompting students to compose such ditties as the following (to the tune of the Merry Widow):

> Was für eine Sorge hat der Student der Chemie
> Gatterman zu übersetzen, Bielstein auch muss sehen,
> Und wenn nach Etwas in L. und B.'s Tabellen
> Sucht er einen schlusslichen Fluch,
> Denkt er Gott sei bedanken wir sprechen
> Englisch auf MIT.

Duncan MacRae, memorandum to John W. Servos, 28 September 1976. The lyrics were composed during the 1913–1914 academic year by R. E. Zimmerman, later a vice president of U.S. Steel, but at the time an instructor in chemistry at the Tech.

6. Faculty rosters are available in the MIT *Catalogue*. Full-length biographies exist for Coolidge, Whitney, and G. N. Lewis: John Anderson Miller, *Yankee Scientist: William David Coolidge* (Schenectady, N.Y., 1963); George Wise, *Willis R. Whitney, General Electric, and the Origins of U.S. Industrial Research* (New York, 1985); and Arthur Lachman, *Borderland of the Unknown: The Life Story of Gilbert Newton Lewis* (New York, 1955). On Noyes see Linus Pauling, "Arthur Amos Noyes," *BMNAS 31* (1958): 322–346; on Walker, "William H. Walker," *National Cyclopedia of American Biography*, vol. A, p. 167; on W. K. Lewis, *Warren Kendall Lewis, John Fritz Medalist for 1966* (Philadelphia, December 1965).

7. Enrollment figures are drawn from the *Report of the President and Treasurer* (Boston, 1905), p. 105; see also Prescott, *When MIT Was "Boston Tech,"* p. 177.

8. Henry P. Talbot and Arthur A. Blanchard, *The Electrolytic Dissociation Theory With Some of Its Applications* (New York, 1905); Arthur A. Noyes and Miles S. Sherrill, *A Course of Instruction in the General Principles of Chemistry. Printed in Preliminary Form for the Classes of the Massachusetts Institute of Technology* (Boston, 1914). The latter served as the basis of *Chemical Principles* (New York, 1922), a widely-used text that introduced the "problem method" of teaching physical chemistry, that is, the technique of leading students to discover basic principles by having them solve a multitude of carefully designed problems.

9. *Newburyport Herald*, quoted in Henry E. Noyes and Harriette E. Noyes, *Genealogical Record of Some of the Noyes Descendants of James, Nicholas and Peter Noyes*, vol. 1, *Descendants of Nicholas Noyes* (Boston, 1904), p. 275; John J. Currier, *History of Newburyport, Massachusetts, 1764–1909*, vol. 2 (Newburyport, 1909), pp. 178 and 287.

10. Stephan Thernstrom, *Poverty and Progress: Social Mobility in a Nineteenth Century City* (Cambridge, Mass., 1964; New York, 1974), pp. 10–11, 27, 30, 36, and 167. See also W. Lloyd Warner, *Yankee City*, 5 vols. (New Haven, 1941–1959). Turn-of-the-century Newburyport appears, under various pseudonyms, in many of John P. Marquand's novels. Marquand, a generation younger than Noyes, spent much of his youth in the town; on Marquand and Newburyport, see Millicent Bell, *Marquand: An American Life* (Boston, 1979).

11. Noyes, "Autobiographical Notes," NAS; "Life Sketch of Arthur A. Noyes," Box 49, Folder 2, CIT.

12. "Life Sketch of Arthur A. Noyes," CIT; Miles S. Sherrill, "Arthur Amos Noyes," *PAAAS 74* (1940): 153; Noyes to George Ellery Hale, 21 January 1918, Adm. EX. BD.: General, NAS.

13. A case for the congenial fit between the values of Puritanism and experimental science was first made by Robert Merton, *Science and Society in Seventeenth Century England* (Bruges, 1938; New York, 1970). On the relevance of this connection to late-nineteenth-century America, see Owen Hannaway, "The German Model of Chemical Education in America: Ira Remsen at Johns Hopkins," *Ambix 23* (1976): 145–164; Robert H. Kargon, *The Rise of Robert Millikan: Portrait of a Life in American Science* (Ithaca, N.Y., 1982), pp. 31–32 and 43–44; David A. Hollinger, "Inquiry and Uplift: Late Nineteenth Century American Academics and the Moral Efficacy of Scientific Practice," in *The Authority of Experts: Studies in History and Theory*, ed. Thomas L. Haskell (Bloomington, Ind., 1984), pp. 141–156; and Larry Owens, "Pure and Sound Government: Laboratories, Playing Fields, and Gymnasia in the Late-Nineteenth-Century Search for Order," *Isis 76* (1985): 182–194.

14. Noyes to Whitney, 1 February 1908, WRW. I thank George Wise for this reference and Larry Owens for awakening me to the spiritual overtones of engineering education in the late nineteenth century.

15. The words are those of MIT's president, Francis Amasa Walker, quoted in Wise, *Whitney*, p. 24.

16. Noyes, "Life Sketch"; Wise, *Whitney*, p. 25.

17. "Transcript of Arthur Amos Noyes," courtesy of the Office of the Registrar, MIT; quote from F. J. Moore, *History of Chemistry*, 3rd ed., rev. William T. Hall (New York, 1939), p. 422.

18. "On the Action of Heat upon Ethylene," *ACJ 8* (1886): 362–363.

19. Noyes, "Life Sketch"; T. L. Davis, "Samuel P. Mulliken," *PAAAS 70* (1935–1936): 562–563.

20. Noyes, "A Talk on Teaching," *Science 27* (1908): 659.

21. Wise, *Whitney*, p. 31; Helen Wright, *Explorer of the Universe: A Biography of George Ellery Hale* (New York, 1966), esp. pp. 48–72.

22. Noyes to Harry M. Goodwin, 5 January 1890, 20 February 1890, and 6 August 1890, HMG.

23. Frederick G. Keyes, "Arthur Amos Noyes," *Nucleus 14* (1936): 13.

24. *Notes on Qualitative Chemical Analysis* (Boston, 1892). This book went through nine subsequent editions and was among the two or three most popular textbooks in its field prior to World War I; see Noyes to George Ellery Hale, 7 August 1914, Roll 27, GEH, and James S. Thompson to W. C. Bray, 6 May 1915, Box 5, CCR. A. A. Noyes and S. P. Mulliken, *Laboratory Experiments on the Class Reactions and Identification of Organic Substances* (Easton, Penn., 1898); A. A. Noyes, *The General Principles of Physical Science* (New York, 1902).

25. On *Chemical Abstracts* see Charles A. Browne, *A History of the American Chemical Society* (Washington, D.C., 1952), p. 65. On Noyes's summers see: "General Institute News," *TR 3* (1901): 416; Noyes to Goodwin, 3 August 1891, 12 September 1892, 12 July 1893, and 13 July 1901, HMG. On ACS presidents see Arnold Thackray, Jeffrey Sturchio, P. Thomas Carroll, and Robert Bud, *Chemistry in America, 1876–1976: Historical Indicators* (Dordrecht, 1985), pp. 456 and 470.

26. Henry P. Talbot, "Report of the Department of Chemistry and Chemical Engineering," in Massachusetts Institute of Technology, *Annual Report of the President and Treasurer, December 10, 1902* (Boston, 1902), pp. 37–38.

27. Wise, *Whitney*, pp. 24 and 41. The depression of the 1890s moderated MIT's growth in enrollments, but did not stop it; the student body expanded from 937 in 1890–1891 to 1,277 in 1900–1901, before leaping to 1,608 in 1902–1903. Enrollments in chemistry and chemical engineering, however, grew at a much faster pace; the combined totals of bachelors degrees granted in these courses grew from 38 in the period 1885–1889 to 147 in the period 1895–1899. See *Report of the President and Treasurer* (Boston, 1905), pp. 105 and 116.

28. Ostwald to Richards, 30 November 1896, Personal Correspondence (1896–1902) Box, TWR; see also Nernst to Richards, 2 March 1897, in same Box.

29. Massachusetts Institute of Technology, *Annual Report of the President and Treasurer, December 14, 1898*, p. 55. Descriptions of apparatus in this laboratory are provided in the *Annual Catalogue*.

30. Noyes to Goodwin, 13 July 1901, HMG.

31. Noyes to Goodwin, 3 August 1891 and 12 July 1893, HMG.

32. Noyes to Goodwin, 15 January 1893, HMG.

33. Ibid.

34. Noyes to Goodwin, 17 March 1893, HMG; Noyes and Clement, "Electrolytic Reduction of Nitrobenzene in Sulphuric Acid Solution," *TQ 6* (1893): 62–64.

35. Noyes and Rolfe M. Ellis, "Synthesis of Diphenylbiphenyl and its Identification as Benzerythrene," *TQ 8* (1895): 178–180; and Noyes and Willard H. Watkins, "The Occurrence of Trimethylene Glycol as a By-Product in Glycerine Manufacture," *TQ 8* (1895): 261–262.

36. Wise, *Whitney*, pp. 59–62. The figure of $1,000 a month comes from Henry G. Pearson to Vannevar Bush, 22 March 1939, AAN, CIW.

37. On Agassiz, Marsh, Cope, Crafts, and Hale see *DSB*; on Crafts's background see also Prescott, *When MIT Was "Boston Tech,"* pp. 152–154; for other examples, see Kenneth M. Ludmerer, *Learning to Heal: The Development of American Medical Education* (New York, 1985), p. 37, and Jane Maienschein, "Physiology, Biology, and the Advent of Physiological Morphology," in *Physiology in the American Context,*

1850–1940, ed. Gerald L. Geison (Bethesda, Maryland, 1987), p. 188. On Hale's first telescope see Helen Wright, *Explorer of the Universe*, pp. 75–76.

38. Noyes to Henry S. Pritchett, 25 November 1901, Roll 27, GEH.

39. Wise, *Whitney*, p. 61.

40. Noyes to Goodwin, 13 July 1901, HMG.

41. Ibid.

42. Noyes to Hale, 19 December 1901, Roll 27, GEH; Prescott, *When MIT Was "Boston Tech,"* pp. 185 and 192.

43. Howard S. Miller, *Dollars for Research: Science and Its Patrons in Nineteenth Century America* (Seattle, 1970), pp. 166–181; Nathan Reingold, "National Science Policy in a Private Foundation: The Carnegie Institution of Washington, in *The Organization of Knowledge in Modern America*, ed. Alexandra Oleson and John Voss (Baltimore, 1979), pp. 313–341.

44. John W. Servos, "To Explore the Borderland: The Foundation of the Geophysical Laboratory of the Carnegie Institution of Washington," *HSPS 14* (1983): 147–185.

45. Based on information in Carnegie Institution of Washington *Year Book 2–16* (1903–1917).

46. Noyes to Gilman, 24 March 1902, AAN, CIW; see also Noyes to Miles Sherrill, 12 October 1902, quoted by Henry G. Pearson in a letter to Vannevar Bush, 24 February 1939, AAN, CIW.

47. Noyes to Charles D. Walcott, 22 January 1903 and 11 October 1904; Walcott to Noyes, 27 January 1903; Noyes to Pritchett, 12 January 1903; all in AAN, CIW.

48. Massachusetts Institute of Technology, *Catalogue 40* (1904–1905): 14. Additional grants of $1,000 were made to the Laboratory by William E. Hale (George Ellery Hale's father) and by Willis R. Whitney. See Noyes to Walcott, 22 January 1903, AAN, CIW.

49. F. G. Keyes to G. N. Lewis, 10 February 1933, Box 3, CCR.

50. Noyes, "The New Research Laboratories at the Institute," *TR 5* (1903): 305–306; on Noyes's earlier troubles in obtaining equipment see Noyes to Goodwin, 17 March 1893, HMG; Whitney to Noyes, 25 May 1910, Box 6, CCR.

51. Noyes, "The New Research Laboratories at the Institute," p. 307.

52. On the Yerkes Observatory see Wright, *Explorer of the Universe*, pp. 108–110.

53. Ivy L. Lee, "How Can the Colleges and Industry Cooperate?" and H. P. Hammond, "Promotion of Engineering Education in the Past Forty Years," *Proceedings of the Society for the Promotion of Engineering Education 21* (1913): 57–58 and *41* (1933): 47; David F. Noble, *America by Design: Science, Technology, and the Rise of Corporate Capitalism* (New York, 1977), pp. 28–32; Monte A. Calvert, *The Mechanical Engineer in America: Professional Cultures in Conflict* (Baltimore, 1967), pp. 232–234; Terry S. Reynolds, *Seventy-Five Years of Progress—A History of the American Institute of Chemical Engineers, 1908–1983* (New York, 1983), pp. 9–15.

54. Noyes, "Talk to First Year Students," *TR 9* (1907): 5; see also Noyes, "Advanced Courses for Specialization," in Massachusetts Institute of Technology *President's Report*, January 1908 (Boston, 1908), p. 18; and "Discussion on the Training of Technical Chemists," *Science 19* (1904): 572.

55. Noyes, "Research Laboratory of Physical Chemistry," *Report of the President and Treasurer* (Boston, 1905), p. 58.

56. Noyes to Robert S. Woodward, 12 October 1905, AAN, CIW. Rosters of Noyes's staff members may be found in relevant issues of the MIT *Catalogue*.

57. Noyes to Hale, 7 October 1903, Roll 27, GEH. On Hudson see Lyndon F. Small and Melville L. Wolfrom, "Claude Silbert Hudson," *BMNAS 32* (1958): 181–220, and Claude S. Hudson, "Autobiography," in *The Collected Papers of C. S. Hudson*, ed. R. M. Hann and N. K. Richtmyer (New York, 1946), pp. xi–xxvii.

58. W. D. Coolidge, "Autobiographical Notes," 10 October 1936 and February 1954, NAS. See also Wise, *Whitney*, passim, and John Anderson Miller, *Yankee Scientist: William David Coolidge*.

59. Described in Arthur A. Noyes, *The Electrical Conductivity of Aqueous Solutions* (Washington, D.C., 1907), pp. 9–23.

60. Joel Henry Hildebrand, "William Crowell Bray," *BMNAS 26* (1951): 13–24; W. C. Bray to William T. Goldsborough, 17 June 1914, Box 5, CCR.

61. Raymond M. Fuoss, "Charles August Kraus," *BMNAS 42* (1971): 119–151; Robert Taft, "The Beginning of Liquid Ammonia Research in the United States," in *Selected Readings in the History of Chemistry*, ed. Aaron J. Ihde and William F. Kieffer (Easton, Penn., 1965), pp. 196–201, esp. p. 200. Kraus's work with Franklin was published in the *ACJ 20* (1898): 820–853; *21* (1899): 1–14; *23* (1900): 277–313; and *24* (1900): 83–93.

62. Richard N. Lewis, "A Pioneer Spirit from a Pioneer Family," *JCE 61* (1984): 3; Gerald E. K. Branch, "Gilbert Newton Lewis," *History of Science Dinner Club Papers*, July 13, 1953, reprinted in *JCE 61* (1984): 18–21; Melvin Calvin, "Gilbert Newton Lewis," in *Proceedings of the Robert A. Welch Foundation Conferences on Chemical Research. XX. American Chemistry—Bicentennial*, ed. W. O. Milligan (Houston, 1977), pp. 116–150; Joel Henry Hildebrand, "Gilbert Newton Lewis," *BMNAS 31* (1958): 210–224.

63. Richards to Julius Stieglitz, 17 November 1916, Professional and Personal Correspondence, 1916 Box, TWR.

64. Compare G. N. Lewis, "The Development and Application of a General Equation for Free Energy and Physico-chemical Equilibrium," *PAAAS 35* (1899): 3–38, and "The Law of Physico-Chemical Change," *PAAAS 37* (1901): 49–69 with T. W. Richards; "The Driving Energy of Physico-Chemical Reaction and Its Temperature Coefficient," *PAAAS 35* (1900): 471–480; "The Possible Significance of Changing Atomic Volume," *PAAAS 37* (1902): 3–17; and "The Significance of Changing Atomic Volume. III.," *PAAAS 38* (1902): 291–317.

65. Richards, "The Possible Significance of Changing Atomic Volume," p. 5 and pp. 15–16.

66. Richards, "Chemistry 8 Notebook," TWR, quoted by Sheldon J. Kopperl, "The Scientific Work of Theodore William Richards," Ph.D. dissertation, University of Wisconsin, 1970, p. 226.

67. Robert F. Kohler, Jr., "The Origin of G. N. Lewis's Theory of the Shared Pair Bond," *HSPS 3* (1971), pp. 350–351; Lewis, *The Anatomy of Science* (New Haven, 1926), p. 6.

68. Lewis to Robert A. Millikan, 28 October 1919, quoted by Robert F. Kohler, Jr., in "The Origin of Lewis's Theory," pp. 351–352.

69. On Noyes's fondness for Tennyson, see Noyes to Goodwin, 20 February 1890; for the signature "Arturo" see Noyes's letters to Hale, Roll 27, GEH; on Noyes's nickname, interview with Duncan MacRae, 12 October 1976.

70. Thomas S. Kuhn, *The Structure of Scientific Revolutions*, 2nd ed. (Chicago, 1970), pp. 52–65.

71. For some testimonials to the intellectual leadership attained by Noyes's laboratory see Wilhelm Boettger, *Amerikanisches Hochschulwesen: Eindrücke und Betrachtungen* (Leipzig, 1906), p. 5; Hugh Taylor, "Fifty Years of Chemical Kinetics," *ARPC 13* (1962): 2; and David Ridgway, "Interview with Joel Hildebrand," *JCE 52* (1975): 48. Among the foreign students and research associates were Yogoro Kato from the Technical College of Tokyo, Ming Chow from China, John Johnston from Scotland via Breslau, Wilhelm Boettger from Leipzig, and Eugene W. Posnjak from Moscow via Leipzig.

72. For the original statement of the solubility-product principle, see Nernst, "Über gegenseitige Beeinflussung der Löslichkeit von Salzen," *ZPC 4* (1889): 372–383. In that article Nernst emphasized the similarities between the behavior of mixtures of partially dissociated gases and solutions. His derivation of the principle in *Theoretical Chemistry* gave greater weight to the law of mass action. Recapitulating his argument: if m_0 is the solubility of the solid electrolyte in water and α_0 its degree of dissociation at the given concentration, then $m_0(1 - \alpha_0)$ will represent the undissociated and $m_0\alpha_0$ the dissociated quantity of the electrolyte. If the solubility is m and the degree of dissociation is α in the presence of another electrolyte, the free ions of which have a concentration x, then

$$m_0(1 - \alpha_0) = m(1 - \alpha),$$

since the solubility of the undissociated molecules is constant, and

$$Km_0(1 - \alpha_0) = (m_0\alpha_0)^2, \text{ and}$$
$$Km(1 - \alpha) = m\alpha(m\alpha + x),$$

since the law of mass action states that the concentration of the products of a reaction is proportional to the square of the concentrations of the reactants. Therefore,

$$(m_0\alpha_0)^2 = m\alpha(m\alpha + x),$$

from which may be obtained the following equation for finding the solubility after the addition:

$$m = -x/2\alpha + [m_0^2(\alpha_0/\alpha)^2 + x^2/4\alpha^2]^{1/2}$$

73. Noyes, "Über die gegenseitige Beeinflussung der Löslichkeit von dissociierten Körpern," *ZPC 6* (1890): 241–267.

74. Noyes and E. H. Woodworth, "Investigation of the Theory of Solubility Effect in the Case of Tri-Ionic Salts," *JACS 20* (1898): 194–201; Noyes, "Die Theorie der Löslichkeitsbeeinflussung bei zweiionigen Elektrolyten mit lauter verschiedenen Ionen," *ZPC 27* (1898): 267–278; Noyes and Max Le Blanc, "Über vermehrte Löslichkeit. Anwendung der Gefrierpuncktsbestimmungen zur Ermittelung der Vörgange in

Lösung," *ZPC 6* (1890): 385–402; Noyes and Willis R. Whitney, "Cryoscopic Experiments with the Aluminates and Borates of the Alkali Metals," *TQ 7* (1894): 70–75.

75. John H. Wolfenden, "The Anomaly of Strong Electrolytes," *Ambix 19* (1972): 175–196.

76. J. R. Partington, "Ionic Equilibrium in Solutions of Electrolytes," *JCS 97* (1910): 1162.

77. Wolfenden, "The Anomaly of Strong Electrolytes," pp. 180–184; Charles A. Kraus and William C. Bray, "A General Relation between the Concentration and the Conductance of Ionized Substances in Various Solvents," *JACS 35* (1913): 1315–1434, esp. pp. 1316–1317; 1391–1392; 1423–1430.

78. Noyes, "Über die gegenseitige Beeinflussung der Löslichkeit von dissociierten Körpern," p. 259; "On the Determination of the Electrolytic Dissociation of Salts by Means of Solubility Experiments," *TQ 4* (1891): 262–264; Noyes and William C. Bray, "The Effect of Salts on the Solubility of Other Salts," *JACS 33* (1911): 1643.

79. Wilhelm Biltz, "Zur Kenntnis der Lösungen anorganischer Salze in Wasser," *ZPC 40* (1902): 185–221, esp. p. 218; Partington, "Ionic Equilibrium in Solutions of Electrolytes," pp. 1163–1164.

80. Noyes, "Über die gegenseitige Beeinflussung der Löslichkeit von dissociierten Körpern," pp. 266–267; "On the Determination of the Electrolytic Dissociation of Salts by Means of Solubility Experiments," pp. 263–264; 272; 290.

81. Between 1890 and 1895, Noyes devoted heroic efforts to calculating ionization constants from solubility data. Nernst had developed the solubility-product principle as a means of calculating solubilities from dissociation values; Noyes reversed the procedure, using the principle to move from observed solubilities to dissociation values. See Noyes, "On the Determination of the Electrolytic Dissociation of Salts by Means of Solubility Experiments"; Noyes, "The Determination of the Electrolytic Dissociation of Salts by Solubility Experiments," *TQ 6* (1893): 237–240; Noyes and Charles G. Abbot, "A Comparison of the Dissociation Values Calculated from Solubility Experiments and from the Electrical Conductivity"; and W. J. Humphreys, "The Dissociation of Electrolytes as Determined by Solubility Experiments," *ACJ 17* (1895): 708–712. After a heated exchange of papers with Arrhenius, Noyes realized that ionization values obtained by this method fit Ostwald's dilution law no better than those derived from conductivity measurements.

82. Noyes and Charles G. Abbot, "A Comparison of the Dissociation Values Calculated from Solubility Experiments and the Electrical Conductivity," p. 62; Noyes, "The Physical Properties of Aqueous Salt Solutions in Relation to the Ionic Theory," *Science 20* (1904): 577–587.

83. Noyes, "The Physical Properties of Aqueous Salt Solutions in Relation to the Ionic Theory," p. 582.

84. Ibid., pp. 586–587.

85. Lewis, "The Use and Abuse of the Ionic Theory," *Science 30* (1909): 3.

86. Peter Debye and Erich Hückel, "Zur Theorie der Elektrolyte. I. Gefrierpunktserniedrigung und verwandte Erscheinungen," *PZ 24* (1923): 185–206; and "Zur Theorie der Elektrolyte. II. Das Grenzgesetz für die elektrische Leitfähigkeit," *PZ 24* (1923): 305–325.

87. Noyes, *The Electrical Conductivity of Aqueous Solutions*, p. 4; Noyes to R. S. Woodward, 3 October 1908, AAN, CIW.

88. Noyes, *The Electrical Conductivity of Aqueous Solutions*, pp. 3–4; Noyes, "Statement of Proposed Investigation for which Grants are Requested," accompanying Noyes to D. C. Gilman, 24 March 1902, AAN, CIW.

89. Noyes, *The Electrical Conductivity of Aqueous Solutions*, pp. 48–49, 333–352; Noyes to Robert S. Woodward, 30 September 1907 and 1 October 1914, AAN, CIW.

90. Noyes and K. G. Falk, "The Properties of Salt Solutions in Relation to the Ionic Theory. I. Mol-Numbers Derived from the Freezing-Point Lowering," *JACS 32* (1910): 1011–1030, on p. 1012.

91. Noyes, *The Electrical Conductivity of Aqueous Solutions*, p. 351.

92. Kraus, "Solutions of Metals in Non-Metallic Solvents. I. General Properties of Solutions of Metals in Liquid Ammonia," *JACS 29* (1907): 1558.

93. Lewis, "A Revision of the Fundamental Laws of Matter and Energy," *TQ 21* (1908): 212–225; Lewis and Tolman, "The Principle of Relativity and Non-Newtonian Mechanics," *PAAAS 44* (1909): 711–726. Stanley Goldberg provides a valuable discussion of these papers in *Understanding Relativity: Origin and Impact of a Scientific Revolution* (Boston, 1984), pp. 252–259.

94. Lewis, *The Anatomy of Science*, pp. 84–85.

95. Goldberg, *Understanding Relativity*, p. 256.

96. Basil Schonland, *The Atomists (1805–1933)* (Oxford, 1968), pp. 75–98.

97. Ibid., pp. 101–104; J. J. Thomson, "The Structure of the Atom," *PM 7* (1904): 237 and *The Corpuscular Theory of Matter* (New York, 1907); Robert E. Kohler, Jr., "The Origin of G. N. Lewis's Theory of the Shared Pair Bond," *HSPS 3* (1971): 348–350 and 353–355.

98. Kraus's papers on solutions of metals in liquid ammonia were published in *JACS 29* (1907): 1557–1571; *30* (1908): 653–658; 1197–1219; and 1324–1344. His work was described as a search for a missing link in anon. (probably G. N. Lewis), "General Institute News," *TR 10* (1908): 59.

99. Tolman, "The Electromotive Force Produced in Solutions by Centrifugal Action," *PAAAS 46* (1910): 109–146; Tolman and T. Dale Stewart, "The Electromotive Force Produced by the Acceleration of Metals," *PR 8* (1916): 97–116. In the first article, Tolman established that electrolytic solutions, under the influence of centrifugal force, produce a measurable electromotive force due to the effect of the force on the mass of the ions; in the second, he showed that an accelerative force will likewise generate an electromotive force in metals, permitting the measurement of the inertia of the free electrons.

100. Wise, *Whitney*, pp. 102–103 and 152–153.

101. "General Institute News—Research Laboratory of Physical Chemistry," *TR 10* (1908): 175; W. C. Bray to C. A. Kraus, 30 June 1913, Box 5, CCR; William C. Bray and Gerald E. K. Branch, "Valence and Tautomerism," *JACS 35* (1913): 1440; G. N. Lewis, *Valence and the Structure of Atoms and Molecules* (New York, 1923), p. 30.

102. K. George Falk and John M. Nelson, "The Electron Conception of Valence," *JACS 32* (1910): 1637–1654; Bray and Branch, "Valence and Tautomerism"; William

C. Arsem, "A Theory of Valency and Molecular Structure," *JACS 36* (1914): 1655–1675, quote from p. 1656.

103. Lewis, "Valence and Tautomerism," *JACS 35* (1913): 1448–1455; "The Atom and the Molecule," *JACS 38* (1916): 762–785.

104. Kohler, "The Origins of Lewis's Theory," p. 344.

105. Ibid. See also Robert E. Kohler, Jr., "Irving Langmuir and the 'Octet' Theory of Valence," *HSPS 4* (1972), p. 40 and pp. 49–50.

106. Lewis, *Valence and the Structure of Atoms and Molecules*, p. 29. Kohler gives a sensitive and convincing account of the role of the cubical atom in the genesis of the theory of the shared-pair bond in "The Origins of Lewis's Theory."

107. Noyes and K. George Falk, "The Properties of Salt Solutions in Relation to the Ionic Theory," *JACS 32* (1910): 1011–1030; *33* (1911): 1436–1460; *34* (1913): 454–485 and 485–489. Noyes and Bray, "The Effect of Salts on the Solubility of Other Salts," pp. 1643–1649; 1673–1686. Charles A. Kraus and William C. Bray, "A General Relation between the Concentration and the Conductance of Ionized Substances in Various Solvents," *JACS 35* (1913): 1315–1434. G. N. Lewis, "A Review of Recent Progress in Physical Chemistry," *JACS 28* (1906): 893–910; "The Use and Abuse of the Ionic Theory," *Science 30* (1909): 1–6; "The Activity of the Ions and the Degree of Dissociation of Strong Electrolytes," *JACS 34* (1912): 1631–1644.

108. Falk and Nelson, "The Electron Conception of Valence," p. 1651; Arsem, "A Theory of Valency and Molecular Structure," p. 1659.

109. Bray to Charles A. Kraus, 30 June 1913, Box 5, CCR.

110. Walden, "The Dielectric Constants of Dissolved Salts," *JACS 35* (1913): 1661.

111. Lewis, "The Atom and the Molecule," pp. 764–767, 775–776, 781–782. Lewis and Merle Randall, "The Activity Coefficient of Strong Electrolytes," *JACS 43* (1921): 1153; reproduced verbatim in Lewis and Randall, *Thermodynamics and the Free Energy of Chemical Substances* (New York, 1923), pp. 324–325. The organic chemist Howard J. Lucas made this point in a interestingly different way: "Lewis advances a reason for both ion formation and molecule formation, namely the tendency of atoms to surround themselves with stable groupings of electrons, which in the case of the smaller atoms is almost always eight. This tendency may be so pronounced that the atom goes over into an ionic condition, a process opposite to what one would expect from the known behavior of positive and negative charges. In other words ion formation and molecule formation are simply different results of a fundamental process, namely the tendency of atoms to take on electron configurations characteristic of the members of the helium family." Lucas to W. A. Noyes, 8 September 1926, Box 6, WAN.

112. W. H. Bragg and W. L. Bragg, *X Rays and Crystal Structure* (London, 1915), esp. pp. 4–5 and 88–97.

113. Noyes to R. S. Woodward, 26 September 1916, AAN, CIW.

114. C. Lalor Burdick, "The Genesis and Beginnings of X-ray Crystallography at Caltech," in *Fifty Years of X-ray Diffraction*, ed. Peter Paul Ewald (Utrecht, 1962), p. 557; Noyes, "Research Laboratory of Physical Chemistry," in *Reports of the President and Treasurer for 1917* (Cambridge, Mass., 1918), p. 82.

115. Duncan MacRae to John W. Servos, 28 September 1976.

116. There is little secondary literature on the development of chemical thermodynamics during the years after 1890. Erwin Hiebert, "Hermann Walther Nernst," *DSB, Supplement I*, pp. 432–453 provides the best discussion of Nernst's work and valuable insights into the concerns of others working on chemical thermodynamics around the turn of the century.

117. Lewis, "A New Conception of Thermal Pressure and a Theory of Solutions," *PAAAS 36* (1900): 145–168; "The Law of Physico-Chemical Change," *PAAAS 37* (1901): 49–69, quote from p. 49. Lewis realized that exact thermodynamic methods had been developed by Gibbs, Duhem, and Planck but believed that their approach carried the liabilities of being based on an unsatisfying analogy between chemical and mechanical systems and of being expressed in "the rather abstruse" equations of entropy and the thermodynamic potential. Rather than adopting the method of Gibbs, Lewis preferred to modify the methods of van't Hoff, Nernst, and Arrhenius, already familiar to chemists.

118. Lewis, "The Law of Physico-Chemical Change," p. 54.

119. Ibid.; see also Lewis, "A New Conception of Thermal Pressure," pp. 148–149 and 156.

120. Lewis, "The Law of Physico-Chemical Change," pp. 60 and 66; quote from p. 60.

121. Ibid., p. 55.

122. Bancroft, review of "Deduction and Application of a General Equation for Free Energy and Physical Equilibrium" (by G. N. Lewis), in *JPC 5* (1901): 405.

123. Lewis, "Outlines of a New System of Thermodynamic Chemistry," *PAAAS 43* (1907): 259–293; quote from p. 284.

124. Walther Nernst, *Theoretical Chemistry from the Standpoint of Avogadro's Rule and Thermodynamics*, 4th ed., trans. Robert A. Lehfeldt (London, 1904), p. 532.

125. Lewis, "Outlines of a New System of Thermodynamic Chemistry," pp. 284–288.

126. Ibid., pp. 269–270 and p. 287.

127. Ibid., pp. 289–292.

128. These techniques are summarized in Lewis and Randall, *Thermodynamics*, pp. 259–275.

129. See Lewis, "Outlines of a New System of Thermodynamic Chemistry," p. 287.

130. In 1907, Lewis would have denied this. At that time he believed activity was as fundamental a quantity as free energy or chemical potential, related expressions of escaping tendency. While Lewis never publicly retracted this interpretation, activity gradually moved from the center toward the periphery of Lewis's thermodynamics; in *Thermodynamics*, for instance, he gave much greater attention to the use of activity as a tool in the calculation of equilibrium constants and free energies than to its ontological significance or its foundational role in the equations of thermodynamics. In a letter written in 1928, he refers to having "invented the somewhat trivial but very useful 'activity.' " Lewis to J. R. Partington, 7 December 1928, Box 3, CCR.

131. J.R.P., review of *Thermodynamics and the Free Energy of Chemical Substances* in *TFS 19* (1923–1924), pp. 231–232. Partington, it should be said, also praised Lewis's book as "the most important contribution to the literature of thermo-

dynamics which has appeared in recent times." Later, Partington would nominate Lewis for a Nobel Prize; see Lewis to Partington, 7 December 1928.

132. Noyes and Bray, "The Effect of Salts on the Solubility of Other Salts. I," p. 1646. Some physical chemists had earlier used "activity coefficient" as a synonym for the degree of dissociation; see Nernst, "Über gegenseitige Beeinflussung der Löslichkeit von Salzen," p. 379. On Noyes's subsequent use of activity coefficients in the analysis of the anomaly of strong electrolytes, see Noyes and Duncan MacInnes, "The Ionization and Activity of Largely Ionized Substances," *JACS 42* (1920): 239–245.

133. The term free energy was coined by Helmholtz in 1882 to describe that portion of the energy of a reaction that was not bound in the form of heat but could be freely converted to other forms of energy. A similar entity, the chemical potential, had been independently defined by Gibbs in 1878. Gibbs's chemical potential differed from Helmholtz's change in free energy insofar as the former represented the maximum work exclusive of work done against a uniform pressure by the change in volume resulting from the reaction, whereas Helmholtz's change in free energy included this form of work. The chemical potential or Gibbs free energy therefore represents the maximum useful work of a reaction, whereas the Helmholtz free energy represents the maximum work of a reaction. After Gibbs's paper was publicized by Ostwald in the early 1890s, it was common for physical chemists to use the version of free energy that fit their immediate purposes better. The resulting ambiguities were part of the reason that Lewis, in 1901, described chemical thermodynamics as "bewildering to the beginner and confusing even to the initiated" (Lewis, "The Law of Physico-Chemical Change," p. 49). Lewis was far more conscious than most of his contemporaries were of the need for consistency.

134. There was another method of determining free-energy values, that of calculating free-energy values directly from calorimetric data. By integrating the Gibbs-Helmholtz equation,

$$\Delta F - \Delta H = T \, \partial(\Delta F)/\partial T,$$

it was possible to obtain a formula expressing the change in free energy in a reaction as a series of terms that could all be evaluated by the measurement of heats of reaction or specific heats, with the exception of a single integration constant. The presence of this indeterminate constant rendered the formula useless. Between 1888 and 1905, Le Châtelier, Richards, Lewis, van't Hoff, and Haber all sought the physical meaning of this constant without success. In a paper published in 1906, Nernst proposed a theorem, often called Nernst's heat theorem or, more simply, the third law of thermodynamics, that showed that the value of this constant could be computed from empirically determined values. The calculation of free-energy values from calorimetric data, however, was a method that rested upon Nernst's theorem; in fact one means of testing that theorem was by comparing free-energy values derived from equilibrium or electromotive-force measurements with those calculated from calorimetric data. Hence, Nernst's heat theorem in no way freed chemists from the task of perfecting noncalorimetric techniques of measuring free-energy changes. On the development of the heat theorem see Hiebert, "Hermann Walther Nernst"; on its role in the determination of free-energy values see Lewis and Randall, *Thermodynamics*, pp. 448–454, and Neil K. Adam, *Physical Chemistry* (Oxford, 1956), pp. 270–273.

135. Lewis and Randall, *Thermodynamics*, p. 291; see also pp. 292–293; Lewis and

George H. Burrows, "The Free Energy of Organic Compounds. I. The Reversible Synthesis of Urea and of Ammonium Cyanate," *JACS 34* (1912): 1515–1529; and Lewis, "The Free Energy of Chemical Substances," *JACS 35* (1913): 1–30.

136. Lewis, "The Potential of the Oxygen Electrode," *JACS 28* (1906): 158–170; Noyes to George Ellery Hale, 5 June 1905, Roll 27, GEH; Noyes, *et al.*, "Report of the Committee on Cooperative Research in Chemistry," *Report of the National Academy of Sciences for the Year 1906* (Washington, D.C., 1907), pp. 18–20; Noyes, *et al.*, "Report of the Committee on Cooperative Research in Chemistry," *Report of the National Academy of Sciences for the Year 1907* (Washington, D.C., 1908), pp. 15–17.

137. Noyes to C. D. Walcott, 13 October 1914, AAN, CIW.

138. Ibid.

139. Lewis to Partington, 7 December 1928.

140. Lewis to Wheeler, 11 November 1911, 9 December 1912, 5 January 1912, 13 March 1912, 18 April 1912, CU-5, PPUC; Robert Wayne Seidel, "Physics Research in California: The Rise of a Leading Sector in American Physics," Ph.D. dissertation, University of California, Berkeley, 1979, pp. 23–24. Lewis's salary in his final year at MIT was $2,500; in his first year at California he received $5,000; see minutes for the meeting of 12 May 1911, ECM; and Lewis to Wheeler, 9 December 1911.

141. Noyes to Hale, 23 May 1914, Roll 27, GEH; see also Bray to Noyes, 3 June 1914, and Noyes to Bray, 10 June 1914, both in Box 5, CCR.

142. H. P. Talbot to Willis R. Whitney, 20 September 1915, 1915 Box, WRW; see also W. D. Harkins to G. N. Lewis, 5 February 1916, Box 2, CCR.

143. On MacInnes, see the revealing "Autobiographical Sketch," NAS; on Burdick, see his "The Genesis and Beginnings of X-Ray Crystallography at Caltech''; on Dickinson and MacRae, "Interview with Duncan MacRae''; on Keyes and Beattie, see E. Bright Wilson and John Ross, "Physical Chemistry in Cambridge, Massachusetts," *ARPC 24* (1973): 5–8, and *American Men of Science*, ed. J. McKeen Cattell and Jaques Cattell (New York, 1933), 5th ed.

144. Statistics on the productivity of Noyes's laboratory are derived from information in Massachusetts Institute of Technology, *Annual Report of the President and Treasurer* for the years 1908 to 1917.

145. The successes of Noyes's laboratory spawned imitators at MIT and helped nudge the Institute toward a policy of being more tolerant of—if not solicitous of—basic research. See Prescott, *When MIT Was "Boston Tech,"* pp. 186–187, 220–221; and John W. Servos, "The Industrial Relations of Science: Chemical Engineering at MIT, 1900–1939," *Isis 71* (1980): 531–549.

146. For Pauling's assessment of Noyes, see Pauling, "Arthur Amos Noyes," *BMNAS 31* (1958): 321; on Lewis as laboratory director see Joel H. Hildebrand, "Gilbert Newton Lewis," *BMNAS 31* (1958): 212–215.

147. Charles A. Kraus, "Arthur Amos Noyes," *SM 43* (1936): 180–181; those elected to the NAS were W. R. Whitney, W. D. Coolidge, C. S. Hudson, C. A. Kraus, W. C. Bray, G. N. Lewis, R. C. Tolman, W. D. Harkins, D. A. MacInnes, F. G. Keyes, and E. W. Washburn.

148. I am indebted to Duncan MacRae for information about the role of Noyes's

students in industrial research. On the annual budget of Noyes's laboratory, see Servos, "The Industrial Relations of Science," p. 541.

149. Characteristically, Noyes found a way to express his attitude toward physics through some lines from Tennyson:

> Who loves not Physics? Who shall rail
> Against her beauty? May she mix
> With men and prosper. Who shall fix
> Her pillars? Let her work prevail.
>
> . . . .
>
> But she is a child fiery-hot to burst
> All barriers in her onward race
> For power. Let her know her place,
> She is the second, not the first.
> A higher hand must make her mild,
> If all be not in vain; and guide
> Her footsteps, moving side by side
> With Chemia, like the younger child.

Quoted in Noyes to H. M. Goodwin, 20 February 1890, HMG.

CHAPTER 4
The Phase Ruler: Wilder D. Bancroft and His Agenda for Physical Chemistry

1. Noyes and Miles S. Sherrill, *A Course of Studies in Chemical Principles*, 2nd ed. (New York, 1938); Noyes to Hale, 23 May 1914, Roll 27, GEH.

2. Harvard University, *Twenty-fifth Anniversary Report of the Class of 1888* (Cambridge, Mass., 1913), p. 12.

3. Harvard University, *Fiftieth Anniversary Report of the Class of 1888* (Cambridge, Mass., 1938), p. 19.

4. M. A. DeWolfe Howe, *The Life and Letters of George Bancroft*, 2 vols. (New York, 1908) is a rich source of information about the Bancroft family as is Lillian Handlin, *George Bancroft: The Intellectual as Democrat* (New York, 1984); see also Russel B. Nye, *George Bancroft: Brahmin Rebel* (New York, 1944).

5. Edward Everett Hale to M. A. DeWolfe Howe, 20 January 1906, quoted in Howe, *Life and Letters of George Bancroft*, vol. 1, p. 8.

6. Howe, *Life and Letters of George Bancroft*, vol. 2, p. 281. On John Chandler Bancroft see Handlin, *George Bancroft*, pp. 201–202, 251–254, 312, and 337. The best biographical sketch of Wilder Bancroft is Alexander Findlay's in *JCS* (1953): 2506–2514; reprinted in *Great Chemists*, ed. Eduard Farber (New York, 1961), pp. 1245–1261.

7. "Undergraduate Record Card of Wilder Dwight Bancroft," Harvard University Archives.

8. Cooke, "Scientific Culture," in *Scientific Culture and Other Essays* (New York, 1881), pp. 28 and 32. On Cooke see George S. Forbes, "Josiah Parsons Cooke, Jr.," in *DSB*, vol. 3, pp. 397–399; and addresses in commemoration of Josiah Parsons Cooke by his former students Charles Loring Jackson, Henry Barker Hill, Augustus Lowell, Francis Humphreys Storer, and Charles William Eliot in *PAAAS 30* (1895):

513–547. On Cooke drawing Richards to Harvard, see T. W. Richards, "Retrospect," NAS.

9. Cooke, *Religion and Chemistry; or, Proofs of God's Plan in the Atmosphere and Its Elements* (New York, 1864), pp. 6, 268.

10. Cooke, *Elements of Chemical Physics*, Boston, 1860, p. 6.

11. Erwin Hiebert and Hans Günther Körber, "Friedrich Wilhelm Ostwald," *DSB, Supplement I*, New York, 1980, p. 462.

12. Richards, "Brief Biography of T. W. Richards between 1868 and 1917," NAS.

13. On Cornell University see Morris Bishop, *A History of Cornell University* (Ithaca, 1962). Information on the chemistry department may be gleaned from the annual Cornell University *Register* and *President's Report*.

14. *Report of the Commissioner of Education for the Year 1899–1900*, vol. 2 (Washington, D.C., 1901), p. 1874; Bishop, *A History of Cornell*, p. 322.

15. Schurman, *Annual Report of the President of Cornell University, 1892–1893*, p. 26.

16. On Schurman's administration and educational goals, see Bishop, *A History of Cornell*, pp. 303–331, and Laurence Veysey, *The Emergence of the American University* (Chicago, 1965), pp. 360–362.

17. In 1894–1895 the first semester of the elementary chemistry course was attended by 342; see Schurman, *Report of the President, 1894–1895*, p. 104; earlier volumes of the *Report* do not provide course enrollments. On provisions for laboratory work in physical chemistry, see Schurman, *Report of the President, 1899–1900*, pp. 35–36.

18. Schurman, *Report of the President, 1895–1896*, p. xxxviii. Trevor was appointed to this full professorship in 1894, but the appointment was made effective in 1900. No explanation was given for this unusual arrangement. See Schurman, *Report of the President, 1893–1894*, p. 49.

19. Statistics from Schurman, *Report of the President, 1894–1895*, p. 105, and Cornell University, *Register, 1894–1895*, p. 92; quote from Albert Hazen Wright, *The Research Club of Cornell University (1919–1965)*, Pre-Cornell and Early Cornell, Studies in History, no. 35, p. 17.

20. Homer W. Gillett, "Wilder D. Bancroft," *Journal of Industrial and Engineering Chemistry 24* (1932): 1200.

21. On enrollments see Schurman, *Report of the President, 1896–1897*, p. lxxvii; on courses see Cornell University *Register*, 1897–1898 to 1902–1903 and a description of the program in physical chemistry in *JPC 2* (1898), on page preceding p. 77.

22. Quote from Leon Gortler, interview with Thomas Jacobs. I thank Professor Gortler for sharing this anecdote with me. See also Gillett, "Wilder D. Bancroft," and C. W. Mason, "Wilder Dwight Bancroft," *JACS 76* (1954): 2602. Van't Hoff, after a visit to Cornell in 1901, wrote that Bancroft reminded him of an impresario; see his account of this visit quoted in Ernst Cohen, *Jacobus Henricus van't Hoff, Sein Leben und Wirken* (Leipzig, 1912), pp. 454–455.

23. Emile M. Chamot and Fred H. Rhodes, "The Development of the Department of Chemistry and of the School of Chemical Engineering at Cornell," unpublished MS, Cornell University Archives (Ithaca, n.d.), pp. 63, 65, 67.

24. Bancroft, "Analytical Chemistry and the Phase Rule," *JPC 6* (1902): 106–117.

25. Chamot and Rhodes, "Development of the Department of Chemistry," p. 67.

26. Some financial records of the *JPC* are preserved in Box 5, WDB. Trevor resigned as coeditor in 1909.

27. Bishop, *A History of Cornell*, pp. 327, 330.

28. Edgar Fahs Smith to Bancroft, 31 July 1905, 1905–1906 Box, Professional Correspondence, WDB.

29. Quote from Bancroft to E. C. Bingham, 8 February 1922, 1921–1922 Box, Professional Correspondence, WDB; Bancroft to T. W. Richards, 4 March 1899, Personal Correspondence, 1896–1902 Box, TWR.

30. Bancroft, "The Future in Chemistry," *Science 27* (1908): 979–980.

31. Bancroft, "Analytical Chemistry and the Phase Rule," p. 106.

32. Bancroft, review of *Gesammelte Schriften von Eilhard Mitscherlich* in *JPC 1* (1896–1897): 176.

33. Bancroft, *The Phase Rule* (Ithaca, 1897), p. iii.

34. For Bancroft's views on the structure of his science see *The Phase Rule*, esp. pp. 1–5; his course descriptions in the Cornell *Register*; and his reviews in the *JPC*, esp. his reviews of J. H. van't Hoff's *Vorlesungen über theoretische und physikalische Chemie* in *JPC 2* (1898): 256–257; of J.L.R. Morgan's *The Elements of Physical Chemistry* in *JPC 3* (1899): 234–236; and of W. Nernst's *Theoretische Chemie vom Standpunkte der Avogadro'schen Regel und der Thermodynamik* in *JPC 3* (1899): 337–338.

35. Bancroft, quoted in H. S. van Klooster, "Metallography in America," *Chemical Age 31* (1923): 291–292.

36. For two recent and critical evaluations of Le Châtelier's principle, see J. de Heer, "The Principle of Le Châtelier and Braun," *JCE 34* (1957): 375–380, and Richard S. Treptow, "Le Châtelier's Principle," *JCE 57* (1980): 417–420.

37. J. Willard Gibbs, "On the Equilibrium of Heterogeneous Substances," in *The Scientific Papers of J. Willard Gibbs*, vol. 1 (London, 1906), p. 96. For a valuable discussion of the history of derivations of the phase rule, see Edward E. Daub, "Gibbs Phase Rule: A Centenary Retrospect," *JCE 53* (1976): 750.

38. M. M. Pattison Muir to J. Willard Gibbs, 14 February 1880, in Lynde P. Wheeler, *Josiah Willard Gibbs: The History of a Great Mind* (New Haven, 1951; Hamden, Conn., 1969), pp. 86–87; on the factors contributing to Gibbs's neglect see pp. 93–105. Albert E. Moyer has recently challenged the idea that Gibbs was neglected, at least among his countrymen, by citing his election to the National Academy of Sciences and other marks of honor that were bestowed upon him. See his *American Physics in Transition: A History of Conceptual Change in the Late Nineteenth Century* (Los Angeles, 1983), p. 99. There is, however, a difference between being honored and being understood. Gibbs's work was unknown to American chemists until introduced to them by Ostwald; I have seen no evidence that American physicists were significantly better acquainted with Gibbs's papers than their brethren in chemistry.

39. See above, Chapter 2, note 21.

40. On Roozeboom's and Ostwald's discovery of Gibbs's phase rule, see Edward E. Daub, "Gibbs Phase Rule: A Centenary Retrospect," pp. 747–751; J.M.A. van Bemmelen, W. P. Jorissen, and W. E. Ringer, "H.W.B. Roozeboom," *BDCG 40* (1907): 5153–5154; and Wilhelm Ostwald, *Lebenslinien, Eine Selbstbiographie*, 3 vols. (Berlin, 1926–1927), vol. 2, p. 61. The relevant portion of Gibbs's mailing list

for reprints is reproduced in Wheeler, *Josiah Willard Gibbs*, pp. 242–243. Ostwald first cites Gibbs's work in the second volume of his *Lehrbuch der allgemeinen Chemie*, the preface of which is dated December 1886; H. W. Bakhuis Roozeboom, "Sur les differentes formes de l'équilibre chimique hétérogène," *Recueil des travaux chimiques des Pays-Bas et de la Belgique 6* (1887): 262–303.

41. J. Willard Gibbs, *Thermodynamische Studien*, trans. Wilhelm Ostwald (Leipzig, 1892); H. W. Bakhuis Roozeboom, "Studien über chemisches Gleichgewicht," *ZPC 12* (1888): 262–303. On the Dutch school of physical chemists who made use of the phase rule, see J. R. Partington, *A History of Chemistry*, vol. 4 (London, 1964), pp. 638–640.

42. Bancroft, "Das chemische Potential der Metalle," *ZPC 12* (1893): 289–297; quotation from Bancroft, "Inorganic Chemistry and the Phase Rule," pp. 40–41.

43. He first uses "solute" in "On Ternary Mixtures," *PR 3* (1895–1896): 21–33; see note p. 21. He introduced the term for the sake of convenience, but also because he was convinced that there was a physico-chemical difference between solvent and solute: a point that he would later give up only with great reluctance. Other papers in this series were published in the same volume, pp. 114–136 and 193–209; and in *JPC 1* (1896–1897): 34–50.

44. Noyes, review in *RACR 3* (1897): 18; see also his review of an earlier paper by Bancroft in *RACR 1* (1895), appearing in *TQ 8* (1895): 306–307.

45. Bancroft, *The Phase Rule*. A German, Wilhelm Meyerhoffer, had earlier published a short monograph on the rule: *Die Phasenregel und ihre Anwendungen* (Leipzig, 1893). See also Bancroft, "Analytical Chemistry and the Phase Rule Classification," *JPC 6* (1902): 106–117; "Inorganic Chemistry and the Phase Rule"; and many of Bancroft's reviews in the *JPC*. Hector R. Carveth used the term "Phase Ruler" in Carveth to Bancroft, 25 February 1910, 1910 Box, Professional Correspondence, WDB.

46. On Kahlenberg see R.G.A. Dolby, "Debates over the Theory of Solution," *HSPS 7* (1976): 354–373.

47. Bancroft, *The Phase Rule*, p. 5.

48. See Bancroft's review of Jones's work in *JPC 4* (1900): 307–308; *9* (1905): 602; and his review of Noyes's *The General Principles of Physical Science* in *JPC 7* (1903): 217–218.

49. Bancroft, review of *Der energetische Imperativ* by Wilhelm Ostwald, *JPC 16* (1912): 789.

50. Bancroft, "Chemical Potential and Electromotive Force," *JPC 7* (1903): 416–427; the text of Gibbs's letter to Bancroft appears in Gibbs, *Collected Works*, vol. 1, pp. 419–434. Roozeboom, in a generally appreciative review of Bancroft's book, pointedly criticized the absence of any derivation of the phase rule: see *JPC 1* (1896–1897): 559.

51. Bancroft, review of *Traité elementaire de mecanique chimique, fondee sur la thermodynamique* in *Science 10* (1899): 82.

52. Bancroft, "Future Developments in Physical Chemistry," *JPC 9* (1905): 225–226; Bancroft to Richards, 9 March 1905, Correspondence on Chemistry, 1895–1928 Box, TWR.

53. Bancroft, "The Relation of Physical Chemistry to Technical Chemistry," *JACS 21* (1899): 1107.

54. See, for instance, Bancroft, "Future Developments in Physical Chemistry," p. 217; and Bancroft, "Physical Chemistry," in *American Chemical Society, A Half-Century of Chemistry in America, 1876–1926*, ed. Charles A. Browne (Easton, Penn., 1926), p. 94.

55. Bancroft, review of *Theoretische Chemie vom Standpunkte der Avogadro'schen Regel und der Thermodynamik* (by Walther Nernst) in *JPC 3* (1899): 337.

56. Nernst, *Theoretical Chemistry from the Standpoint of Avogadro's Rule and Thermodynamics*, trans. C. S. Palmer (London, 1895), p. 567; Bancroft, "Wilhelm Ostwald, The Great Protagonist," *JCE 10* (1933): 541–542.

57. Cooke, *The Credentials of Science. The Warrant of Faith*, 2nd ed. (New York, 1893), pp. 30–31; quote from p. 30.

58. George Bancroft quoted by Richard Hofstadter, *The Progressive Historians: Turner, Beard, Parrington* (New York, 1970), pp. 15–16. On the development of research in New York's industries see Reese Jenkins, *Images and Enterprise: Technology and the American Photographic Industry, 1839–1925* (Baltimore, 1975); George Wise, *Willis R. Whitney, General Electric, and the Origins of U.S. Industrial Research* (New York, 1985); and Martha Moore Trescott, *The Rise of the American Electrochemicals Industry, 1880–1910: Studies in the American Technological Environment* (Westport, Conn., 1981).

59. Cooke, *The Credentials of Science*, p. 94.

60. Bancroft completed a single course in calculus, for which he received the grade of C; Record Card of Wilder Dwight Bancroft, Harvard University Archives.

61. For more evidence on this point see John W. Servos, "Mathematics and the Physical Sciences in America, 1880–1930," *Isis 77* (1986): 611–629.

62. Cooke, *The Credentials of Science*, pp. 119–120. On the cultural value of laboratory science see Chapter 3, note 13.

63. Arthur Gordon Webster, "Some Practical Aspects of the Relations between Physics and Mathematics," *PR 18* (1904): 316. On the development of mathematics in America, see David Eugene Smith and Jekuthiel Ginsburg, *A History of Mathematics in America before 1900* (Chicago, 1934); for evidence on the mathematical preparation of American scientists see Servos, "Mathematics and the Physical Sciences in America, 1880–1930."

64. Kahlenberg, quoted in Farrington Daniels, "Physical Chemistry," in *Chemistry . . . Key to Better Living* (Washington, D.C., 1951), p. 50.

65. Neil K. Adam, *Physical Chemistry* (Oxford, 1956), p. 326.

66. Roozeboom, "Erstarrungspunkte der Mischkrystalle zweier Stoffe," *ZPC 30* (1899): 385–412.

67. There is a voluminous literature on the phase rule and its uses; I have found the following especially helpful: Alexander Findlay, *The Phase Rule and Its Applications* (London, 1903; 5th ed., 1923); Cecil H. Desch, *Metallography* (London, 1910; 4th ed., 1939); Frederick N. Rhines, *Phase Diagrams in Metallurgy: Their Development and Application* (New York, 1956); Francis J. Turner and John Verhoogen, *Igneous and Metamorphic Petrology*, 2nd ed. (New York, 1960); and E. C. Rollason, *Metallurgy for Engineers* (Norwich, 1973).

68. On the history of metallurgy, see Cyril Stanley Smith, A *History of Metallography* (Chicago, 1960); and R. F. Mehl, *A Brief History of the Science of Metals* (New York, 1948). The question of how and why the place of metals in chemistry changed during the eighteenth and nineteenth centuries deserves study.

69. An example would be Frederick W. Taylor, whose empirical studies of steels led him to the discovery of "high speed steel" in 1898; see Daniel Nelson, *Frederick W. Taylor and the Rise of Scientific Management* (Madison, Wisconsin, 1980), pp. 86–87; Thomas Jay Misa, "Science, Technology and Industrial Structure: Steelmaking in America, 1870–1925," Ph.D. dissertation, University of Pennsylvania, 1987, pp. 164–179, 197–201. Even Andrew Carnegie, who liked to boast of having introduced science into the steel industry, really used his chemists to monitor the quality of raw materials and finished metals rather than to develop new processes or products; see Rossiter W. Raymond to Charles D. Walcott, 8 February 1915, Box 33, CDW.

70. Important here was the development of the thermocouple pyrometer by Le Châtelier and Barus in 1888 and Moissan's invention of the electric furnace in 1892.

71. On the work of these scientists, see Desch, *Metallography*, pp. 4–8, where the slant is decidedly British, and the relevant articles in the *DSB*.

72. Results of these investigations were published in a lengthy series of articles in the *JPC* between 1902 and 1914. All appeared under the names of Bancroft's collaborators, principally, E. S. Shepherd, G. B. Upton, F. E. Gallagher, and B. E. Curry. It is clear from the texts of these papers and from Bancroft's correspondence that their work was closely supervised by Bancroft.

73. Bancroft's application for aid to the Carnegie Institution of Washington accompanied his letter to C. D. Walcott of 9 August 1902, WDB, CIW. See also Bancroft's annual reports in Carnegie Institution of Washington, *Year Book 3–10* (1904–1911). On Bancroft's personal finances, see H. W. Gillett, "Wilder D. Bancroft"; C. W. Mason, "Wilder Dwight Bancroft"; and Bancroft to Louis Kahlenberg, 8 September 1914, Box 4, LK.

74. Bancroft to C. D. Walcott, 9 August 1902, WDB, CIW; and Bancroft, "Report on a Systematic Chemical Study of Alloys," in Carnegie Institution of Washington, *Report of the Executive Committee to the Board of Trustees, December 13, 1904* (Washington, 1904), p. 69.

75. Shepherd, "The Constitution of Copper-Zinc Alloys," *JPC 8* (1904): 422.

76. Cecil H. Desch, *Metallography* (London, 1910; 4th ed., 1937), p. 21.

77. Shepherd, "The Constitution of Copper-Zinc Alloys," p. 423, William C. Roberts-Austen had earlier established many of the major features of the copper-zinc diagram; see his "Fourth Report to the Alloys Research Committee," *Proceedings of the Institute of Mechanical Engineers* (1897): 31.

78. See, for example, E. S. Shepherd and G. B. Upton, "The Tensile Strength of Copper-Tin Alloys," *JPC 9* (1905): 441–476; A. T. Lincoln, David Klein, and Paul E. Howe, "The Electrolytic Corrosion of Brasses," *JPC 11* (1907): 501–536. More recent workers have found the task of relating states of phase equilibrium to the physical properties of alloys no easier; see, for instance, Frederick N. Rhines, *Phase Diagrams in Metallurgy*, p. 7.

79. Shepherd and Upton, "The Tensile Strength of Copper-Tin Alloys," p. 443; Bancroft to R. S. Woodward, 3 April 1905, WDB, CIW.

80. Alfred Sang, "The Industrial Outlook for Physical Chemistry," *Proceedings Engineers' Society of Western Pennsylvania 23* (1907): 45. On the conservatism of research and development work at U.S. Steel, see Gertrude G. Schroeder, *The Growth of Major Steel Companies, 1900–1950* (Baltimore, 1953), pp. 111–113. Firms using electrochemical methods to refine metals appear to have been quicker to invest in research and more receptive to Bancroft's students. A number of Bancroft's pupils found jobs with firms like the Pittsburgh Reduction Co., the Aluminum Castings Co., and the Niagara Electrochemical Co. Martha Moore Trescott, *The Rise of the American Electrochemicals Industry* contains much information on the beginnings of research and development at such firms. On the evolution of links between the scientific community and the steel industry after World War I, see Misa, "Science, Technology and Industrial Structure," pp. 324ff.

81. Rudolph Ruer, *The Elements of Metallography*, trans. C. H. Mathewson (New York, 1909), p. vii. George Marion Howe, the dean of American metallurgists, gave the phase rule greater prominence in his work; see his *Iron, Steel and Other Alloys*, 2nd ed. (Cambridge, Mass., 1906).

82. John Johnston, "Applications of Science to the Metallurgical Industry," *SM 48* (1939): 494–495, 498.

83. Lewis and Merle Randall, *Thermodynamics and the Free Energy of Chemical Substances* (New York, 1923), p. 186; Noyes to Stuart J. Bates, 17 March 1919, Box 2, JABS. See also Noyes, review of "Analytical Chemistry and the Phase Rule Classification" (by Wilder D. Bancroft), in *RACR 8* (1902): 440, where Noyes writes: "The Phase Rule is considered to be all important; but the far more useful Law of Mass Action is not referred to."

84. Ostwald, review of *Die Phasenregel und ihre Anwendungen* (by W. Meyerhoffer), in *ZPC 12* (1893): 398–399; Nernst, "Development of General and Physical Chemistry During the Last Forty Years," *Annual Report of the Smithsonian Institution, 1908* (Washington, D.C., 1909), pp. 250–251; W. A. Noyes to Bancroft, 8 November 1907, 1907–1909 (A–C) Box, Professional Correspondence, WDB. Bancroft did not withdraw the sentence. See Bancroft, "H.W.B. Roozeboom," *Proceedings of the American Chemical Society* (1907): 92.

85. Jones, *A New Era in Chemistry* (New York, 1913), pp. 72–73.

86. See Kraus, "Josiah Willard Gibbs," *Science 89* (1939): 281; see also John Johnston, "Applications of Science to the Metallurgical Industry," *SM 48* (1939): 493–503.

87. Merton, *Science, Technology and Society in Seventeenth Century England* (Bruges, 1938; New York, 1970).

88. In addition to the work on alloys, Bancroft's laboratory also made important contributions during the period prior to World War I to the study of dynamic isomerides and the technology of chromium plating: see Findlay, "Wilder Dwight Bancroft," pp. 2511–2512 and George Dubpernell, "The Development of Chromium Plating," *Plating 47* (1960): 37–39.

89. Thomas Jacobs, interview with Leon Gortler. I thank Professor Gortler for bringing this interview to my attention.

90. E. S. Shepherd and George A. Rankine joined the staff of the Geophysical Lab-

oratory. Going to the USDA between 1897 and 1911 were Frank K. Cameron, A. E. Taylor, J. M. Bell, J. W. Turrentine, J. K. Haywood, and L. F. Hawley.

91. There is an interesting literature on Teeple. See William Haynes, "John Edgar Teeple," in *Great Chemists*, ed. Eduard Farber, pp. 1411–1420; and Robert P. Multhauf, *Neptune's Gift: A History of Common Salt* (Baltimore, 1978), pp. 207–209. Teeple's *The Industrial Development of Searles Lake Brines* (New York, 1929) is a unique and entertaining mixture of personal history and multi-component phase diagrams. Among Bancroft's students who entered teaching careers were: G. A. Perley (New Hampshire), O. W. Brown (Indiana), John A. Wilkinson (Iowa State), and Harry B. Weiser (Rice).

92. Bancroft, "Physical Chemistry," in *American Chemical Society, A Half-Century of Chemistry in America, 1876–1926*, ed. Charles A. Browne, p. 94.

93. Bancroft considered the electrolytic dissociation theory a useful working hypothesis but repeatedly expressed reservations about its validity: see his review of recent literature on the electrolytic dissociation theory in *JPC 1* (1896–1897): 449; review of *The Theory of Electrolytic Dissociation and Some of Its Applications* (by Harry C. Jones), *JPC 4* (1900): 307; "Some Applications of the Electrolytic Dissociation Theory to Medicine and Biology," *Electrochemical Industry 1* (1902–1903): 305; and review of *Qualitative Analysis as a Laboratory Basis for the Study of General Inorganic Chemistry* (by William Conger Morgan), in *JPC 11* (1907): 421. Bancroft himself proposed a modified form of Ostwald's dilution law in 1895, but later despaired of finding a valid equation: "Until we have some method of determining the percentage dissociation of a strong electrolyte, it is rather a waste of time to speculate as to the form of the dilution law." See Bancroft, review of "Ionic Sizes in Relation to the Conductivity of Electrolytes" (by W. R. Bousfield), in *JPC 10* (1906): 71.

94. Eschenbach Printing Co. to Bancroft, 1 April 1918, 1918 Box, Professional Correspondence, WDB. A circulation of 500 was quite respectable for a scientific journal at the time. In 1919 the *American Journal of Physiology* had 460 subscribers, and the *Journal of General Physiology*, 327. See Jacques Loeb to Simon Flexner, 4 August 1919, SF.

95. "Subscriptions to Journal of Physical Chemistry," appended to W. W. Buffum to Williams and Wilkins Co., 22 December 1932, Folder D-6-7, File 29, Drawer 1, FG; Linus Pauling, "Arthur Amos Noyes," in *Proceedings of the Robert A. Welch Foundation Conferences on Chemical Research, XX. American Chemistry—Bicentennial*, edited by W. O. Milligan (Houston, 1977), p. 93.

96. Frank K. Cameron to B. B. Boltwood, 15 January 1909; Cameron to Bancroft, 1 February 1909 and 25 February 1909, all in 1907–1909 (A–C) Box, Professional Correspondence, WDB; Charles H. Herty to Bancroft, 12 June 1909 (D–Z) Box, Professional Correspondence, WDB.

97. Frank K. Cameron to Bancroft, 8 May 1909, 1907–1909 (A–C) Box, Professional Correspondence, WDB; on the "publishing trust," see Louis Kahlenberg to Willis R. Whitney, 16 December 1909, Box 1, LK. On W. A. Noyes's efforts see Charles Albert Browne and Mary Elvira Weeks, *A History of the American Chemical Society* (Washington, D.C., 1952), pp. 68–96; W. A. Noyes, "The Next Step in Publication for the American Chemical Society," *Journal of Industrial and Engineering*

Chemistry 1 (1909): 148–149. William Albert Noyes was a distant cousin of Arthur Amos Noyes.

98. Cameron to Bancroft, 1 February 1909.

99. Bancroft, review of *Laboratory Manual of Physical Chemistry* (by A. W. Davison and H. S. van Klooster), in *JPC 27* (1923): 99.

100. Bancroft, review of *Physical Chemistry and Biophysics for Students of Biology and Medicine* (by Matthew Steel), in *JPC 32* (1928): 794.

CHAPTER 5
Physical Chemistry in the "New World of Science"

1. Van't Hoff toured the United States in 1901 and returned in the fall of 1904 for the International Congress of Arts and Sciences at St. Louis; Ostwald lectured at Berkeley in the summer of 1903, at the St. Louis Congress, and was an exchange professor at Harvard in the fall of 1905; Arrhenius lectured at Berkeley in the summer of 1904 and gave the Benjamin Silliman Lectures at Yale in 1912; Nernst was the Silliman Lecturer at Yale in 1906. American physical chemists on the program at St. Louis in 1904 included A. A. Noyes, Louis Kahlenberg, and W. D. Bancroft.

2. Among the basic books on these developments are Alfred D. Chandler, Jr., *The Visible Hand: The Managerial Revolution in American Business* (Cambridge, Mass., 1977); Robert H. Wiebe, *The Search for Order, 1877–1920* (New York, 1967); Thomas L. Haskell, *The Emergence of Professional Social Science: The American Social Science Association and the 19th Century Crisis of Authority* (Urbana, 1977); Laurence R. Veysey, *The Emergence of the American University* (Chicago, 1965).

3. Arnold Thackray, Jeffrey L. Sturchio, P. Thomas Carroll, and Robert Bud, *Chemistry in America, 1876–1976: Historical Indicators* (Dordrecht, 1985), pp. 256, 313, 268; Department of Commerce, Bureaus of Foreign and Domestic Commerce, *Statistical Abstract of the United States 1919* (Washington, D.C., 1920), p. 788; U.S. Bureau of the Census, *Census of Population: 1960*, vol. 1, *Characteristics of the Population* (Washington, D.C., 1961), Part A, pp. 1-14, 1-15, Table 8.

4. Monte A. Calvert offers an especially rich treatment of such conflicts in *The Mechanical Engineer in America, 1830–1910: Professional Cultures in Conflict* (Baltimore, 1967), esp. pp. 43–107.

5. See, for example, Charles Eliot, "The New Education," *Atlantic Monthly 23* (1869): 202–220 and 365–366; reprinted in *American Higher Education: A Documentary History*, vol. 2, ed. Richard Hofstadter and Wilson Smith (Chicago, 1961), pp. 624–641.

6. On the roles of scientists in Edison's work see Thomas P. Hughes, *Networks of Power: Electrification in Western Society, 1880–1930* (Baltimore, 1983), pp. 23–27; David A. Hounshell, "Edison and the Pure Science Ideal in Nineteenth-Century America," *Science 207* (1980): 104–112.

7. For an especially sensitive reading of the motives of John D. Rockefeller, see John Ettling, *The Germ of Laziness: Rockefeller Philanthropy and Public Health in the New South* (Cambridge, Mass., 1981). On Carnegie, see Joseph Frazier Wall, *Andrew Carnegie* (New York, 1979), and Howard S. Miller, *Dollars for Research: Science and Its Patrons in Nineteenth-Century America* (Seattle, 1970), pp. 166–181.

8. The board of advisors in chemistry, composed of Ira Remsen, T. W. Richards,

and Edgar Fahs Smith, was not as aggressive as similar boards in other sciences; their reticence no doubt encouraged the trustees of the Carnegie Institution of Washington to think that chemistry was a science that could fend for itself. In 1913 and again in 1914, G. N. Lewis solicited funds from the Carnegie Institution for a laboratory of chemical energetics, the aim of which would be to determine the free energies of formation of chemical substances. By then, however, the Carnegie Institution had locked itself into the support of so many projects that it lacked the funds to consider a major new undertaking. See R. S. Woodward to G. N. Lewis, 18 October 1913, 8 June 1914; Lewis to Woodward, 6 June 1914; Lewis to C. D. Walcott, 6 October 1914; and Woodward to Walcott, 16 October 1914, all in Box 33, CDW. The best account of strategies of giving at the CIW is Nathan Reingold, "National Science Policy in a Private Foundation: The Carnegie Institution of Washington," in *The Organization of Knowledge in Modern America, 1860–1920*, ed. Alexandra Oleson and John Voss (Baltimore, 1979), pp. 313–341.

9. Thackray, Sturchio, Carroll, and Bud estimate that 3,830 chemists were working in industrial research in 1921 and that 866 chemists were teaching in colleges and universities in 1920. See *Chemistry in America, 1876–1976*, pp. 347 and 367. Since many industrial chemists lacked Ph.D.s and most faculty members in chemistry departments possessed doctorates, it seems reasonable to assume that the ratio of chemists with doctorates in the two categories was much closer to equality.

10. Louis Galambos, "The American Economy and the Reorganization of the Sources of Knowledge," in *The Organization of Knowledge in Modern America, 1860–1920*, pp. 269–282; Leonard S. Reich, "Industrial Research and the Pursuit of Corporate Security: The Early Years of Bell Laboratories," *Business History Review* 54 (1980): 503–529.

11. America's profligacy was among Little's favorite themes during the decade prior to World War I: "A new competitor is even now girding up his loins and training for the race, and that competitor is strangely enough the United States—that prodigal among nations, still justly stigmatized as the most wasteful, careless and improvident of them all." See Little, "Industrial Research in America," *Journal of Industrial and Engineering Chemistry 5* (1913): 793, and E. J. Kahn, Jr., *The Problem Solvers: A History of Arthur D. Little, Inc.* (Boston, 1986), esp. pp. 41–42. See also David J. Rhees, "The Chemists' Crusade: The Rise of an Industrial Science in Modern America, 1907–1922," Ph.D. dissertation, University of Pennsylvania, 1987, pp. 34–51.

12. See Table 2.5.

13. Johnston to W. C. Bray, 26 May 1911, Box 5, CCR.

14. Warren C. Scoville, *Revolution in Glassmaking: Entrepreneurship and Technological Change in the American Industry, 1880–1920* (Cambridge, Mass., 1948), pp. 297–298; John Lawrence Enos, *Petroleum Progress and Profits: A History of Process Innovation* (Cambridge, Mass., 1962), p. 35 and pp. 101–102.

15. George Wise, "Ionists in Industry: Physical Chemistry at General Electric, 1900–1915," *Isis 74* (1983): 7–21; "A New Role for Professional Scientists in Industry: Industrial Research at General Electric, 1900–1916," *TC 21* (1980): 408–429.

16. William D. Coolidge, "The Development of Ductile Tungsten," in *The Sorby Centennial Symposium on the History of Metallurgy*, ed. Cyril S. Smith (New York, 1965), p. 449.

17. By far the most detailed and comprehensive account of the development of the American chemical industry is Williams Haynes, *American Chemical Industry, A History*, 6 vols. (New York, 1945–1954); volumes 2 and 3 deal with the era of World War I.

18. On the history of chemical engineering during this era, see Terry Reynolds, "Defining Professional Boundaries: Chemical Engineering in the Early Twentieth Century," *TC 27* (1986): 694–716.

19. Arthur D. Little, "Chemical Engineering Research: Lifeblood of American Industry," and Charles O. Brown, "High Pressure Synthesis—Basis of New Chemical Engineering Industries," in *Twenty-five Years of Chemical Engineering Progress: Silver Anniversary Volume, American Institute of Chemical Engineers*, ed. Sidney D. Kirkpatrick (New York, 1933), pp. 1–14 and 152–168; Margaret Jackson Clarke, O.S.B., "The Federal Government and the Fixed Nitrogen Industry, 1915–1926," Ph.D. dissertation, Oregon State University, 1977.

20. Clarke, "The Federal Government and the Fixed Nitrogen Industry," pp. 79–108.

21. Olive Bell Daniels, "Farrington Daniels: Chemist and Prophet of the Solar Age: A Biography," typescript (Madison, Wisconsin, 1978), pp. 93–94. Kraus and Daniels did come up with a design so successful that it became standard issue for the American Army during the interwar years. Lewis to Frederick P. Gay, 15 April 1935, Box 2, CCR. See also Lewis to H. R. Hatfield, 27 November 1917, and Lewis to Morse Cartwright, 4 December 1918, CCR.

22. See Robert H. Kargon, "The New Era: Science and American Individualism in the 1920's," in *The Maturing of American Science* (Washington, D.C., 1974), pp. 1–29; "Temple to Science: Cooperative Research and the Birth of the California Institute of Technology," *HSPS 8* (1977): 3–31.

23. Lewis to Charles H. Warren, 6 May 1927, Box 4, CCR.

24. Keyes to S. W. Stratton, 12 May 1925, Folder 475, PPMIT.

25. McIntosh to G. A. Hulett, 12 August 1922, Box 3, GAH.

26. Roger L. Geiger, *To Advance Knowledge: The Growth of American Research Universities 1900–1940* (New York, 1986), pp. 125–126 and p. 229.

27. Rodebush to Lewis, 16 December 1918, Box 4, CCR. On the U.S. Industrial Alcohol Co. see Haynes, *American Chemical Industry*, vol. 3, pp. 126–127.

28. Lewis to Rodebush, 5 April 1919, Box 4, CCR.

29. William H. Walker, Warren K. Lewis, and William H. McAdams, *Principles of Chemical Engineering* (New York, 1923); Olaf A. Hougen, "Seven Decades of Chemical Engineering," *Chemical Engineering Progress 73* (1977): 94 and 96; F. W. Frerichs, *et al.*, "Report of Committee on Chemical Engineering Education," *TAICE 3* (1910): 127–128, 130. This infusion of physical chemistry into chemical engineering was met by considerable resistance; see below, chapter 6, and Edward E. Daub, "Chemical Engineering at the University of Wisconsin: The Early Years," in *A Century of Chemical Engineering*, ed. William F. Furter (New York, 1982), esp. p. 192.

30. Robert M. Yerkes, *The New World of Science: Its Development During the War* (New York, 1920).

31. Systematic botany and paleontology are examples of disciplines that were less successful than physical chemistry in establishing vital ties with other sciences and that

as a result languished in the early twentieth century. On paleontology, see H. W. Menard, *Science: Growth and Change* (Cambridge, Mass., 1971).

32. Robert Kohler, *From Medical Chemistry to Biochemistry: The Making of a Biomedical Discipline* (Cambridge, 1982). On the origins of chemical engineering, see Jean-Claude Guédon, "Conceptual and Institutional Obstacles to the Emergence of Unit Operations in Europe," in *History of Chemical Engineering*, ed. William F. Furter (Washington, D.C., 1980), pp. 45–75; and Terry Reynolds, "Defining Professional Boundaries."

33. Bunsen, "Über die Prozesse der vulkanischen Gesteinsbildung Islands," *Poggendorff's Annalen Physik und Chemie 83* (1851): 197–272; *Zeitschrift der deutschen geologischen Gesellschaft 13* (1861): 61–63; Durocher, "Essai de pétrologie comparée ou recherches sur la composition chimique et mineralogique des roches ignée, sur les phénomènes de leur émission et sur leur classification," *Annales des Mines* v *11* (1857): 220. On the history of petrology see F. Y. Loewinson-Lessing, *A Historical Survey of Petrology*, trans. S. I. Tomkeieff (Edinburgh, 1954); Louis V. Pirsson, "The Rise of Petrology as a Science," in *A Century of Science in America*, ed. Edward S. Dana, *et al.* (New Haven, 1918), pp. 248–267; and Adolph Knopf, "Petrology," in *Geology 1888–1938: Fiftieth Anniversary Volume of the Geological Society of America* (New York, 1941), pp. 335–363.

34. On van't Hoff's work see Hans P. Eugster, "The Beginnings of Experimental Petrology," *Science 173* (1971): 481–489. Van't Hoff reviewed some of his results in his *Physical Chemistry in the Service of the Sciences* (Chicago, 1903), pp. 97–123. On Arrhenius's geological interests, see Ernest H. Riesenfeld, *Svante Arrhenius* (Leipzig, 1931).

35. F. Bascom, "Fifty Years of Progress in Petrography and Petrology, 1876–1926," *Johns Hopkins Studies in Geology*, no. 8 (1927): 33–82; James Vincent Elsden, *Principles of Chemical Geology: A Review of the Application of the Equilibrium Theory to Geological Problems* (London, 1910), p. v. On Vogt, Lagorio, Doelter, and Loewinson-Lessing see Loewinson-Lessing, *Historical Survey of Petrology*; Christoffer Oftedahl, "Johan Hermann Lie Vogt," *DSB*, vol. 14, pp. 58–59; and Walther Fischer, "Cornelio August Severinus Doelter," in *DSB*, vol. 4, pp. 140–142. Some of the earliest experimental studies of rock formation were undertaken by Scotland's James Hall in the first decade of the nineteenth century and by the Frenchmen Ferdinand André Fouqué and Auguste Michel-Lévy in the late 1870s. Their initiatives, however, did not lead to the development of enduring research traditions.

36. On American petrographers educated in Germany, see Servos, "To Explore the Borderland," pp. 180–181. See also Edward B. Mathews, "Memorial of Joseph Paxson Iddings," *BGSA 44* (1933): 352–374; E. S. Larsen, "Charles Whitman Cross," *BMNAS 32* (1958): 100–112; Adolph Knopf, "Louis V. Pirsson, 1860–1919," *BMNAS 34* (1960): 228–248; and Francis Birch, "Reginald A. Daly," *BMNAS* (1960): 30–64.

37. Loewinson-Lessing, *Historical Survey of Petrology*, pp. 71–77.

38. Mathews, "Joseph Paxson Iddings"; L. M. Dennis, "Frank Wigglesworth Clarke, 1847–1931," *BMNAS 15* (1934): 139–165.

39. See Iddings, "Origin of Igneous Rocks," *BPSW 12* (1892): 89–214. This paper contains an extensive historical introduction.

40. Iddings, "On the Crystallization of Igneous Rocks," *BPSW 11* (1889): 65–114.

41. Iddings, "Origin of Igneous Rocks," pp. 154–156.

42. Iddings, "The Mineral Composition and Geologic Occurrence of Certain Igneous Rocks in the Yellowstone National Park," *BPSW 11* (1890): 212–213 and 217.

43. Iddings, "Origin of Igneous Rocks," pp. 154–156.

44. Barus and Iddings, "Note on the Change of Electric Conductivity Observed in Rock Magmas of Different Composition on Passing from Liquid to Solid," *AJS* iii *44* (1892): 249.

45. Iddings, "Origin of Igneous Rocks," pp. 158–160; Iddings acknowledged that he here was following suggestions made earlier by Lagorio; see A. Lagorio, "Über die Natur der Glasbasis, sowie der Krystallisationsvorgänge im eruptiven Magma," *Tschermak's mineralogische und petrographische Mittheilungen 8* (1887): 421–529.

46. N. L. Bowen, "The Later Stages of the Evolution of Igneous Rocks," *JG 23* (1915), supplement, pp. 4–5.

47. Iddings, "Origin of Igneous Rocks," p. 194.

48. On Van Hise see Thomas C. Chamberlin, "Charles Richard Van Hise, 1857–1918," *JG 26* (1918): 690–697, and Maurice Vance, *Charles Richard Van Hise: Scientist Progressive* (Madison, 1960). Van Hise discusses his own intellectual development in "The Problems of Geology," *JG 12* (1904): 589–616.

49. Van Hise, "The Training and Work of a Geologist," *Science 16* (1902): 326.

50. Van Hise, *A Treatise on Metamorphism* (Washington, D.C., 1904). Benjamin LeRoy Miller discusses the influence of Van Hise's work in "Progress in Ore Genesis Studies, 1876–1926," *Johns Hopkins Studies in Geology*, no. 8 (1927): 121–135. See also Adolph Knopf, "Petrology." On Van Hise and Lincoln see Lincoln to Van Hise, 19 October 1899, US MSS S, July 1899–June 1900, Box 6, USGSLS; and Van Hise to Lincoln, 24 October 1899, US MSS S, USGSVH. Lincoln took his Ph.D. under Wisconsin's Louis Kahlenberg in 1899.

51. Van Hise, "Some Principles Controlling the Deposition of Ores," *JG 8* (1900): 730–770; "Metamorphism of Rocks and Rock Flowage," *BGSA 9* (1898): 269–328; "The Problems of Geology"; and *A Treatise on Metamorphism*. The distinction between zones of fracture and flow was not original to Van Hise. See G. K. Gilbert, *Report on the Geology of the Henry Mountains* (Washington, D.C., 1877), p. 82.

52. Van Hise, "The Problems of Geology," p. 609.

53. Ibid., pp. 610–611; Van Hise, "The Training and Work of a Geologist," p. 326.

54. E. S. Larsen, "Geochemistry," in *Geology, 1888–1938. Fiftieth Anniversary Volume of the Geological Society of America*, pp. 395–398; F. Y. Loewinson-Lessing, *Historical Survey of Petrology*, pp. 42–48.

55. Walcott to Van Hise, 24 February 1900, Box 6, USGSLD.

56. Servos, "To Explore the Borderland," pp. 152–158.

57. Van Hise, J. S. Diller, and S. F. Emmons to Walcott, 13 March 1900, Box 17, GFB.

58. On Becker, see George Merrill, "George Ferdinand Becker," *National Academy of Sciences Memoirs 21*, Second Memoir (1924); and Arthur L. Day, "George Ferdinand Becker," *BGSA 31* (1919): 14–25. Also useful are Charles A. Anderson, "Eugene Thomas Allen," *BMNAS 40* (1969): 1–17; and L. M. Dennis, "Frank Wigglesworth Clarke." The work of Becker's laboratory may be followed in the *Annual Report*

of the U.S.G.S. On Day, see Robert B. Sosman, "Arthur Louis Day," *American Philosophical Society Year Book* (1960): 134–140; Philip H. Abelson, "Arthur Louis Day," *BMNAS 47* (1975): 27–47; and Servos, "To Explore the Borderland," pp. 173–175.

59. Walcott to Becker, 15 June 1901, and Day to Becker, 23 September 1901, both in Box 18, GFB.

60. Arthur L. Day and E. S. Shepherd, "The Lime-Silica Series of Minerals," *AJS* iv *22* (1906): 265.

61. Day and Allen, *Isomorphism and the Thermal Properties of the Feldspars* (Washington, D.C., 1905), p. 17.

62. Ibid., pp. 27–28 and 68.

63. Their melting points were based upon the inspection of heating and cooling curves and were, especially for mixtures rich in albite, acknowledged to be imprecise; in such extremely viscous mixtures the heat of fusion was liberated only slowly, making it difficult to assign an unambiguous melting point. See Day and Allen, *Isomorphism and the Thermal Properties of the Feldspars*, pp. 41–42.

64. Ibid., p. 68. As their co-worker N. L. Bowen later observed, Day and Allen had no conclusive evidence that their melting-point curve corresponded to the liquidus; their melting points could have fallen anywhere within the range between the liquidus and solidus curves. Nevertheless, Bowen's more precise measurements confirmed that their melting points came very close to lying on the liquidus; see Bowen, "The Melting Phenomena of the Plagioclase Feldspars," *AJS* iv *35* (1913): 577, 585.

65. Day and Allen, *Isomorphism and the Thermal Properties of the Feldspars*, pp. 66–70. Subsequent workers, especially N. L. Bowen, gave much more attention to the mechanisms by which early-forming crystals interacted with the liquid melt. Bowen suggested that if the early-forming crystals maintained contact with the liquid, they would undergo a continuous reaction, steadily approximating the composition of the original liquid, and thereby obliterating, as it were, their earlier history. But if early-forming crystals were removed from contact with the liquid, by gravitational settling or other physical processes, they would retain their original composition; see Bowen, "The Reaction Principle in Petrogenesis," *JG 30* (1922): 177–198.

66. Day and Allen, *Isomorphism and the Thermal Properties of the Feldspars*, pp. 69–70.

67. Walcott, "Work of the Year," in USGS, *Annual Report, 1904–1905 26* (1905): 99.

68. On the beginnings of the Geophysical Laboratory, see Servos, "To Explore the Borderland"; there is no history of the laboratory, but on its work see Robert B. Sosman, "The Work of the Geophysical Laboratory of the Carnegie Institution of Washington," in *A Century of Science in America*, pp. 284–287; Hatten S. Yoder, "Experimental Mineralogy: Achievements and Prospects," *Bulletin minéralogique 103* (1980): 5–26.

69. Robert S. Woodward, *et al.*, "Report of the Advisory Committee on Geophysics," in CIW, *Year Book 1* (1902): 27; the language is taken from Van Hise, "Proposal for Laboratories of Geo-physics for the Carnegie Institution of Washington," 31 July 1902, GL, CIW.

70. Of the twenty-four scientists who held staff positions between 1907 and 1916,

sixteen held degrees in chemistry or physical chemistry, six in geology or petrography, and two in physics. On the staff of the Geophysical Laboratory, see Servos, "To Explore the Borderland," p. 151.

71. The most important of Bowen's theoretical works are "The Later Stages of the Evolution of Igneous Rocks," *JG 30* (1922): 177–198, and *The Evolution of Igneous Rocks* (Princeton, 1928). Bowen's importance was in developing a generally persuasive account of how igneous rocks of the crust could all be derived from a single basaltic magma through the process of fractional crystallization. His work stands in need of historical study, as does the entire development of thought on igneous petrogenesis.

72. Bancroft, "H.W.B. Roozeboom," *Proceedings of the American Chemical Society* (1907): 92; "Physical Chemistry," in *A Half Century of Chemistry in America, 1876–1926*, ed. Charles A. Browne (Easton, Penn., 1926), p. 100.

73. On Bowen, see the biographical memoirs by Hans P. Eugster in *BMNAS 52* (1980): 34–79; and C. E. Tilley in *Biographical Memoirs of Fellows of the Royal Society 3* (1957): 6–22. Bowen expresses his debt to Shepherd in "The Binary System: $Na_2Al_2Si_2O_8$ (Nephelite—Carnegieite)—$CaAl_2Si_2O_8$ (Anorthite)," *AJS* iv *33* (1912): 573.

74. Bowen, "The Melting Phenomena of the Plagioclase Feldspars," pp. 577–579 and 586–590.

75. Ibid., pp. 593–594. Bowen himself expressed some surprise that the laws of perfect solutions should apply to his water-free, viscous, and concentrated solutions of silicates while they failed when applied to many dilute aqueous solutions; but, as he pointed out, the result was to be expected in cases of pairs of liquids, for all concentrations, when the liquids were miscible in all proportions and showed neither volume change nor heat effect on mixing.

76. In his excellent memoir on Bowen, Hans P. Eugster calls attention to several such instances. For example, when Bowen evaluated the role of the Soret effect in differentiation, he studied the literature on diffusion and heat flow, and acquainted himself with Gaussian distributions and error functions. "Very few petrologists of his generation," Eugster observes, "had that much courage." See Eugster, "N. L. Bowen," p. 46.

77. Examples of textbooks include Alfred Harker, *The Natural History of Igneous Rocks* (New York, 1909); James Vincent Elsden, *Principles of Chemical Geology: A Review of the Application of Equilibrium Theory to Geological Problems* (London, 1910); H. E. Boeke, *Grundlagen der physikalisch-chemischen Petrographie* (Berlin, 1915); and R. H. Rastall, *Physico-Chemical Geology* (London, 1927). The 1924 conference was held under the auspices of the Faraday Society, the Geological Society, and the Mineralogical Society; papers and comments were published in J. S. Flett, G. Tyrrell, P. Niggli, *et al.*, "The Physical Chemistry of Igneous Rock Formation," *TFS 20* part 3, no. 60 (1925): 414–501.

78. G. W. Tyrell, "Review of Recent Work on the Origin and Differentiation of Igneous Rocks," *TFS 20* part 3, no. 60 (1925): 418; F. Y. Loewinson-Lessing, *A Historical Survey of Petrology*, trans. S. I. Tomkeieff (Edinburgh, 1954), p. 37.

79. Day and Shepherd, "Quartz Glass," *Science 23* (1906): 670–672; Shepherd and G. A. Rankin, "The Binary System of Alumina with Silica, Lime, and Magnesia," *AJS* iv *28* (1909): 292–333. Harrison E. Howe, "Optical Glass for War Needs," in

The New World of Science, ed. Robert M. Yerkes, p. 119; see also F. E. Wright, *The Manufacture of Optical Glass and of Optical Systems: A Wartime Problem* (Washington, D.C., 1921), pp. 10–13; Haynes, *American Chemical Industry*, vol. 3, pp. 361–363.

80. Edward W. Washburn, "Physical Chemistry and Ceramics," and P. H. Bates, "The Application of the Fundamental Knowledge of Portland Cement to Its Manufacture and Use," *Journal of the Franklin Institute 193* (1922): 747–774 and 289–294; Robert B. Sosman, *The Properties of Silica* (New York, 1927), pp. 809–812. Washburn, a former pupil of A. A. Noyes, became head of the Department of Ceramics at the University of Illinois; Bates worked on Portland cement at the National Bureau of Standards. Producers of both ceramics and Portland cement funded research fellowships during the 1920s; see C. J. West and Callie Hull, "Research Scholarships and Fellowships Supported by Industry," *IECNE 7* (20 July 1929): 2, and "The Industrial Fellowships of Mellon Institute during 1928–1929," *IECNE* (10 April 1929): 6–7.

81. Kohler, *From Medical Chemistry to Biochemistry*, pp. 314–315; John T. Edsall, "Edwin Joseph Cohn, 1892–1953," *BMNAS 35* (1961): 46–84, and "Physical Chemistry at Harvard Medical School," in *The Chemistry and Physiology of Human Proteins*, ed. D. H. Bing (New York, 1979), pp. 1–10; George Corner, *A History of the Rockefeller Institute, 1901–1953* (New York, 1964).

82. The first academic facilities for experimental petrology appear to have been built at the University of Chicago; see C. E. Tilley, "Norman Levi Bowen," p. 15; D. Jerome Fisher, *The Seventy Years of the Department of Geology, University of Chicago, 1892–1961* (Chicago, 1963), pp. 47–48.

83. The American Medical Association did not include physical chemistry in its roster of required premedical courses, but by 1932 a majority of American medical colleges recommended that prospective applicants have some preparation in the subject. See Roy I. Grady, "Chemistry in the College Curriculum of the Pre-medical Student," *JCE 9* (1932): 111–113; D. L. Randall, "Premedical Physical Chemistry," *JCE 9* (1932): 1096; Jack P. Montgomery, *et al.*, "Premedical Requirements in Chemistry," *JCE 9* (1932): 1118. Most medical schools attempted to convey some basic understanding of the subject in their biochemistry courses (see Kohler, *From Medical Chemistry to Biochemistry*, p. 195); a few required their first-year students to complete courses in physical chemistry at affiliated universities.

84. Geiger, *To Advance Knowledge*, pp. 270–271; J. McKeen Cattell and Jacques Cattell, *American Men of Science*, 5th ed. (New York, 1933), p. 1267.

85. Aaron J. Ihde and H. A. Schuette, "The Early Days of Chemistry at the University of Wisconsin," *JCE 29* (1952): 65–72; Merle Curti and Vernon Carstensen, *The University of Wisconsin; 1848–1925: A History*, vol. 2 (Madison, 1949), pp. 347–352. Kahlenberg held the chair until 1918; he was succeeded by Joseph Howard Mathews (chair, 1919–1952) and Farrington Daniels (chair, 1952–1959). Mathews and Daniels were both students of T. W. Richards. On Minnesota, see Samuel Colville Lind, "Autobiographical Notes," 1 September 1943, NAS. Illinois made overtures to T. W. Richards, A. A. Noyes, and Louis Kahlenberg in 1905–1906; see Herbert S. Bailey to Edward Bartow, 2 January 1906, and E. A. Birge to S. A. Forbes, 11 December 1905, both in Box 1, LAS, 1900–1913; Margaret E. Mateby to Edward Bartow, 4 January 1906, Box 2, LAS, 1900–1913. Michigan sought a physical chemist to reor-

ganize its department of chemistry in 1926; see G. N. Lewis to John R. Effinger, 4 January 1926, Box 2, CCR.

86. W. F. Giauque, "Gilbert Newton Lewis (1875–1946)," *American Philosophical Society Yearbook* (1946): 321; the Nobel laureates were Harold C. Urey (1933), William F. Giauque (1949), Glenn T. Seaborg (1951), Willard F. Libby (1960), and Melvin Calvin (1961). Twenty-one students from California received NRC fellowships in chemistry between 1919 and 1936 compared with seventeen from Caltech, thirteen from Princeton, and a dozen each from Illinois, Yale, and Wisconsin. Next to Harvard and Caltech, California was the most popular destination of NRC fellows during this same period. See National Research Council, *National Research Fellowships, 1919–1938* (Washington, D.C., 1938).

87. Geiger, *To Advance Knowledge*, pp. 270, 276. On Wheeler, see Laurence R. Veysey, *The Emergence of the American University*, pp. 362–365.

88. Ludwig Boltzmann, "Summer in Berkeley—1904," trans. Irene Jerison, *Westways 68* (1976): 34, 36, and 78.

89. Bancroft to Wheeler, 1 February 1900, Box 4A, CU-5, PPUC. Wheeler would enter a bidding war with the University of Chicago for Loeb in 1902–1903; see Philip J. Pauly, *Controlling Life: Jacques Loeb and the Engineering Ideal in Biology* (New York, 1987), pp. 106–108. Stanford's courses were taught by S. W. Young, who spent a sabbatical year at Leipzig in 1899–1900.

90. W. B. Rising to Wheeler, 18 May 1901, and Rising to John U. Nef, 3 April 1901, Box 4A, CU-5, PPUC. Edmund O'Neill, undated MS on Frederick Gardner Cottrell, Box 1, CCR. On Cottrell see Frank Cameron, *Cottrell, Samaritan of Science* (Garden City, N.Y., 1952).

91. In the 1909–1910 academic year, Cottrell taught no less than seven courses, several of them time-consuming laboratory classes. See University of California, *President's Biennial Report, 1908–1910*, p. 295.

92. On Cottrell's scrubber, see Harry J. White, "Centenary of Frederick Gardner Cottrell," *Journal of Electrostatics 4* (1977–1978): 1–34.

93. The Research Corporation was chartered in 1912. It grew, though not perhaps as quickly as Cottrell had hoped. By the mid-1920s, the Corporation was making annual grants of roughly $50,000 per year, enough to do some good, but a fairly small sum by comparison with the amounts spent by the Rockefeller and Carnegie philanthropies. Grant recipients were placed under no legal obligation to sign patent rights over to the Corporation, but many appear to have felt a moral obligation to do so. On the Research Corporation, see Cameron, *Cottrell, Samaritan of Science*, passim.

94. Total enrollment at Berkeley increased from 1,532 to 2,699 during the decade, yet majors in chemistry declined from 87 to 51; Edwin E. Slossen, *Great American Universities* (New York, 1910), p. 181. Wheeler had sought to bring A. A. Noyes to Berkeley in 1908, without success. See Noyes to George Ellery Hale, 17 December 1908 and 1 January 1909, Roll 27, GEH.

95. On Hulett see E. C. Sullivan, "George Augustus Hulett," *BMNAS 34* (1960): 82–105.

96. Wheeler to Hulett, 11 September 1911. This and other letters relating to the negotiations between Wheeler and Hulett may be found in Box 3, GAH. See especially Hulett to Edmund O'Neill, 12 April 1911; Cottrell to Hulett, 17 May 1911; O'Neill to

Hulett, 11 July, 3 August, and 12 September 1911; Wheeler to Hulett, 17 August and 24 August 1911; and Hulett to Wheeler, 19 September 1911.

97. Hulett to F. K. Cameron, undated, but reply to letter from Cameron to Hulett of 26 January 1911, Box 2, GAH. See also Hulett to Willis R. Whitney, 11 March 1911, Box 2, GAH; Hulett to H. B. Fine, 11 April 1911, Box 3, GAH; and Hulett to O'Neill, 12 April 1911.

98. Hulett to Wheeler, 19 September 1911; see also O'Neill to Hulett, 12 September 1911.

99. Lewis to Wheeler, 11 November 1911, Box 61, CU-5, PPUC. Hulett was among those Lewis consulted; see Hulett to Lewis, 23 November 1911, Box 2, GAH. On attitudes toward Noyes and Lewis at MIT, see below, chapter 6.

100. Lewis to Wheeler, 9 December 1911 and 5 January 1912, Box 61, CU-5, PPUC.

101. Holden to George Ferdinand Becker, 12 November 1896, Box 17, GFB; Holden moved to New York the following year. Jacques Loeb to Osterhout, 9 January 1912, WJO; on Loeb's unhappiness at Berkeley see Pauly, *Controlling Life*, pp. 132–136.

102. Lewis to Wheeler, 18 April 1912, Box 61, CU-5, PPUC. On the construction of a staff see Lewis to Wheeler, 5 January 1912, 8 February 1912, and 13 March 1912, all in Box 61, CU-5, PPUC; and 5 April 1913, Box 68, CU-5, PPUC. On the new laboratory see Lewis to Wheeler, 10 September 1915, Box 68, CU-5, PPUC.

103. University of California, *Report of the President, 1910–1912*, pp. 33–34; ironically Wheeler cited Edmund O'Neill's report on chemistry and California's industrial development to buttress these points. Lewis, "Report from the Chairman of the Department of Chemistry to the President, 1914–1915," 28 May 1915, Box 4, CCR.

104. Lewis, "Report for the Year 1915–16 from the Department of Chemistry to the President," 8 June 1916, Box 4, CCR.

105. California graduated 52 Ph.D.s in chemistry between 1914 and 1923. Of the 42 for whom biographical information was readily available, 34 took their first jobs in universities, 4 in industry, and 4 in government. Ten years after receiving their Ph.D.s, 29 remained in academic life, 11 were in industrial work, and 2 in government. Melvin Calvin provides a list of Berkeley's Ph.D. recipients in chemistry during Lewis's era in "Gilbert Newton Lewis," in *Proceedings of the Robert A. Welch Foundation. Conferences on Chemical Research. XX. American Chemistry—Bicentennial*, ed. W. O. Milligan (Houston, 1977), pp. 146–149.

106. Lewis, "Report from the Chairman of the Department of Chemistry to the President, 1914–1915," 28 May 1915, Box 4, CCR; on majors see University of California, *President's Biennial Report, 1910–1912* and *1918–1919*. On the curriculum see Lewis to M. A. Hines, 3 March 1919, Box 2, CCR. Slossen, *Great American Universities*, p. 168.

107. W. C. Bray to E. B. Spear, 12 June 1913, Box 6, CCR.

108. Ibid.

109. Bray to W. A. Noyes, 16 June 1916, Box 5, CCR; Lewis to Wheeler, 26 July 1916 and 16 August 1916, Box 89, CU-5, PPUC; Wheeler to Lewis, 7 August 1916, Box 4, CCR.

110. Lewis to M. A. Hines, 3 March 1919, Box 2, CCR; Lewis to Wheeler, 5 January 1912, Box 61, CU-5, PPUC.

111. Seidel, "Physics Research in California"; Lewis to Wheeler, 18 January 1916, Box 89, CU-5, PPUC; 17 February 1916 and 24 January 1917, both in Box 4, CCR; "Report Concerning the Activities of the Department of Chemistry, July 1, 1925 to June 30, 1926," Box 8, CCR.

112. Edmund O'Neill to Osterhout, 1 February 1910, WJO. Lewis's new laboratory at Berkeley was approximately ten times the size of Noyes's research laboratory at MIT. Whereas Noyes's laboratory had room for about fifteen research workers, Lewis's new laboratory had space for a hundred and fifty; whereas Noyes installed less than $10,000 worth of equipment in his laboratory, Lewis obtained $60,000 to equip his. See Lewis to Wheeler, 1 March 1915 and 10 September 1915, Box 68, CU-5, PPUC; Lewis to J. C. Merriam, 25 August 1919, Box 3, CCR.

113. In 1904–1905, Pritchett and several members of the MIT Corporation sought to effect a merger of MIT and Harvard's Lawrence Scientific School. Their apparent lack of confidence in MIT's future precipitated a revolt among MIT's graduates and faculty and Pritchett's resignation. See Samuel C. Prescott, *When M.I.T. Was "Boston Tech"* (Cambridge, Mass., 1954), pp. 193–203.

114. Cameron, *Cottrell, Samaritan of Science*, p. 55. Joseph LeConte (1823–1901), professor of geology and natural history and the first president of the University of California, was well known for his books on geology, comparative physiology, and vision.

115. The average number of students per year majoring in Wisconsin's programs in chemical engineering, industrial chemistry, and mining and metallurgy rose from 20 during the period 1910–1914 to 62 in the period 1920–1924. Annual enrollments in physical chemistry increased from approximately 40 to 80 during the same period. See Daub, "Chemical Engineering at the University of Wisconsin," pp. 183–186, and Farrington Daniels, "Physical Chemistry," in *Chemistry . . . Key to Better Living* (Washington, D.C., 1951), p. 43. Physical chemistry was required of majors in chemical engineering.

116. Lauder W. Jones, "The School of Chemistry," in University of Minnesota, *President's Report for the Year 1919–1920*, p. 171.

117. See, for example, James W. Neckers, *The Building of a Department: Chemistry at Southern Illinois University, 1927–1967* (Carbondale, 1979), pp. 27–29.

CHAPTER 6
From Physical Chemistry to Chemical Physics

1. Walter F. Rittman, "The Application of Physical Chemistry to Industrial Processes," *TAICE 7* (1914): 48.

2. William H. Walker, "Some Present Problems in Technical Chemistry," *PSM 66* (1905): 447.

3. Charles F. Burgess, letter to *Electrical World* of 7 May 1901, quoted in Alexander McQueen, *A Romance in Research: The Life of Charles F. Burgess* (Pittsburgh, 1951), p. 59.

4. Edward E. Daub, "Chemical Engineering at the University of Wisconsin: The Early Years," in *A Century of Chemical Engineering*, ed. William F. Furter (New

York, 1982), pp. 160, 164, 180, 192. See also L. Kermit Herndon, James O. Pence, and James R. Sithrow, "The Teaching of Chemical Technology," in *Collected Papers on the Teaching of Chemical Engineering: Proceedings of the Chemical Engineering Division of the Society for the Promotion of Engineering Education, Conference and Second Chemical Engineering Summer School, Pennsylvania State College, June 20–30, 1939* (New York, 1940), p. 247.

5. Noyes, "Advanced Courses for Specialization," in Massachusetts Institute of Technology, *President's Report, January 1908* (Boston, 1908), p. 18.

6. Noyes, "Discussion on the Training of Technical Chemists," *Science 19* (1904): 572. See also Noyes, "Talk to First Year Students," *TR 9* (1907): 5; Noyes, "Research Laboratory of Physical Chemistry," in Massachusetts Institute of Technology, *Reports of the President and Treasurer* (Boston, 1912), p. 89; William C. Bray to W. Lash Miller, 7 January 1907, Box 5, CRR.

7. Walker to Noyes, 20 January 1919, Folder 555, PPMIT. See also Walker to Noyes, 14 May 1916, and Walker to Richard C. Maclaurin, 16 May 1916, both in Folder 555, PPMIT. On Walker, see "William H. Walker," *National Cyclopedia of American Biography*, vol. A, p. 167; and Warren K. Lewis, "Reminiscences of William H. Walker, 'Father of Chemical Engineering,' " *Chemical Engineering* (July 1952): 158–159, 178. Among those who shared many of Walker's views was Arthur D. Little, the influential consulting chemist who became a member of the Visiting Committee in Chemistry in 1912. Walker had been Little's partner before joining MIT's faculty in 1905. See E. J. Kahn, Jr., *The Problem Solvers: A History of Arthur D. Little, Inc.* (Boston, 1986).

8. Walker, "A Laboratory Course in Industrial Chemistry," *TR 6* (1904): 163–174.

9. Walker to C. H. Warren, 9 January 1919, Folder 555, PPMIT. Arthur D. Little expressed a similar view in "Industrial Research in America," *Journal of Industrial and Engineering Chemistry 5* (1913): 796, although his language was more temperate.

10. Arthur D. Little, *et al.*, "Report of the Visiting Committee of the Department of Chemistry and Chemical Engineering," 6 December 1915, Folder 1259, PPMIT. See also Alfred H. White, "Chemical Engineering Education," in *Twenty-Five Years of Chemical Engineering Progress*, ed. Sidney D. Kirkpatrick (Philadelphia, 1933), pp. 355–356, and M. C. Whitaker, "The New Chemical Engineering Course and Laboratories at Columbia University," *TAICE 5* (1912): 162. For an example of a course of study organized around unit operations, see William H. Walker, Warren K. Lewis, and William H. McAdams, *Principles of Chemical Engineering* (New York, 1923).

11. Plans for the School of Chemical Engineering Practice are discussed in Little, *et al.*, "Report of the Visiting Committee," 6 December 1915. See also R. T. Haslam, "The School of Chemical Engineering Practice of the Massachusetts Institute of Technology," *Journal of Industrial and Engineering Chemistry 13* (1921): 465–466.

12. On the creation and history of the Research Laboratory of Applied Chemistry, see the MIT Executive Committee to Walker, 28 April 1908, Folder 555, PPMIT; Little, "A Laboratory for Public Service," *TR 11* (1909): 16–24; Walker, "Cooperation in Industrial Research: The University," *Transactions of the American Electrochemical Society 29* (1916): 30–31; Warren K. Lewis to S. W. Stratton, 21 April 1925, Folder 495, PPMIT; and the reports of the director of the Laboratory in the annual MIT *President's Report*.

13. Little, "A Laboratory for Public Service," pp. 16–21; Walker, "Cooperation in Industrial Research: The University," p. 30; Walker, "The University and Industry," *Journal of Industrial and Engineering Chemistry 8* (1916): 63–65.

14. Compare, for example, Maclaurin's statements in "Universities and Industries," *Journal of Industrial and Engineering Chemistry 8* (1916): 60–61 and his "Report of the President," in Massachusetts Institute of Technology, *President's Report, January 1917* (Cambridge, Mass., 1917), p. 17. There is a book-length biography of Maclaurin: Henry G. Pearson, *Richard Cockburn Maclaurin, President of the Massachusetts Institute of Technology* (New York, 1937).

15. Talbot, "Relation of Educational Institutions to the Industries," *Journal of Industrial and Engineering Chemistry 12* (1920): 946; E. B. Wilson to Edwin S. Webster, 21 April 1920, Folder 73, PPMIT. On Talbot, see James F. Norris, "Henry Paul Talbot," *TR 29* (1927): 479–480.

16. "M.I.T.: Resources—Capital Gifts—Income—Teachers' Salaries, 1910–1948," undated memorandum in MIT History Folder, VB.

17. C. D. Walcott to Henry S. Pritchett, 29 March 1919, Box 33, CDW. A $300,000 gift from George Eastman made the School of Chemical Engineering Practice possible. See R. T. Haslam to S. W. Stratton, 9 May 1924, Folder 94, PPMIT. Walker, while working with Little, had developed a process for making artificial silk (lustron silk) from cellulose acetate. He also developed methods for waterproofing fabrics using esters of cellulose and made valuable contributions to the technology of reducing the corrosion of iron and steel. On his work at Edgewood Arsenal, see L. F. Haber, *The Poisonous Cloud: Chemical Warfare in the First World War* (Oxford, 1986), pp. 167–168.

18. Noyes to Hale, 8 May 1915 and 13 May 1915, and Noyes to James A. B. Scherer, 13 May 1915, all on Roll 27, GEH. On the Throop Institute and Hale's plans for it, see Helen Wright, *Explorer of the Universe: A Biography of George Ellery Hale*, pp. 239–251.

19. Noyes to Maclaurin, 31 January 1916, Roll 27, GEH.

20. Walker to Maclaurin, 21 March 1919, Folder 555, PPMIT; Noyes to Maclaurin, 5 April 1919 and 7 April 1919, and Noyes to Hale, 9 April 1919, Roll 28, GEH.

21. On the Division of Industrial Cooperation and Research, see John W. Servos, "The Industrial Relations of Science: Chemical Engineering at MIT, 1900–1939," *Isis 71* (1980): 531–549.

22. E. B. Wilson to Maclaurin, 12 April 1919, Folder 46, PPMIT. There is an intriguing possibility that Hale planted doubts about Noyes in order to secure him for Throop.

23. Maclaurin to Noyes, 12 April 1919, and Noyes to Maclaurin, 23 April 1919, Folder 149, PPMIT; and Noyes to Hale, 20 November 1919, Roll 28, GEH. A copy of the letter of resignation, dated 17 November 1919, is on Roll 28, GEH. Noyes had told Maclaurin that his resignation was "inevitable" in April; see Noyes to Hale, 21 April 1919, Roll 28, GEH.

24. Roger L. Geiger, *To Advance Knowledge: The Growth of American Research Universities* (New York, 1986), p. 178; Noyes to James A. B. Scherer, 20 November 1919, Box 2, JABS; Noyes to Hale, 21 April 1919, Roll 28, GEH. Noyes did not accept the nonresident professorship.

25. The work of the Research Laboratory of Physical Chemistry in the 1920s may be followed in Frederick G. Keyes's contributions to the annual MIT *President's Report*. For a fuller account of MIT's tribulations during the 1920s, see Servos, "The Industrial Relations of Science," pp. 540–545.

26. Compton to Max Mason, 7 April 1930, Folder 84, PPMIT; Servos, "Industrial Relations of Science," p. 541 and pp. 545–548.

27. Noyes to James A. B. Scherer, 20 November 1919, Box 2, JABS. See also Noyes to George Ellery Hale, 18 November 1919 and 20 November 1919, Roll 28, GEH. Before moving to California Noyes had been careful to obtain assurances of continued support from the Carnegie Institution of Washington.

28. Hale to Noyes, 2 January 1909, Roll 27, GEH; Hale to Henry S. Pritchett, 28 October 1921, CITCC.

29. Noyes and James E. Bell, "Honor Students in Chemistry," *JCE 3* (1926): 888; see also A. A. Ashdown, "Arthur Amos Noyes," *TR 38* (1936): 420; Noyes, "Notes for an Address on Educational Opportunities at C.I.T.," undated MS, Box 49, CIT; Noyes to R. A. Millikan, 13 September 1928, Roll 28, GEH; and Noyes, "Proposed Development of the Work in Chemistry," undated MS (circa 1930), Box 25, Folder 15, RAM. On Noyes's views of the relationship between engineers and scientists at Caltech, see Noyes to Willis R. Whitney, 12 April 1920, WRW; and Noyes to Millikan, 14 March 1925, Roll 28, GEH.

30. Noyes to James A. B. Scherer, 5 December 1919; Noyes to Scherer, 3 February 1919; and Noyes to S. J. Bates, 17 March 1919, Box 2, JABS.

31. For a fuller treatment of Millikan's appointment, see Robert H. Kargon, *The Rise of Robert Millikan: Portrait of a Life in American Science* (Ithaca, 1982), pp. 97–103.

32. George Ellery Hale to Charles D. Walcott, 24 July 1914, copy in Box 2, JABS; see also Noyes to E. B. Wilson, 11 October 1920, A. A. Noyes, 1920 folder, EBW.

33. Linus Pauling, "Arthur Amos Noyes," *BMNAS 31* (1958): 327. Especially valuable for understanding the workings of the Executive Council and Noyes's role in it is Noyes to Henry S. Pritchett, 29 October 1921, CITCC.

34. On NRC fellowships, see Kargon, *The Rise of Robert Millikan*, pp. 116–117; John W. Servos, "The Knowledge Corporation: Chemistry at Caltech," *Ambix 23* (1976): 186. I wish to acknowledge my special debt to Robert H. Kargon, in whose seminar I first became acquainted with the history of Caltech.

35. On local support for Caltech, see Kargon, *The Rise of Robert Millikan*, pp. 105–108.

36. Geiger, *To Advance Knowledge*, pp. 185–186.

37. Millikan to King Gillette, 26 May 1929, Box 19, Folder 9, RAM.

38. Millikan to I. C. Copley, 27 August 1929, Box 19, Folder 9, RAM.

39. RAM contains much information about these fund-raising efforts. Essential to the plan was the recruiting of California Institute of Technology Associates, donors who pledged Caltech a thousand dollars a year in return for membership in this exclusive club. The membership quickly grew to over one hundred, although it is worth pointing out that from the outset two or three individuals paid the dues, in whole or in part, of many of these members. Caltech's local support, in other words, was not quite as broadly based as the Institute's Executive Council wished the public and foundation

officials to believe. See George Farrand to Millikan, 3 October 1946, Box 39, Folder 12; "List of Men whose Associateships are Partially Paid by A. C. Balch," Box 19, Folder 15; Millikan to Harry Chandler, 17 July 1935, Box 19, Folder 15; Millikan to A. M. Clifford, 12 June 1925, Box 27, Folder 9, all in RAM; "California Institute Associates Cash Book, June 1927–November 1944," p. 20, California Institute of Technology Archives.

40. "Memorandum Relating to the Application of the California Institute of Technology to the Carnegie Corporation of New York for Aid in Support of Project of Research on the Constitution of Matter and the Development of the Scientific Departments of the Institute," 17 September 1921, Box 6, GEH, CIT.

41. Ibid. Millikan's hopes of disintegrating nuclei with high-voltage electrical discharges and of tracing cosmic rays to a process of atom-building were both disappointed. On Millikan's research, see Kargon, *The Rise of Robert Millikan*, pp. 123–144.

42. On Caltech's relations with philanthropic foundations see Kargon, *The Rise of Robert Millikan*, pp. 109–118; Geiger, *To Advance Knowledge*, pp. 162–163; and Paul A. Hanle, *Bringing Aerodynamics to America* (Cambridge, Mass., 1982), pp. 9–13.

43. Servos, "The Industrial Relations of Science," pp. 544–555. As an engineering school, Caltech did of course support work in the applied sciences. Royal Sorenson, professor of electrical engineering, was active as a consultant to power companies, and Millikan himself, in return for a gift of $105,000 toward a high-tension laboratory and other considerations, gave advice to the Southern California Edison Company. In 1922 a small research laboratory of applied chemistry was organized that undertook work on commercial problems. And, during the 1930s, Caltech's aeronautical engineers and wind tunnel became important resources in the development of an aircraft industry in the Los Angeles region. Nevertheless, basic research enjoyed hegemony over applied research at Caltech, at least until World War II. Professors of applied science were generally paid less than those active in basic research; the research laboratory of applied chemistry was kept small and encouraged to work with trade associations like the American Petroleum Institute rather than with private firms; and both faculty and students were indoctrinated in the belief that there was a hierarchy of knowledge in which physics, mathematics, and chemistry held higher stations than engineering. Noyes, who seems to have perceived greater dangers in commercial work than Millikan, consistently used his influence to keep applied science in a subordinate position; see Noyes to Millikan, 14 March 1925 and 13 September 1928, both on Roll 28, GEH; and Noyes to Millikan, 7 April 1927 and 3 June 1927, both in Box 8, Folder 16, RAM. See also Jeffrey L. Sturchio and Arnold Thackray, "Interview with Arnold O. Beckman," 23 April 1985, Center for the History of Chemistry, Philadelphia. I thank Jeffrey Sturchio for a copy of this interview. On relations with the Southern California Edison Company and the high-voltage laboratory, see Kargon, *The Rise of Robert Millikan*, pp. 132–134, and D. M. Trott to C.I.T. Executive Council, 7 October 1924, Box 2, AF; on the research laboratory of applied chemistry, see "Report of the Chairman of the Executive Council, 10 March 1922," Box 138, GEH, CIT; on aeronautics at Caltech, see Richard P. Hallion, *Legacy of Flight: The Guggenheim Contribution to American Aviation* (Seattle, 1977), pp. 187–206.

44. "C.I.T. Budget, 1923–1924," Box 140, GEH, CIT. Especially valuable in trac-

ing Noyes's plans for the Gates Laboratory are Noyes to Leonard B. Loeb, 8 February 1916, and Noyes to Stuart J. Bates, 17 March 1919, both in Box 2, Folder 14, JABS; Noyes, "Undergraduate Education for Chemical Research," Box 49, Folder 5, CIT; "Proposed Development of the Work in Chemistry," Box 25, Folder 15, RAM; Noyes to Hale, 20 February 1919, Noyes to Millikan, 14 March 1925, and Noyes to Millikan, 13 September 1928, all on Roll 28, GEH. For a description of work in the Gates Laboratory during the 1920s and 1930s, see Linus Pauling, "Fifty Years of Physical Chemistry in the California Institute of Technology," *ARPC 16* (1965): 1–14; on Noyes's special interest in able undergraduates, see David Ridgway, "Interview with Kenneth Pitzer," *JCE 52* (1975): 219–220.

45. Noyes, for example, served on the governing board of the Mount Wilson Observatory and undertook a study of ionization in stellar atmospheres after moving to Pasadena, and his lieutenant, Richard Chace Tolman, pursued research on relativistic thermodynamics and cosmology that brought him into close touch with both physicists and astrophysicists.

46. Noyes to Leonard B. Loeb, 8 February 1916, Box 2, JABS. Loeb did not accept the position.

47. Hale to Noyes, 18 July 1915, Noyes to Hale, 21 July 1915, and Noyes to Hale, 24 September 1915, all on Roll 27, GEH; Noyes to Hale, 24 June 1918, Roll 28, GEH; Charles Lalor Burdick, "The Genesis and Beginnings of X-ray Crystallography at Caltech," in *Fifty Years of X-ray Diffraction*, ed. Peter Paul Ewald (Utrecht, 1962), pp. 556–558; Linus Pauling, "Roscoe Gilkey Dickinson," unpublished MS, dated 14 July 1945, Box 40, CIT.

48. Noyes to Henry S. Pritchett, 5 December 1921, CITCC.

49. Paul H. Emmet to John W. Servos, 8 April 1975. On Tolman's work see Linus Pauling, "Fifty Years of Physical Chemistry at the California Institute of Technology," p. 3; J. G. Kirkwood, O. R. Wulf, and P. S. Epstein, "Richard Chace Tolman," *BMNAS 27* (1952): 139–153; aspects of Tolman's work on chemical kinetics are ably treated in Christine M. King and Keith J. Laidler, "Chemical Kinetics and the Radiation Hypothesis," *AHES 30* (1984): 45–86.

50. It did not hurt that Noyes was a charter member of the NRC fellowship board in chemistry. Don M. Yost to John W. Servos, 2 March 1977. On Badger, see "Richard McLean Badger," *Engineering and Science 38* (December 1974–January 1975): 24; Pauling, "Fifty Years of Physical Chemistry in the California Institute of Technology."

51. During the late 1920s, X-ray crystallographic work was resumed in the physics department at MIT under a student of W. L. Bragg; see Ralph W. G. Wyckoff, "The Development of X-ray Diffraction in the U.S.A.: The Years Before 1940," in *Fifty Years of X-ray Diffraction*, p. 433.

52. Albert W. Hull, "Autobiography," in *Fifty Years of X-ray Diffraction*, pp. 582–587.

53. Ralph W. G. Wyckoff, "Reminiscences," in *Fifty Years of X-ray Diffraction*, pp. 691–694.

54. Ralph W. G. Wyckoff, "The Development of X-ray Diffraction in the U.S.A.: The Years Before 1940," in *Fifty Years of X-ray Diffraction*, p. 432.

55. On Röntgen's discovery, see Bern Dibner, *Wilhelm Conrad Röntgen and the Discovery of X Rays* (New York, 1968).

56. In addition to *Fifty Years of X-ray Diffraction*, ed. Ewald, valuable sources on the early years of X-ray diffraction include Paul Forman, "The Discovery of the Diffraction of X-rays by Crystals: A Critique of the Myths," and P. P. Ewald, "The Myth of Myths: Comments on P. Forman's Paper," *AHES 6* (1969/1970): 38–71 and 72–81.

57. On the development of X-ray diffraction techniques, see especially W. L. Bragg, "The Growing Power of X-ray Analysis," in *Fifty Years of X-ray Diffraction*, pp. 120–136.

58. C. L. Burdick, "The Genesis and Beginnings of X-ray Crystallography at Caltech," in *Fifty Years of X-ray Diffraction*, p. 557; Linus Pauling, "Early Work on X-ray Diffraction in the California Institute of Technology," in *Fifty Years of X-ray Diffraction*, p. 624.

59. Linus Pauling, "Problems in Inorganic Structures," in *Fifty Years of X-ray Diffraction*, pp. 141–142; Robert J. Paradowski, "The Structural Chemistry of Linus Pauling," Ph.D. dissertation, University of Wisconsin, 1972, pp. 183–203; J. H. Sturdivant, "The Scientific Work of Linus Pauling," in *Structural Chemistry and Molecular Biology*, ed. Alexander Rich and Norman Davidson (San Francisco, 1968), pp. 3–5.

60. Details regarding Pauling's early years are drawn largely from his own reminiscences, especially "Fifty Years of Physical Chemistry in the California Institute of Technology," esp. pp. 1–3, and from his interview with John L. Heilbron of 27 March 1964. I thank Professor Heilbron for a transcript of that interview. Some additional detail is available in Robert J. Paradowski, "Structural Chemistry of Linus Pauling," pp. 33–44. I am also indebted to Judith Goodstein's lively sketch of Pauling, "Atoms, Molecules, and Linus Pauling," *Social Research 51* (1984): 691–708. For a bibliography of Pauling's writings to 1967, see Gustav Albrecht, "Scientific Publications of Linus Pauling," in *Structural Chemistry and Molecular Biology*, pp. 887–907.

61. For an account of Pauling's introduction to X-ray analysis, see his "Early Work on X-ray Diffraction in the California Institute of Technology," pp. 625–628.

62. Goodstein, "Atoms, Molecules, and Linus Pauling," pp. 697–698; Pauling also completed a longer manuscript on the theory of concentrated solutions, which Noyes and Debye discouraged him from publishing; see Heilbron, Interview with Pauling, Session I, pp. 14–15. For more on Pauling's program of studies, see Paradowski, "Structural Chemistry of Linus Pauling," pp. 49–56.

63. Goodstein, "Atoms, Molecules, and Linus Pauling," pp. 699–701.

64. An especially rich collection of letters from American students in Europe is in CCR. See, especially, Harold C. Urey to Lewis, 9 September 1923, 1 January 1924, 4 May 1924, and W. H. Williams to Lewis, 1 February 1927, all in Box 4; Linus Pauling to Lewis, 10 June 1926, 8 January 1927, and Wendell Latimer to Lewis, 9 June 1930, all in Box 3.

65. There is strong evidence that Lewis had seconded Noyes's advice that Pauling visit Europe; see Pauling to Lewis, 10 June 1926 and 7 March 1928, both in Box 3, CCR.

66. Niels Bohr, "On the Constitution of Atoms and Molecules," *PM* vi *26* (1913): 1–25, 476–502, and 857–875; for a more detailed description of the physical issues

confronted by Bohr, Sommerfeld, and their colleagues, see Max Jammer, *The Conceptual Development of Quantum Mechanics* (New York, 1966), pp. 69–133.

67. No one expected precision in the case of multielectron systems, since exact solution of equations involving the interactions of three or more bodies was impossible; nevertheless, by use of perturbation theory it was possible to obtain approximate solutions, and these were disappointingly distant from observed data. For a succinct itemization of the Bohr atom's shortcomings, see John H. Van Vleck, "The New Quantum Mechanics," *CR 5* (1928): 489–497.

68. Bohr briefly discussed the tetrahedral symmetry of carbon compounds in the final installment of his 1913 paper. He suggested that the bonds uniting the carbon and hydrogen atoms in methane consisted of two electrons rotating in a ring perpendicular to the line joining the two nuclei, but he gave no attention to the questions that would most interest a chemist, such as why a carbon atom should preferentially bond with four hydrogen atoms. His theory was, in its original form, incompetent to tell chemists anything about the chemical bond they did not already know. See Bohr, "On the Constitution of Atoms and Molecules," p. 874.

69. Lewis, "The Atom and the Molecule," *JACS 38* (1916): 768. In his papers of 1913, Bohr supposed that the number of electrons in an atom was approximately equal to half an element's atomic weight; by 1916 he, like most physicists, had accepted Moseley's demonstration that neutral atoms had electrons equal in number to their place in the periodic table.

70. Ibid., p. 778.

71. Robert E. Kohler, Jr., "The Origin of G. N. Lewis's Theory of the Shared Pair Bond," *HSPS 3* (1971): 343–376.

72. Lewis, "The Atom and the Molecule," pp. 763–767.

73. Lewis, "The Atom and the Molecule," pp. 768–771, 777–778. See also Robert E. Kohler, Jr., "Irving Langmuir and the Octet Theory of Valence," *HSPS 4* (1972): 39–87, and "The Lewis-Langmuir Theory of Valence and the Chemical Community, 1920–1928," *HSPS 6* (1975): 431–468. Lewis's model also had its chemical shortcomings. A few molecules, like PCl_5 or B_2H_6, were inexplicable according to its principles, and it did not explain why some atoms have several valences.

74. Joe D. Burchfield, *Lord Kelvin and the Age of the Earth* (New York, 1975).

75. Bohr, "The Structure of the Atom and the Physical and Chemical Properties of the Elements," in *The Theory of Spectra and Atomic Constitution* (Cambridge, 1922), pp. 61–126; Lewis, *Valence and the Structure of Atoms and Molecules* (New York, 1923), pp. 56–57. Helge Kragh, "Niels Bohr's Second Atomic Theory," *HSPS 10* (1979): 123–186 provides more detail on chemists' responses to Bohr's work; see esp. pp. 165–167.

76. See Maurice L. Huggins, "Evidence from Crystal Structures in Regard to Atomic Structures," *PR 27* (1926): 286–297, esp. 296–297.

77. Lewis, "Valence and the Electron," *TFS 19* (1923): 453–454; Van Vleck, "The New Quantum Mechanics," p. 504.

78. Jammer's *Conceptual Development of Quantum Mechanics* is the best history of this era; Barbara Lovett Cline's *Men Who Made a New Physics: Physicists and the Quantum Theory* (New York, 1965) is an excellent nonmathematical account of the origins of quantum mechanics; especially valuable because of its focus on chemistry is

W. G. Palmer, *A History of the Concept of Valency to 1930* (Cambridge, 1965), pp. 125–171. For translations of Schrödinger's papers, see Erwin Schrödinger, *Collected Papers on Wave Mechanics*, translated by J. F. Shearer, from the 2nd German edition of 1928 (London, 1928).

79. Heilbron, Interview with Pauling, Session I, pp. 17–19 and 24–25; Session II, pp. 8–9. Linus Pauling and Sterling B. Hendricks, "The Prediction of the Relative Stabilities of Isosteric Isomeric Ions and Molecules," *JACS 48* (1926): 641–651; and Pauling, "The Dynamic Model of the Chemical Bond and Its Application to the Structure of Benzene," *JACS 48* (1926): 1132–1143. In these papers, Pauling assumed that the shared electron pair in a valence bond moved in elliptical orbits about two atomic nuclei and that the electronic structure of molecules was similar, in certain cases, to the electronic structure of atoms whose outer shells contained the same number of electrons. These ideas were fairly common in the literature between 1923 and 1925.

80. Heilbron, Interview with Pauling, Session I, p. 19; Pauling, "The Sizes of Ions and the Structure of Ionic Crystals," *JACS 49* (1927): 765.

81. Pauling, "The Sizes of Ions and the Structure of Ionic Crystals," p. 766.

82. Heilbron, Interview with Pauling, Session I, p. 22.

83. Pauling, "The Theoretical Prediction of the Physical Properties of Many-Electron Atoms and Ions. Mole Refraction, Diamagnetic Susceptibility and Extension in Space," *Proceedings of the Royal Society A114* (1927): 181–211. Pauling's technique involved the assumption that each electron in a many-electron atom would have a distribution in space of a hydrogen-like electron under the influence of a nuclear charge which was distorted by the screening effect of inner-shell electrons. Pauling recognized that ionic radii are not constant but vary to some degree with the environment in which they find themselves.

84. Pauling, "The Sizes of Ions and the Structure of Ionic Crystals"; "The Coordination Theory of the Structure of Ionic Crystals," in *Festschrift zum 60. Geburtstage Arnold Sommerfelds* (Leipzig, 1928), pp. 11–17; "The Principles Determining the Structure of Complex Ionic Crystals," *JACS 51* (1929): 1010–1026. These rules were in part anticipated by the Norwegian geochemist V. M. Goldschmidt.

85. W. L. Bragg, "The Growing Power of X-ray Analysis," in *Fifty Years of X-ray Diffraction*, p. 127. See also J. H. Sturdivant, "The Scientific Work of Linus Pauling," in *Structural Chemistry and Molecular Biology*, pp. 4–5; Linus Pauling, "Fifty Years of Progress in Structural Chemistry and Molecular Biology," in *The Twentieth Century Sciences: Studies in the Biography of Ideas* (New York, 1972), pp. 285–287; Paradowski, "Structural Chemistry of Linus Pauling," pp. 204–236.

86. Heilbron, Interview with Pauling, Session II, pp. 9–10.

87. A few paragraphs cannot begin to do justice to the exceedingly complex history of the various atomic models proposed by Bohr, Sommerfeld, and their colleagues in the period between 1913 and 1925. Readers seeking more detail should begin with Max Jammer's still unsurpassed *The Conceptual Development of Quantum Mechanics*, especially pp. 89–156. Other important sources include John L. Heilbron, "The Kossel-Sommerfeld Theory and the Ring Atom," *Isis 58* (1967): 451–485 and Helge Kragh, "Niels Bohr's Second Atomic Theory."

88. Edmund C. Stoner, "The Distribution of Electrons among Atomic Levels," *PM 48* (1924): 725–726, 734–735. Carbon did represent a problem, since according to

Stoner its outermost shell consisted of two discrete subshells each containing two electrons. Yet its tetrahedral symmetry suggested that its four outer electrons were equivalent.

89. Jammer, *The Conceptual Development of Quantum Mechanics*, pp. 133–156.

90. Rodebush, "The Electron Theory of Valence," *CR 5* (1928): 516.

91. See, for example, Van Vleck, "The New Quantum Mechanics," p. 504.

92. Heilbron, Interview with Pauling, Session II, p. 7 and pp. 16–17; W. Heitler and F. London, "Wechselwirkung neutraler Atome und homöopolare Bindung nach der Quantenmechanik," *Zeitschrift für Physik 44* (1927): 455–472.

93. Especially important were papers by O. Burrau, Edward U. Condon, and Friedrich Hund. On Burrau and Condon's papers see Linus Pauling, *The Nature of the Chemical Bond*, 3rd ed. (Ithaca, 1960), p. 23 and Heilbron, Interview with Pauling, Session II, pp. 7–8. On Hund's work, see Palmer, *A History of the Concept of Valency*, pp. 166–171 and Robert S. Mulliken, "Molecular Scientists and Molecular Science: Some Reminiscences," *JCP 43* (1965): S2–S11, esp. S6–S7.

94. W. Heitler and F. London, "Wechselwirkung neutraler Atome und homöopolare Bindung nach der Quantenmechanik," *Zeitschrift für Physik 44* (1927): 455–472. Valuable accounts of Heitler and London's work include: C.W.F. Everitt and W. M. Fairbank, "Fritz London," *DSB*, vol. 8, pp. 473–475; J. H. Van Vleck, "The New Quantum Mechanics," pp. 500–506; J. H. Van Vleck and Albert Sherman, "The Quantum Theory of Valence," *Reviews of Modern Physics 7* (1935): 167–228; Palmer, *A History of the Concept of Valency*, pp. 154–166; and C. A. Russell, *The History of Valency* (Oxford, 1971), pp. 303–309.

95. Van Vleck, "The New Quantum Mechanics," p. 506.

96. F. London, "Zur Quantentheorie der homöopolaren Valenzzahlen," *Zeitschrift für Physik 46* (1928): 455–477.

97. Pauling, "The Application of the Quantum Mechanics to the Structure of the Hydrogen Molecule and Hydrogen Molecule-Ion and to Related Problems," *CR 5* (1928): 173–213. On this paper and Pauling's other efforts to develop Heitler and London's theory see Paradowski, "The Structural Chemistry of Linus Pauling," pp. 317–332 and 420–432.

98. Pauling, "The Shared-Electron Chemical Bond," *PNAS 14* (1928): 359 and 360.

99. In the new nomenclature, electronic energy levels are labeled by numbers and letters referring to the principal and azimuthal quantum numbers, respectively. The letters applied to the lowest-energy-states, s, p, d, and f, were borrowed from spectroscopy, where they referred to the sharp, principal, diffuse, and fundamental spectral series. An electron in the $2s$ orbital had a principal quantum number of 2 and an azimuthal quantum number of 0; a $2p$ electron had a principal quantum number of 2 and an azimuthal quantum number of 1. The azimuthal quantum number could only assume values 0, 1, 2, . . . , $n-1$, where n is the principal quantum number. The value of the magnetic quantum number (m) can assume values 0, ± 1, . . . , $\pm l$, where l is the azimuthal quantum number. Hence, the $2s$ subshell contains but one orbital ($n = 2$, $l = 0$, $m = 0$) and the $2p$ subshell contains three ($n = 2$, $l = 1$, $m = 0$, $+1$, -1).

100. For Lewis's views on carbon monoxide, see his *Valence and the Structure of Atoms and Molecules*, pp. 127–128. London discussed the valence states of carbon in

his paper of January 1928. There he explained the multiple valencies of carbon (and other atoms such as nitrogen) and explained them by reference to a diagram showing the possible configurations of carbon's outer electron shell:

				Valence
		1		
Azimuthal Quantum Number	0	---		
Magnetic Quantum Number	0	−1	0	1
Number	2	2		0
of	2	1	1	2
Electrons	1	1	1	1	4

While this scheme explained how the valence of carbon might assume several values, it did not explain why the four bonds of quadrivalent carbon were identical to one another or why the quadrivalent state was usually favored over the bivalent state. This is precisely where Pauling's concept of hybridization proved valuable. It should be noted that Pauling later would present strong evidence suggesting that carbon in carbon monoxide is not, strictly speaking, bivalent. Rather, he suggested that the structure of carbon monoxide was best interpreted as a combination of two structures, : C : : Ö : and : C : : : O : , with the latter having greater importance. See Pauling, "The Nature of the Chemical Bond. III. The Transition from One Extreme Bond Type to Another," *JACS 54* (1932): 1000–1001.

101. Pauling, "The Shared-Electron Chemical Bond," p. 361.

102. Heilbron, Interview with Pauling, Session II, p. 16.

103. These papers appeared in *JACS 53* (1931): 1367–1400 and 3225–3237; *54* (1932): 988–1003 and 3570–3582; *JCP 1* (1933): 362–374 (with G. W. Wheland) and 606–617 and 679–686 (both with J. Sherman). See Paradowski, "Structural Chemistry of Linus Pauling," pp. 332–343 and 432–463 for a skillful explication of these papers.

104. Pauling and G. W. Wheland, "The Nature of the Chemical Bond. V. The Quantum-Mechanical Calculation of the Resonance Energy of Benzene and Naphthalene and the Hydrocarbon Free Radicals," *JCP 1* (1933): 362–374.

105. Ibid., p. 363; Pauling, "The Calculation of Matrix Elements for Lewis Electronic Structures of Molecules," *JCP 1* (1933): 280–283.

106. When a molecule forms, according to Hund and Mulliken, the electronic structure of the individual atoms is disrupted and electrons are redistributed into molecular orbitals, which might be viewed as surrounding both nuclei. These molecular electronic structures are built up in accordance with Pauli's exclusion principle and leave their signatures in band spectra much as those of atoms do in line spectra. In the transition from atomic to molecular orbitals, electrons may either gain or lose energy. Electrons in energy states that are common to both an atomic and the molecular structure lose energy, since in the molecule they are bound by a far larger nuclear charge. Those that enter energy levels without a counterpart in the constituent atoms typically gain energy. The balance between these electrons, called bonding and antibonding electrons, respectively, determines whether the energy of the molecule as a whole is higher or lower than that of the individual atoms, i.e., whether the combination is unstable or stable. Robert S. Mulliken, "Bonding Power of Electrons and Theory of Valence," *CR 9* (1931): 347–388 provides an especially clear account of the theory.

For fuller historical treatment see Palmer, *A History of the Concept of Valency to 1930*, pp. 166–171; Russell, *The History of Valency*, pp. 309–312; and Arturo Russo, "Mulliken e Pauling: Le due vie della chimica-fisica in America," *Testi & Contesti 6* (1982): 37–59.

107. For a detailed comparison of molecular-orbital and valence-bond theories, see Van Vleck and Sherman, "The Quantum Theory of Valence." Van Vleck and Sherman maintained that neither method, in its pure form, yielded predictions that were consistent with experimental data. The molecular-orbital method tended to exaggerate the ionic character of bonds and the valence-bond method tended to underestimate ionic contributions. Nevertheless, by adding additional terms to the wave equations generated by these methods, it was possible to correct for these tendencies. Hence, they concluded that it was "meaningless . . . to argue which of the two methods is the better since they ultimately merge. In fact, they may be regarded as simply two different starting points of a perturbation calculation, corresponding to different choices of unperturbed wave functions" (p. 171). This view, expressed earlier by Slater, gradually won widespread acceptance among chemists and physicists.

108. In addition to Heitler, London, Slater, Pauling, Hund, and Mulliken, this group also would include G. Herzberg, E. Hückel, John Van Vleck, John E. Lennard-Jones, and F. Block. Mulliken, the son of Noyes's friend at MIT, Samuel P. Mulliken, took his bachelor's degree at MIT in chemical engineering and his Ph.D. in physical chemistry at Chicago under William D. Harkins. Subsequently, however, Mulliken taught in physics departments and maintained closer contact with physicists than chemists. Mulliken, it should be noted, did his first independent research under Noyes while an undergraduate at MIT. Apparently this was one of the rare occasions when Noyes's instinct for talent failed him, for Noyes expressed disappointment in his work. Mulliken might well be called the one that got away. See Mulliken's Introduction to Part I in *Selected Papers of Robert S. Mulliken*, ed. D. A. Ramsay and J. Hinze (Chicago, 1975), p. 4. On Mulliken see also the excerpts from his interview with Thomas S. Kuhn (February 1964) in *Selected Papers of Robert S. Mulliken*, pp. 5–10; and "Molecular Scientists and Molecular Science: Some Reminiscences," *JCP 43* (1965): S2–S11, and in *Selected Papers of Robert S. Mulliken*, pp. 22–31. Through the courtesy of the late Professor Mulliken I have also benefited from reading his unpublished autobiography, "Life of a Scientist."

109. Mulliken, "Life of a Scientist," p. 91.

110. Pauling, "Fifty Years of Progress in Structural Chemistry and Molecular Biology," pp. 290–291.

111. Noyes, "The Constitution of Benzol," *TQ 1* (1887–1888): 79–90.

112. On this point see Pauling, "The Nature of the Chemical Bond. IV. The Energy of Single Bonds and the Relative Electronegativity of Atoms," *JACS 54* (1932): 3573; and *The Nature of the Chemical Bond*, 3rd ed. (Ithaca, 1960), p. 67.

113. Noyes and Arnold O. Beckman, "A Periodic Table of the Structure of Atoms and Its Relation to Ion Formation and Valence," *PNAS 13* (1927): 737–743; "The Structure of Atoms as a Periodic Property and Its Relation to Valence and Ion-Formation," *CR 5* (1928): 85–107.

114. Pauling, "The Theoretical Prediction of the Physical Properties of Many-Electron Atoms and Ions."

115. Noyes to Hale, 18 February 1923, Roll 28, GEH Papers; Noyes to E. B. Wilson, 14 October 1927, 1926–27 A. A. Noyes folder, EBW; Heilbron, Interview with Pauling, Session I, p. 31; Warren Weaver Diaries, October 23–25, 1933, quoted by Goodstein, "Atoms, Molecules, and Linus Pauling," p. 707.

116. Lewis to Pauling, 25 August 1939, Box 3, CCR; Heilbron, Interview with Pauling, Session II, pp. 20–21; ECM, CIT, 23 March 1929.

117. G. B. Kistiakowsky, review of *The Nature of the Chemical Bond* in *JACS 62* (1940): 457. See also Mulliken's comments in his interview with Thomas S. Kuhn of February 1964, excerpted in *Selected Papers of Robert S. Mulliken*, pp. 9–10, and his review of *The Nature of the Chemical Bond* in *JPC 44* (1940): 827–828.

118. Pauling, "Fifty Years of Progress in Structural Chemistry and Molecular Biology," p. 997. On Lucas see W. G. Young and Saul Winstein, "Howard J. Lucas," *BMNAS 43* (1973): 162; Dean Stanley Tarbell and Ann Tracy Tarbell, *Essays on the History of Organic Chemistry in the United States, 1875–1955* (Nashville, 1986), esp. pp. 243–282; and Leon Gortler, "The Physical Organic Community in the United States, 1925–50: An Emerging Network," *JCE 62* (1985): 753–757. On Badger, see "Richard McLean Badger," *Engineering and Science 38* (1974–1975): 24, and his "A Relation Between Internuclear Distances and Bond Force Constants," *JCP 2* (1934): 128–131. There is a large literature on Pauling's role in the development of molecular biology; especially valuable is Robert Olby, *The Path to the Double Helix* (Seattle, 1974), pp. 272–289.

119. Following Noyes's death in 1936, Millikan split authority in the chemistry division between Pauling, who was made director of the new Crellin Chemical Laboratory, and Tolman, who was made Chairman of the Division. Pauling's own colleagues, Tolman, Dickinson, and Lacey, had lobbied against entrusting Pauling with sole authority over department affairs. See Richard C. Tolman, Roscoe G. Dickinson, and William N. Lacey to Millikan, 12 June 1936; Pauling to Millikan, 17 October 1936; and Dickinson to Millikan, 7 April 1937; all in Box 20, folder 26, RAM.

CHAPTER 7
A Dissenter's Decline

1. Thomas Graham, "Liquid Diffusion Applied to Analysis," *Philosophical Transactions of the Royal Society 151* (1861): 183–224. On Graham see R. A. Smith, *The Life and Work of Thomas Graham* (Glasgow, 1884). From the outset, the nomenclature of colloid chemistry was ambiguous, in large part because colloids were defined negatively, that is, as substances that did not diffuse readily. Over time, the roster of substances that could be put into this state grew longer and longer. Many of these materials, such as finely divided gold, chromium, and silver, had little in common with glue. For a succinct and critical discussion of the nomenclature of colloid chemistry, see Neil K. Adam, *Physical Chemistry* (Oxford, 1956), pp. 577–601.

2. For a recapitulation of this research, see J. R. Partington, *A History of Chemistry*, vol. 4 (London, 1964), pp. 729–743.

3. Partington, *A History of Chemistry*, vol. 4, p. 729 and pp. 733–734.

4. On colloids and biochemistry see Robert E. Kohler, Jr., "The History of Biochemistry: A Survey," *Journal of the History of Biology 8* (1975): 275–318, esp. pp. 290–291. Other valuable sources include Joseph S. Fruton, *Molecules and Life: His-*

torical Essays on the Interplay of Chemistry and Biology (New York, 1972), pp. 131–148; Marcel Florkin, *A History of Biochemistry (Comprehensive Biochemistry, 30)* (Amsterdam, 1972), pp. 279–283; and John T. Edsall, "Proteins as Macromolecules," *Archives of Biochemistry and Biophysics, Supplement 1* (1962): 12–20.

5. Grete Ostwald, *Wilhelm Ostwald, Mein Vater* (Stuttgart, 1953), p. 61.

6. Biographical sketches include A. Lottermoser, "Wolfgang Ostwald 60 Jahre alt," *KZ 103* (1943): 89–94 (with bibliography); H. Erbring, "Wolfgang Ostwald, 1883–1943," *KZ 115* (1949): 3–5; Gustav F. Hüttig, "Wolfgang Ostwald," *Forschungen und Fortschritte 20* (1944): 118–119; and Martin H. Fischer, "Wolfgang Ostwalds Weg zur Kolloidchemie," *KZ 145* (1956): 1–2. On Ostwald and Loeb, see Philip J. Pauly, *Controlling Life: Jacques Loeb and the Engineering Ideal in Biology* (New York, 1987), p. 113 and pp. 151–152.

7. Grete Ostwald, *Wilhelm Ostwald*, pp. 61–62.

8. Wolfgang Ostwald, *An Introduction to Theoretical and Applied Colloid Chemistry*, trans. Martin H. Fischer (New York, 1917), pp. 218–219. Ostwald describes his itinerary on p. ix.

9. Ibid., p. 127.

10. Ibid., p. xi. See also p. 3 and p. 76.

11. Ibid., p. 180.

12. Harry N. Holmes, "The Growth of Colloid Chemistry in the United States," in *Twenty Years of Colloid and Surface Chemistry: The Kendall Award Addresses*, ed. K. J. Mysels, C. M. Samour, and J. H. Hollister (Washington, D.C., 1973), p. 7; W. A. Noyes to W. J. Hale, 14 January 1914, Box 4, WAN.

13. Kohler, "The History of Biochemistry: A Survey," pp. 290–291. After World War I, textbooks of physical chemistry usually included a chapter on colloids, and textbooks of colloid chemistry typically began by defining the study of colloids as a part or outgrowth of physical chemistry. See, for example, The Svedberg, *Colloid Chemistry* (New York, 1928), p. 15; H. R. Kruyt, *Colloids: A Textbook*, trans. H. S. van Klooster (New York, 1927), p. vii; Emil Hatschek, *An Introduction to the Physics and Chemistry of Colloids*, 2nd ed. (Philadelphia, 1916), preface to first edition (of 1913).

14. My survey of the articles in the general and physical section of the *JACS* shows the following growth:

Five-year Period	Number of Articles on Colloid & Surface Chemistry	Total number of Articles	% on Colloid & Surface Chemistry
1904–1908	7	155	5
1909–1913	9	201	4
1914–1918	38	274	14
1919–1923	82	438	19

15. Ross Aiken Gortner, later a leading American colloid chemist, traced his first contact with the subject to 1910, when Bancroft invited him to contribute a paper on "Colloids in Biology" to a symposium; see *Selected Topics in Colloid Chemistry with Especial Reference to Biochemical Problems* (Ithaca, 1937), p. 24. Bancroft's papers

on the photographic plate and the theory of emulsification appeared in series form in the *JPC* between 1910 and 1912.

16. Bancroft, "Chemical Activity at Princeton," *Princeton Alumni Weekly 27* (13 May 1927): 918. On the fire, see also Bancroft to T. W. Richards, 26 February 1916, Professional and Personal Correspondence 1916, YWR; and Bancroft to Louis Kahlenberg, 26 March 1916, Box 4, LK. On Bancroft's move into the new laboratory, see Bancroft to Francis Garvan, 28 October 1921, File 29, Drawer 1, Folder D-6-7, FG.

17. Bancroft, *Applied Colloid Chemistry: General Theory* (New York, 1921), p. 2.

18. Bancroft, *Applied Colloid Chemistry*, quote from 3rd ed. (1932), p. v. Bancroft to R. H. Treman, 15 September 1925, 1923–1925 Box, Professional Correspondence, WDB; Bancroft to Herbert Hoover, 23 February 1926, and Bancroft to Emerson, 20 February 1928, both in 1926–1930 Box, Professional Correspondence, WDB.

19. George Scatchard, "Half a Century as a Part-time Colloid Chemist," in *Twenty Years of Colloid and Surface Chemistry*, p. 103. The terms used to describe colloid chemists are almost as confusing as the nomenclature of the science itself. Kohler divides colloid chemists between those with reductionist and vitalist leanings; Furukawa speaks of physicalists and organic-structuralists; Svedberg divided his colleagues into quantitative and qualitative camps. I prefer Scatchard's division between unionists and isolationists. Most unionists were reductionists, but few colloid chemists, either isolationists or unionists, really qualify as vitalists. Neither Bancroft nor Ostwald, for example, viewed living matter as being exempt from physical and chemical laws. Indeed, they displayed a remarkable readiness to equate *in vitro* and *in vivo* phenomena.

Most isolationists were "physicalists" in the sense that Furukawa uses the term. That is, they tended to think of colloids as agglomerates rather than compounds of definite composition and to emphasize the role of physical forces in adsorption and minimize the importance of valence bonds. The methods of "physicalists" like Bancroft and Ostwald, however, were anything but those of the physicist. "Organic-structuralists," like The Svedberg, often displayed a far greater respect for the value of mathematical physics than their "physicalist" colleagues. So, someone like Bancroft would be placed in the "physicalist" camp by Furukawa and the "qualitative" camp by Svedberg. Scatchard's categories manage to avoid most of these ambiguities. See Kohler, "The History of Biochemistry: A Survey," pp. 290–291; Yasu Furukawa, "Hermann Staudinger and the Emergence of the Macromolecular Concept," *Historia Scientiarum*, no. 22 (1982), p. 1; and Svedberg, *Colloid Chemistry*, p. 7 and p. 16.

20. Bancroft, *Applied Colloid Chemistry*, 2nd ed. (New York, 1926), pp. 42–43, 299–300. The physical model was applied to charcoal in part because it was known to adsorb noble gases such as argon, in which case no valence bonds could be involved. Irving Langmuir was a leading advocate of the idea that the short-range forces acting at the surface of colloids sometimes were weak residual valences.

21. Ibid., pp. 216–226. See also John T. Edsall, "Proteins as Macromolecules," *Archives of Biochemistry and Biophysics, Supplement 1* (1962): 15–18.

22. On Staudinger, see Furukawa, "Hermann Staudinger and the Emergence of the Macromolecular Concept"; on the work of Einstein, Perrin, and Svedberg, see Mary Jo Nye, *Molecular Reality: A Perspective on the Scientific Work of Jean Perrin* (New York, 1972), pp. 97–142.

23. Mathews, "Adsorption," *Physiological Reviews 1* (1921): 560.

24. Ibid., p. 579; see also p. 558.

25. Ibid., pp. 588–589.

26. Loeb, *Proteins and the Theory of Colloidal Behavior* (New York, 1922), esp. pp. 19–36, 40–64, and 169–188. Pauly has persuasively argued that Loeb's attitude toward colloid chemistry was colored by his distaste for Wolfgang Ostwald's metaphysical romanticism and nationalism; see Pauly, *Controlling Life*, pp. 151–153.

27. Bancroft, *Applied Colloid Chemistry*, p. 170 and pp. 225–226.

28. Ibid., pp. 271–273, 339–341.

29. Loeb to Svante Arrhenius, 14 December 1923, Box 1, JL.

30. Fruton, *Molecules and Life*, pp. 133–137.

31. L. A. Munro, "The Place of Colloid Chemistry in Chemical Education," *JCE 13* (1936): 462–464; Edsall, "Proteins as Macromolecules," p. 17; Fruton, *Molecules and Life*, p. 137 and pp. 143–165.

32. Bancroft to Francis P. Garvan, 16 September 1921, File 29, Drawer 1, Folder D-6-7, FG; A. B. Lamb to J. H. Hildebrand, 15 December 1921, Correspondence 1917–1923 (Fries to Larson) Box, ABL.

33. C. L. Parsons to William J. Pope, 20 September 1921; Bancroft to E. C. Bingham, 8 February 1922; Parsons to Bancroft, 2 August 1922, all in 1921–1922 Box, Professional Correspondence, WDB.

34. A copy of the 1932 subscription list of the *JPC* is in File 29, Drawer 1, Folder D-6-7 of the FG. Francis P. Garvan to Bancroft, 25 October 1921, 1921–1922 Box, Professional Correspondence, WDB; Bancroft to Garvan, 11 December 1927; "Journal of Physical Chemistry, Statement of Cash Receipts and Disbursements for Period September 2, 1929 to August 15, 1933"; William Buffum to Bancroft, 9 September 1929; all in File 29, Drawer 1, Folder D-6-7, FG.

35. E. C. Bingham to Bancroft, 5 January 1922, 1921–1922 Box, Professional Correspondence, WDB.

36. Lewis to E. C. Bingham, 21 February 1922, 1921–1922 Box, Professional Correspondence, WDB. Edward W. Washburn, Lewis's former student at MIT, makes a nearly identical statement in "Physical Chemistry and Ceramics," *Journal of the Franklin Institute 193* (1922): 750. Both seem to be derived from Arthur B. Lamb to Joel H. Hildebrand, 15 December 1921, Correspondence 1917–1923 Box, ABL. In addition to Lewis and Washburn, members of the ACS committee on the future of the *JPC* included E. C. Bingham (chair), Bancroft, John Johnston, Harry N. Holmes, E. C. Franklin, James Kendall, A. A. Noyes, and Hugh Taylor.

37. Bancroft to E. C. Bingham, 8 February 1922, 1912–1922 Box, Professional Correspondence, WDB.

38. E. C. Bingham to Committee on Publications, undated but certainly written in 1922, 1933–1953 and undated Box, Professional Correspondence, WDB. See also Bingham to Bancroft, 5 January and 2 February 1922, 1921–1922 Box, Professional Correspondence, WDB.

39. G. N. Lewis to E. C. Bingham, 21 February 1922; John Johnston to W. D. Bancroft, 28 November 1921; E. W. Washburn to E. C. Bingham, 15 March 1922; C. L. Parsons to W. D. Bancroft, 26 September 1922; all in 1921–1922 Box, Professional Correspondence, WDB.

40. Bancroft to S. A. Tucker, 28 March 1921, and Bancroft to Francis Garvan, 24

June 1922, both in File 29, Drawer 4, Folder D-6-96, FG; Bancroft to Garvan, 8 October 1923, 11 December 1927, 2 May 1928, and 5 July 1929, all in File 29, Drawer 1, Folder D-6-7, FG.

41. Bancroft, "How to Ripen Time," *JPC 35* (1931): 1921.

42. Bancroft to Orlando F. Scott, 11 February 1933, 1933–1953 and undated Box, Professional Correspondence, WDB.

43. Bernard presented his views in *Leçons sur les anesthésiques et sur l'asphyxie* (Paris, 1875), quoted in Bancroft and George H. Richter, "The Chemistry of Anesthesia," *JPC 35* (1931): 224.

44. Bancroft to Scott, 11 February 1933. The most serious of the objections Bancroft here alludes to was the question of how anesthetics could produce their effect when present in nervous tissue in minute concentrations. Bancroft thought he had the answer: a slightly acidified albumin sol, treated with sodium sulphate until it was on the verge of precipitating, would begin to flocculate with the addition of one drop of alcohol or chloral hydrate. In much the same way, he supposed, the electrolytes normally present in nerve cells kept protoplasmic colloids in a "critical state," ready to coagulate in the presence of extremely small amounts of a flocculating agent. See Bancroft and Richter, "The Chemistry of Anesthesia," pp. 226–227.

45. Fourteen articles by Bancroft and his associates appeared in *PNAS 16–20* (1930–1934); eleven others were published in *JPC 35–36* (1931–1932). Richter left Cornell in 1932 and later became professor and dean of chemistry at the Rice Institute; Rutzler became professor of chemistry at the Case Institute of Technology.

46. Bancroft and G. H. Richter, "Reversible Coagulation in Living Tissue," *PNAS 17* (1931): 294. Bancroft preferred to call sodium thiocyanide by its German name, sodium rhodanate, to avoid associations with the well-known poison, potassium cyanide.

47. Favorable accounts appeared in the *New York American*, 4 March 1931, 26 April 1932, and 22 May 1932; *New York Evening Journal*, *New York Evening Post*, and *New York World Telegram*, 25 April 1932 and 14 November 1932; *New York Sun*, 15 November 1932 and 9 February 1933; *New York Times*, 6 February 1933; *New York Herald Tribune*, 10 February 1933; and *Time*, 3 December 1934. Critical reviews of Bancroft's work include Walter Freeman, "Psychochemistry," *JAMA 97* (1931): 295; T. M. Burkholder, "The Effect of Sodium Thiocyanate on the Action of Anesthetic and Narcotic Drugs," *Journal of Laboratory and Clinical Medicine 18* (1932): 29–41; V. E. Henderson and G.H.W. Lucas, "Claude Bernard's Theory of Narcosis," *Journal of Pharmacology and Experimental Therapy 44* (1932): 253–267; and Henry E. Guerlac, "Combined Action of Ethyl Urethane and Sodium Thiocyanate on the Living Cell," *Proceedings of the Society for Experimental Biology and Medicine 30* (1932): 265–268. Bancroft's procedures were also roundly condemned by the Tompkins County Medical Society; see H. B. Sutton, N. S. Moore, and B. F. Hauenstein, Special Narcotic Committee of the Tompkins County Medical Society, to the Editor of the *JAMA*, 10 June 1932, File 29, Drawer 1, Folder D-6-7, FG.

48. *New York Times*, 6 February 1933, p. 17.

49. "Awards and Recognitions in Chemistry and Medicine," *JAMA 100* (4 March 1933): 667. See also "Sodium Thiocyanate (Rhodanate) and the Theory of Agglom-

eration,'' *JAMA 99* (31 December 1932): 2270–2271, and ''Bancroft's Thiocyanate Therapy,'' *JAMA 100* (28 January 1933): 262–263.

50. Chauncey D. Leake, ''Sodium Thiocyanate (Rhodanate) and the Theory of Agglomeration,'' *JAMA 100* (4 March 1933): 682.

51. Ibid., p. 683.

52. D. P. Morgan to Bancroft, 18 February 1933 and 2 March 1933; Bancroft to Morgan, 22 February 1933; all in 1933–1953 and undated Box, Professional Correspondence, WDB.

53. Bancroft, Esther C. Farnham, and John E. Rutzler, Jr., ''One Aspect of the Longevity Problem,'' *Science 81* (1935): 152. Bancroft's theory was given a full and uncritical treatment in Robert J. Hartman, *Colloid Chemistry* (Boston, 1939), pp. 527–530.

54. A. B. Lamb, ''Memorandum: Conference with Mr. Buffum,'' 11 August 1932, Correspondence 1929–1933 (A-Ba) Box, ABL; Henry A. Barton to John T. Tate, 19 May 1932; Barton to Karl T. Compton, 27 March 1932; Compton to Bancroft, 2 June 1932; Barton, ''Memorandum for Mr. Buffum,'' 22 June 1932; ''Report of Progress Submitted to the Chemical Foundation,'' 1 August 1932; all in HAB; Lamb to Farrington Daniels, 2 July 1932, Correspondence, Selected Files, 1928–1932, *Journal of Chemical Physics* Folder, FD; C. L. Parsons to Bancroft, 19 August 1932, 1931–1932 Box, Professional Correspondence, WDB.

55. ''Memorandum of conference between Mr. Buffum, Professor Bancroft and Dr. Barton concerning the attitude of the Institute of Physics toward the Journal of Physical Chemistry,'' 18 February 1932, Box 5, WDB.

56. Karl T. Compton to The Chemical Foundation, 1 October 1931, File 29, Drawer 3, Folder D-6-82, FG. See also ''William W. Buffum, The Chemical Foundation, and Physics,'' *Physics Forum 9* (1938): 211–212. The founding societies of the American Institute of Physics were The American Physical Society, Optical Society of America, Acoustical Society of America, Society of Rheology, and American Association of Physics Teachers.

57. ''Memorandum of conference between Mr. Buffum, Professor Bancroft and Dr. Barton,'' 18 February 1932.

58. Barton, letter to leading physical chemists and physicists, 12 April 1932, HAB.

59. Birge to Barton, 3 May 1932, HAB (despite his objections, Birge later joined the editorial board of the *JCP*); Urey to Barton, 17 April 1932, HAB.

60. Barton, letter to leading physical chemists and physicists, 5 May 1932, HAB.

61. Barton knew that Bancroft opposed the immediate creation of a journal of chemical physics before this letter was sent; see Bancroft to Barton, 22 April 1932, HAB. It appears that Barton sought to present Bancroft with a *fait accompli* that would make it embarrassing for Bancroft to give public voice to his objections.

62. Bancroft to Garvan, 14 May 1932, 1931–1932 Box, Professional Correspondence, WDB.

63. Gortner to Barton, 18 May 1932, HAB; Gortner to C. L. Parsons, 12 May 1932, 1931–1932 Box, Professional Correspondence, WDB.

64. Barton to John Tate, 19 May 1932; Barton to Compton, 27 May 1932; Barton, ''Memorandum for Mr. Buffum,'' 22 June 1932; all in HAB.

65. Farrington Daniels to Urey, 5 August 1932, Correspondence, Selected Files 1928–1932, *Journal of Chemical Physics* folder, FD. Daniels was quoting an unnamed scientist.

66. Barton to Ross A. Gortner, 28 May 1932; Barton to Bancroft, 2 June 1932; Barton to Compton, 4 June 1932; John T. Tate, George B. Pegram, and Karl T. Compton, ''Report of the American Institute of Physics to the Council of the American Physical Society,'' 25 November 1932; all in HAB; Compton to Bancroft, 6 August 1932, 1931–1932 Box, Professional Correspondence, WDB; Barton to Daniels, 20 June 1932 and 9 August 1932, both in Correspondence, Selected Files 1928–1932, *Journal of Chemical Physics* folder, FD.

67. Tate, Pegram, and Compton, ''Report of the American Institute of Physics to the Council of the American Physical Society,'' 25 November 1932; Urey, ''Statement of the Proposed Subject Matter of the Journal of Chemical Physics,'' no date, but probably composed in late June or July 1932, HAB.

68. Daniels to Urey, 30 August 1932, Correspondence, Selected Files 1928–1932, *Journal of Chemical Physics* folder, FD.

69. Ibid. See also Urey to Daniels, 9 September 1932, Correspondence, Selected Files 1928–1932, *Journal of Chemical Physics* folder, FD.

70. Buffum to Bancroft, 6 August [1932], 1933 and Undated Box, Professional Correspondence, WDB; C. L. Parsons to Bancroft, 19 August 1932, 1931–1932 Box, Professional Correspondence, WDB; Bancroft to Compton, 20 August 1932, HAB.

71. Lind to Bancroft, 18 May 1933, 1933–1953 and Undated Box, Professional Correspondence, WDB.

72. Of the 133 contributors to the first volume of the *JCP*, 84 identified themselves as chemists or physical chemists in *American Men of Science*. On the subsequent history of the *Journal of Chemical Physics*, see J. W. Stout, ''*The Journal of Chemical Physics*: The First 50 Years,'' *ARPC 37* (1986): 1–23.

73. C. W. Mason, ''Wilder Dwight Bancroft,'' *JACS 76* (1954): 2601. Among the firms Bancroft worked for were the General Chemical Company, Butte and Superior Mining Co., General Motors Corp., Douglas Packing Co., Douglas Pectin Corp., General Foods Corp., Air Reduction Co., Eastman Kodak, Norton Co., Technicolor, and the Silica-Gel Corp. On his consulting activities, see Boxes 1 and 3, WDB.

74. Bancroft, ''The Ramifications of a Research Problem,'' *Transactions of the American Electrochemical Society 50* (1927): 9–10.

75. Bancroft to Charles A. Browne, 7 May 1926, Box 2, CAB.

76. Bancroft, review of *Elementary Physical Chemistry* (by Hugh S. Taylor) in *JPC 31* (1927): 1584.

Index

thermodynamics: and chemical affinity, 18–
19; difficulties in applying, 31; in Perrin's
work, 333–334n.22; third law of, 81,
352n.134. *See also* chemical thermodynamics
Thomsen, Julius, 17–19, 22, 24, 29–30, 47
Thomson, J. J., 80, 130–134, 137, 202,
337n.68
Thorpe, Frank H., 101
Throop Polytechnic Institute, 261–262, 264–
265, 269
Tolman, Robert Chace, 151, 153; and Lewis,
130, 245, 247; Noyes and, 198, 254, 270–
271, 298; and Pauling, 275, 283; research
of, 131, 272, 297, 349n.99, 377n.45; in
World War I, 215
Toronto, University of, 165
Tower, O. F., 62, 67, 94
Trevor, Joseph E., 62, 69, 82, 95, 163–166,
169

Uhlenbeck, G., 287
ultramicroscope, 300, 306
unit operations, 258
U.S. Bureau of Mines, 69, 213, 218, 238,
242, 262
U.S. Bureau of Soils, 67, 90, 195, 199
U.S. Geological Survey, 98, 223–224, 227,
229–234
U.S. Steel Corporation, 153, 189–190, 216,
360n.80
universities: conservatism of, in the 1920s,
238–239; and industrial culture, 205–206,
209–210, 218–220; and physical chemistry
in America, 53–55, 73–75, 87, 90, 92–99,
153, 163, 204, 216–219, 239; and physical
chemistry in Germany, 50–51; and provisions for research, 110
University College London, 51–52
Uppsala, University of, 20, 34, 44
Urey, Harold C., 252, 315, 317, 319, 324

Van Hise, Charles R., 224, 227–229, 233–
234, 240
Van Vleck, John, 321, 383n.107
Vogel, H. W., 86
Vogt, J.H.L., 222, 230

Waage, Peter, 15–19, 22, 27–28, 34, 47
Waals, J. D. van der, 80, 172
Wagner, Julius, 48

Walcott, Charles D., 229, 233
Walden, Paul, 49, 135, 295
Walker, James, 41, 49–50, 332n.16,
340n.105
Walker, James Wallace, 332n.16
Walker, William H., 101, 219, 374n.17; con-
flict with Noyes, 259–263; education views
of, 257–258
Washburn, Edward W., 151, 153, 310–311
Weiser, Harry B., 312
Werner, Alfred, 274
Wesleyan University, 62, 94
Wheeler, Benjamin Ide, 150, 240–246, 248–
249
White, J., 99
Whitney, Willis R., 56; on Bancroft, 82–83;
and GE, 69, 114, 117, 210, 213; and indus-
trial research, 92; at MIT, 95, 101, 116; on
Morgan, 70; and Noyes, 62, 106, 109,
111, 113, 120, 151, 153
Wiedemann, Gustav, 44, 48
Wierl, R., 297
Williams, Roger, 153
Williamson, Alexander, 15, 28, 34–35
Willstätter, Richard, 331–332n.12
Wilson, E. B., 262
Wisconsin, University of, 53, 227, 239, 249,
252; chemical facilities at, 94, 96; chemis-
try and chemical engineering at, 255; en-
rollments in physical chemistry at,
372n.115; Kahlenberg at, 98, 240; van
Hise at, 227
Wise, George, 92, 109, 210
Wislicenus, Johannes, 19, 23, 48, 51, 58–59,
61, 88
Wöhler, Friedrich, 22
Wolcott Gibbs Memorial Laboratory, 79
World War I: and colloid chemistry, 303–304;
and industrial research, 208, 211–213, 237,
246; and *Journal of Physical Chemistry*,
309; mission research during, 269; and
Noyes, 259–261; and physical chemists,
72, 91, 213–220
Wright, F. E., 229, 234
Wyckoff, Ralph W. G., 272

X-ray crystallography, 136; at Caltech, 271–
272, 274–275; development of, 272–274;
Pauling on, 284–285